Imagining Ecocatastrophe

This volume examines scholarly perspectives on eco-imaginaries, focusing in particular on how eco-catastrophes have been represented in literature and different visual forms, including film, television and cartoons, among other cultural media. It draws on literary genres such as science fiction, climate fiction, speculative fiction, petrofiction, post-apocalyptic narratives and nuclear fiction to examine the role that literature plays in the dissemination of information about environmental crisis in the Anthropocene and in preparing mankind for a better and sustainable future. Deeply embedded in theoretical conceptualisations, the essays in this volume address issues of natural disasters, deforestation, nuclear disasters and pandemics, among others, which constitute the core subjects of environmental humanities.

A seminal study on the literary and cultural representations of ecodisaster in the global context, and with contributions from across the world, this book, truly interdisciplinary in nature, will be an invaluable read for students, academicians and researchers in literature, film studies, climate change studies, disaster studies, gender studies and cultural studies.

Scott Slovic is a senior scientist at the Oregon Research Institute and a distinguished professor of environmental humanities emeritus at the University of Idaho, USA.

Joyjit Ghosh is Professor in the Department of English Literature, Language and Cultural Studies, Vidyasagar University, Midnapore, West Bengal, India.

Samit Kumar Maiti is Assistant Professor in the Department of English, Seva Bharati Mahavidyalaya, Kapgari, Jhargram, West Bengal, India.

Imagining Ecocatastrophe

Reading Literary and Cultural Texts in the Global Context

Edited by Scott Slovic, Joyjit Ghosh and Samit Kumar Maiti

Routledge
Taylor & Francis Group
LONDON AND NEW YORK

First published 2026
by Routledge
4 Park Square, Milton Park, Abingdon, Oxon OX14 4RN

and by Routledge
605 Third Avenue, New York, NY 10158

Routledge is an imprint of the Taylor & Francis Group, an informa business

British Library Cataloguing-in-Publication Data
A catalogue record for this book is available from the British Library

ISBN: 978-1-032-72419-5 (hbk)
ISBN: 978-1-032-99177-1 (pbk)
ISBN: 978-1-003-60275-0 (ebk)

DOI: 10.4324/9781003602750

Typeset in Sabon
by Apex CoVantage, LLC

Contents

Contributors

Laura Akers, PhD, is an associate scientist at Oregon Research Institute, specialising in cognitive framing, narrative psychology and behaviour change. For more than 30 years, she has been studying the psychology and rhetoric of both social movements and individual behaviour. Her focus on cognitive framing (different ways to organise our understanding) emphasises the roles of narrative, metaphor and cognitive style. Her study of public interest communication has included research on sociopolitical violence (terrorism, genocide and state terror), as well as the language of the environmental movement. Dr Akers is an expert on meta-narratives, the background group-narratives that function as key cultural beliefs about societies. More recently she has been studying the role of rhetorical "salience markers" in drawing public attention to issues of interest, and the relevance of narrative immersion as a way to describe individuals' relationships with culturally promoted worldviews.

Mir Ahammad Ali is Assistant Professor and currently the Head of the Department of English, at Bhatter College, Dantan (affiliated to Vidyasagar University, West Bengal). He has completed his doctoral research on the psychological aspect of partition violence as a PhD scholar in the Department of English, Vidyasagar University. He is the co-editor of the book *The Bleeding Border: Stories of Bengal Partition* (Niyogi Books, New Delhi, 2022). He is the sole editor of *Exploring Filmature: Problems and Prospects of Adaptation Studies* (BCPD, Dantan, Kolkata, 2024). He has currently been associated with the *Bhatter College Journal of Multidisciplinary Studies* ISSN: 2449 3301 as an Associate Editor In-charge of Arts and Humanities. He is the Editor-in-Chief of *The Golden Line: A Magazine of English Literature* ISSN: 2395–1591 [Online] ISSN: 2395–1583 [Print]. At present, he is the Coordinator of the Centre for Performance Studies at Bhatter College, Dantan. He has participated and presented academic papers in many national and international seminars/conferences and published research articles on a wide variety of topics including partition studies, trauma studies, Bengal partition, environmental/ecological studies, performance studies, endangered folk performances of Bengal, indigenous studies and film studies in many reputed journals and books.

Sk Tarik Ali, Assistant Professor in English in the West Bengal Education Service, teaches in the PG department of English, Hooghly Mohsin College, Chinsurah, Hooghly, West Bengal, India. He has completed his PhD on the representation of resource conflicts in Indian eco-fiction with a special focus on the works of Mahasweta Devi and Amitav Ghosh from Vidyasagar University, West Bengal, India. His articles and book reviews have been published in such prestigious journals as *South Asian Review*, *Journal of Postcolonial Writing*, *Ecozon@: European Journal of Literature, Culture and Environment* and *Marg ASIA: Journal of Centre for Asian Studies*.

Nikhilesh Dhar is Assistant Professor of English at Onda Thana Mahavidyalaya, Bankura, West Bengal. He is the editorial Board Member of *Consortium: An International Journal of Literary and Cultural Studies* and reviewer of UGC-listed journal. He has authored a book, *Portrayal of Women in Keats: A Brief Study*, and has some national and international publications to his credit. His research interests include American literature, South Asian literature, Dalit literature, partition literature, diaspora studies, ecocriticism, postcolonialism and posthumanism. He is a PhD Research Supervisor at the Department of English, Bankura University.

Georgios Dimoglou is an external PhD candidate at the Centre for the Arts in Society, Leiden University, Netherlands, with his research focused on interactive films. He graduated from Aristotle University of Thessaloniki (AUTh), Greece, with an MA in Film and Television Studies (2021) and has an integrated master's in film studies (2019), having the highest grades in his class during both graduations. In 2021, he received a best MA thesis award from AUTh. Subjects of interest include the relationship between films and society, film genres, intermediality and new media. He presented his PhD proposal at the 6th Narrative, Media, and Cognition conference in October 2021. As an editor and co-director of short films and videoart, he has participated in various festivals and exhibitions, such as the 14th Audiovisual Arts Festival of the Ionian University, Greece (May–June 2021), the Museum of Modern Art of Thessaloniki (October 2020–June 2021), and the Bibart Biennale Internazionale d'Arte di Bari, Italy (July 2020). Moreover, he has worked in digitisation and photography at the National Library of Greece – Stavros Niarchos Foundation Cultural Center and in videography/editing at the CYENS Centre of Excellence in Nicosia, Cyprus.

Joyjit Ghosh is Professor of English, Vidyasagar University, Midnapore, West Bengal. He was the Deputy Coordinator of the UGC DRS SAP Project on the translation, documentation and digitisation of Adivasi and folk literature and culture of West Bengal from 2015 to 2020. He has a fair number of publications to his credit. His published works include articles/book chapters on D. H. Lawrence, Rabindranath Tagore and Mahasweta Devi, among other great minds. His book *Imaging D. H. Lawrence: His*

Mind and Art in Letters (2012) has been widely acclaimed by Lawrence scholars. His research interests include Indian writing in English, Bangla Dalit literature, environmental studies, animal studies, diasporic literature and translation studies. *The Bleeding Border: Stories of Bengal Partition* (2022), coedited with Mir Ahammad Ali, is one of his leading publications. Along with Prof. Scott Slovic and Dr Samit Kumar Maiti, he edited *Ecodisaster Imaginaries in India: Essays in Critical Perspectives* (2023). Recently his article (co-authored with Dr Sk Tarik Ali) titled "'Can I Become a Tree?': Plant Imagination in Contemporary Indian Poetry in English" has been published in the special issue "Critical Green Theories and Botanical Imaginaries: Exploring Human and More-than-Human World Entanglements" of the journal *Open Critical Studies*, edited by Peggy Karpouzou and Nikoleta Zampaki.

Asha Susan Jacob is the Principal of Bishop Speechly College for Advanced Studies, Pallom, Kottayam, Kerala. Previously, she was Associate Professor in the Postgraduate and Research Department of English at St. Thomas College, Kozhencherry, Kerala. Her extensive experience in academia includes significant contributions to the field of English literature and education. Dr Jacob's leadership and scholarly work continue to impact and advance academic excellence and research in her field.

Elwin Susan John teaches at the Department of English at Sophia College (Autonomous), Mumbai, India. His areas of research interest and publications are mostly located within a cultural study of the body in humanities, literary disability studies, posthumanism, Indian literature and Early Modern English literature.

Tushar Kanti Karmakar is Assistant Professor at the Department of English, Kaliyaganj College, Uttar Dinajpur, West Bengal, India. He has completed his MPhil as Junior Research Fellow at The University of Burdwan, West Bengal, India, and is currently pursuing his PhD at Visva Bharati, Shantiniketan, West Bengal, India. He is writing his dissertation on American climate change fiction. His research interests include climate change literature, eco-criticism, environmental humanities and American literature. He has published a book titled *Narrating Bharatvarsha: A Study of the Role of Women, Religion, and Caste in Rabindranath Tagore's Gora and The Home and the World*. He has a forthcoming book chapter on the intersection of nationalism and indigenousness from ibidem press.

Mahinur Gözde Kasurka completed her PhD in English Literature department at Middle East Technical University with her dissertation entitled "A Posthumanist Study of the Dystopian Novel: Margaret Atwood's *Oryx and Crake*, Jeanette Winterson's *The Stone Gods*, David Mitchell's *Cloud Atlas*." Currently, she works as a full-time Instructor Dr at National Defence University, School of Foreign Languages. Also, she works as part-time Lecturer Dr at İstanbul Bilgi University, in the Department of English Language and Literature. Her research interests are British novels, literary theories, women and writing, psychology and literature, dystopian and post-apocalyptic fiction, posthumanisms and ecocriticisms. She has presented several papers at international conferences. She has attended Rosi Braidotti's summer school on "Posthuman Knowledge(s)."

Anna Frieda Kuhn is a PhD candidate and lecturer at the University of Würzburg's Chair of English Literature and British Cultural Studies. Her research primarily focuses on the intersections of classical reception studies, globalisation studies and postcolonial ecocriticism. Kuhn has also been involved in international academic exchanges and research stays at Cambridge University, Jawaharlal Nehru University in Delhi and the University of Cape Town. Additionally, she has organised and participated in several international conferences and symposia. Anna Frieda Kuhn is also the editor of the multilingual blog Denksport:ART, which showcases diverse cultural and literary discussions.

Bapin Mallick is Assistant Professor at the Department of English, DR B.R. Ambedkar College (Affiliated to the University of Kalyani), Betai, Nadia West Bengal. He has completed MA in English from Rabindra Bharati University, Kolkata. He has done an MPhil on "The Earth Haunted Mind: Exploring Environmental Consciousness of Ruskin Bond in Select Short Stories" from the Department of English, University of Gour Banga. He has completed PhD on the topic entitled "Representing the Non-Human: A Study of Ruskin Bond's Select Writings from Deep Ecological Perspective" from the Department of English, Bankura University, West Bengal. He is the managing editor of *Consortium: An International Journal of Literary and Cultural Studies*. His research interests include deep ecology, posthumanism and ecocriticism.

Amit Mandal is a PhD Research Scholar in the Department of English, University of Delhi. He has presented papers at various national and international conferences. Some of his recent presentations include "Traversing Anthropocentrism and Solastalgia: A British Dystopian Imaginary of 'The Wall' by John Lanchester and the 'Dreamland' by Rosa Rankin Gee," Sheffield Hallam University, UK (6–20 August, 2021); "The Exotic and the Climactic Landscape: A Study of the Fleeting Imaginary of the Sundarbans," Concerned Philosophers for Peace, Pellissippi State Community College and University of Notre Dame, Knoxville, US, October 29–30, 2021; "Narratives of Tentacularity and Debate of Non-human Agency in the Films "Penguin Bloom (2020)" by Glendyn Ivin; and "A Street Cat Named Bob (2016)" by Roger Spotisswoode," The University at Albany, 2 April 2022. His area of interest includes environmental humanities and posthumanism.

Shankha Shubhra Mandal is Junior Research Fellow pursuing his PhD in the Department of English Literature, Language and Cultural Studies, Vidyasagar University, Midnapore, West Bengal. His areas of research include energy humanities, climate literature and ecocriticism.

Asmae Ourkiya (pronounced Es-muh) (They/Them) is the founder of the Ecofeminist Institute, an initiative dedicated to providing consultancy and research services to individuals and organisations by providing holistic approaches towards addressing climate change and environmental challenges. Dr Ourkiya obtained their PhD degree from Mary Immaculate College, University of Limerick, after successfully defending their thesis titled "Queer Ecofeminism: From Binary Environmental Endeavours to Postgender Pursuits," which has been published as a book by Rowman & Littlefield. They have contributed to numerous platforms, books and magazines, disseminating their research and demonstrating a commitment to promoting social justice alongside environmental concerns. Their work is grounded in the belief that climate justice cannot be attained without addressing social justice, a concept that encompasses the interconnected objectives of the Sustainable Development Goals. As an important voice in the global conversation around sustainability, gender and human rights, Asmae continues to bridge different schools of thought to promote lasting, positive change. Asmae has over 10 years of experience as a writer, speaker, researcher, editor, mentor and consultant, and works towards bridging the gap between the humanities, climate politics and climate sciences.

Dimitris Papacharalampous is Head of the Audiovisual Literacy and Research Department of Hellenic Film and Audiovisual Center-Creative Greece. He has taught (a) *Cultural Heritage* and *Cinematic Cultural Representations (City, Body, Gender)* in Athens University at the graduate and post-graduate levels (2007–20022), (b) *Inclusive Education* in the University of Nicosia at the post-graduate level (2020–2021). He teaches *Cultural Management and Media*, *Representation of Media in Cinema* and *Representation of Masculinity in Cinema* in European Communication Institute at the post-graduate level (2020–2024). He is the author of the book *Ο τυφλός ντετέκτιβ. Film Noir και Οιδίπους Τύραννος/The Blind Detective. Film Noir and Oedipus Rex* and of the book articles: (a) "Ιατρική και Κινηματογράφος/Medicine and Cinema" in *Το Σώμα. Τέχνη και Ιατρική/The Body. Art and Medicine* (K. Rigopoulou ed.), (b)"Πολιτισμικές Αναπαραστάσεις της Ελληνικής Κρίσης στις Πολιτικές Γελοιογραφίες/Cultural Representations of Greek Crisis in Political Cartoons" in *Η Εικόνα της Ελλάδας/The Image of Greece* (K. Rigopoulou, ed), (c) "Το Βλέμμα του Αλέξανδρου.Η Αναπαράσταση (και Διεκδίκηση) της Μακεδονίας στον Κινηματογράφο/Alexander's Gaze. The Representation (and Claim) of Macedonia in Cinema" in *Μακεδόνων Χώρα.Πολιτικές της Μνήμης/Land of Macedons. Memory politics* (ed. I. Andreadis), and (d) "Cine-Inclusion in Class. Fostering Inclusive Film Literacy" in *Adapting* *to the Changing Tides. Media and Information Literacy New Trends and Challenges* (ed. K. Karpouzis).

Shifana P. A., a doctoral candidate in the Department of English Language and Literature at St Thomas College, Kozhencherry, delves into the intricate realms of humour studies, cultural studies and post-colonial history. With a keen eye for scholarly discourse, she has contributed significantly to the field through publications such as "Alternate History: Representation of Indo-Pak Partition in Political Cartoons," "Between Religion and Gender: Construction of Muslim Women in Barsa," and "The Incompetent Antagonism of Wit: A Study of Hamlet and Catch–22." Shifana's interdisciplinary scholarship extends to literature, gender studies and climate change communication, as evidenced by her diverse research topics and contributions to esteemed journals like the *Journal of Indian History and Culture*, *Malayalam Literary Survey* and *IAFOR Journal of Arts & Humanities*. Her research explores the multifaceted functions of humour across diverse contexts and its profound impact on social, cultural and political dynamics.

Saumya Priya is working as Postdoctoral Researcher in the Faculty of Social Sciences at Lund University (Sweden). Dr Priya has extensive teaching experience as a Lecturer in the Department of English. She has a keen interest in interdisciplinary research, especially in areas like environmental humanities, digital humanities, gender and film studies, diaspora studies, Francophone Studies and comparative literature. She has organised and attended a number of National/International conferences and has over 35 papers, book chapters, and invited talks to her credit. She recently co-edited a book titled *New Trends and Digital Adoption: Paradigm Shift in Higher Education.* Dr Priya is also on the editorial board of research journals like *ETKI Journal of Literature, Theatre and Culture Studies*, and *Shodh Sankalp Journal.* In addition to this, she is also an active member of the American Comparative Literature Association and the Danish Psychoanalytical Society. Dr Priya also has a degree in creative writing from the University of Kent (Paris) and is currently working on her first short story collection.

Paddaja Roy is Assistant Professor of English at Ratnapith College, Assam, India. She is also a Research Scholar in the Department of English, Bodoland University, Assam, India. Her areas of interest include Northeast writings, media studies, Indian writing in English, translation studies, creative writing and oral literatures. She is an avid reader and has also worked as content curator for Amenta Publications.

Matthew Spencer is a lecturer in the Department of English at Auburn University. His areas of speciality include American literature, Anglophone literature, ecocriticism, Environmentalist writing, and climate fiction. His work has previously appeared in *Postcolonial Interventions*, *Mississippi*

Quarterly, *Medical Humanities*, and the edited collections *Teaching Anglophone South Asian Diasporic Literature* (MLA, 2024), and *In the Crossfire of History: Revisiting Women's War Resistance Discourse* (Rutgers UP, 2022), among others.

Inna Sukhenko, PhD, is a visiting researcher of Helsinki Environmental Humanities Hub, the Department of Cultures, the University of Helsinki. Her current project is about the literary imaginaries of nuclear energy within literary energy narrative studies. After defending her PhD in Literary Studies (Dnipro, Ukraine), she has been a research fellow of Erasmus Mundus (Bologna, 2008; Turku, 2011–2012), Cambridge Colleges Hospitality Scheme (2013), SUSI (Ohio, 2015), Open Society Foundation/Artes Liberales Foundation (Warsaw, 2016–2017), JYU Visiting Fellowship Programme (Jyväskylä, 2021), PIASt Fellowship Programme (Warsaw, 2021) and iASK Fellowship Program (2023). Her general research interests lie within environmental humanities, energy humanities, literary energy narrative studies, ecocriticism, nuclear criticism, petroculture, nuclear fiction and energy ethics. She co-teaches/coordinates courses on nuclear narrative studies and Chernobyl studies at the University of Helsinki.

Will Underland is a PhD student in the English Literature program at the University of Mississippi. He also holds an MA in English from Middle Tennessee State University. His research interests include ecocriticism, film studies and contemporary American literature.

Indrė Žakevičienė started studies at Vilnius University, the faculty of Philology, in 1988. She has studied Lithuanian language and literature. She graduated from the university in 1993 and in the same year started her doctoral studies at Vytautas Magnus University in Kaunas. She maintained her doctoral thesis on the poetry of Vladas Šlaitas in 1998 and went on with her lectures on English-American literature at the same university. She was assigned to take the lead of the Research Center of Comparative Literature at Vytautas Magnus University in 2000 and was the head of the Center till 2012. From 2003 to 2009 she was the head of the Department of Lithuanian Literature. In 2003 she started her master's studies at the Department of Psychology (Vytautas Magnus University), in order to go on with the research in the field of interdisciplinary studies. She is deeply interested in literary translation and comparative studies, interdisciplinary aspects – ecocriticism, ecopsychology, cognitive poetics and ethical criticism.

Acknowledgements

This volume is the result of a Three-Day International Webinar on "Imagining Catastrophe: Literary and Cultural Representations of Environmental Disaster," which was organised by the Department of English and Internal Quality Assurance Cell (IQAC), Seva Bharati Mahavidyalaya, Kapgari, Jhargram, West Bengal, India, from 22 to 24 October 2021. The editors are grateful to the college authority and the academic community of the college for extending generous institutional support. The editors hereby express their cordial thanks to all the resource persons, chairpersons and participants of the Webinar for offering valuable comments, suggestions and insights, and for transforming the Webinar into a vibrant virtual platform for the exchange of knowledge. The present volume contains selected research papers presented at the Webinar and the papers invited from globally acclaimed scholars and experts in environmental humanities. The editors owe a great deal of thanks to all the contributors for their patience, cooperation and scholarly contributions.

As with all publications, this volume could not have been produced without the assistance of many individuals. We are indebted to Maaz Bin Bilal, Grace Cavalieri and Kiran Bhat for giving us cordial permission to quote from their poems in the Introduction to the book. We are also indebted to Prof. K. Satchidanandan, Vinita Agrawal and Kiriti Sengupta for kindly introducing us to these poets.

The editors hereby take an opportunity to express their deep gratitude to their colleagues, fellow academicians and family members for their invaluable feedback and moral support. Special thanks to Dr Sk Tarik Ali, Assistant Professor of English, West Bengal Education Service, Hooghly Mohsin College, Chinsurah, Hooghly, West Bengal, India, for assisting us with thoughtful suggestions and insightful research ideas.

Finally, the editors would like to express their heartfelt thanks and gratitude to the editorial team and all others associated with the Routledge, for their guidance and assistance in bringing the volume to publication.

Introduction

As we write this Introduction to the *Imagining Ecocatastrophe*, the COVID-19 pandemic is no longer a crisis – no longer a catastrophe. The lockdown phase is over along with its "new normal" regulations, and we are back to our pre-COVID lifestyles. The word "catastrophe," when we look at its etymology, implies a sense of "an 'overturning' or 'sudden turn' in individual or cultural consciousness" (Parham 3). COVID-19 was indeed a sudden global turn that baffled us, and we asked questions that had never occurred to us before:

Who spread it? Where'd it come? Is it not trite: who did it?

Where we go from here may scream to the skies
of the bravery of all those who fight who did it.

Nature's self-correction, the sins of the fathers,
Did a god will this to set aright who did it?[1]

The lines borrowed from Maaz Bin Bilal's pandemic poem "Who Did It?" shed light on both "individual" and "cultural consciousness." The reference to the "self-correction" of nature, the allusion to a divine will to "set aright," "the sins of the fathers" are all part of a debate that is very much relevant in today's context. Nothing goes wrong in nature. Nature has her own mechanism to correct the order if it is jeopardised. Does the poet imply this idea when he speaks of "Nature's self-correction"? What are "the sins" that our fathers or our forefathers or people before them committed? Is the word "sins" confined to the religious discourse alone? Or does it have other shades of meaning which are central to the discussion on ecocatastrophe? *The New Shorter Oxford English Dictionary* describes "ecocatastrophe" as "an occurrence of major damage to the natural environment, especially where caused by human activity." This may be a point of beginning for us as we introduce the idea of ecocatastrophe for the present volume.

DOI: 10.4324/9781003602750-1

In the context of climate change and environmental disorder, "Anthropocene" (coined by Paul Crutzen) is indeed a buzzword today. It is not easy to define the concept of "Anthropocene," but one may say that among other things it indicates "the emergence of humanity as a dangerous geological force" (Clark 80). To borrow the words of Dipesh Chakraborty, the "geologic now of the Anthropocene has become entangled with the now of human history" (Chakraborty 36). In the last few hundred years man has destroyed forest cover, polluted the air with tonnes of greenhouse gases, poisoned waterbodies and disturbed biodiversity. In addition, to echo the words of Nnimmo Bassey, a celebrated poet from Nigeria, "Man's greed for natural resources has risen so high that to sate the taste man would need several earths."[2] Man's selfish and reckless ventures in the name of "development" and merciless exploitation of natural resources have made the earth vulnerable. It has made the world warmer "with a higher risk of catastrophes, a reduced ice cover, higher sea levels and a climate out of control" (Nayar, *Vulnerable Earth* 19). We bear witness to occurrences of natural disasters like flash floods, wildfires, cyclones, earthquakes, landslides and mudslides among others almost every day. The "Foreign" page of *The Telegraph* dated 13 July 2024 covers the news of two kinds of disaster in two different parts of the world with the titles: "Las Vegas heat breaks records" and "62 missing after Nepal landslides." For our present purpose, we would like to elaborate on the first event. The news reads:

> Across the West, large parts of California, Nevada, Arizona and Utah were under excessive heat warnings, indicating "extremely dangerous heat conditions". Officials believe that the heat may be to blame for more than 90 deaths reported in the West this month, though it could take months of investigations for authorities to determine an accurate death toll.
>
> More than a dozen high-temperature records were expected to fall, from the West Coast to the High Plains, where temperatures will soar. Some places in the deserts and valleys of California, Arizona and Nevada will once again hit the 43s and 48s Celsius.
>
> The Las Vegas Valley, home to about 2.3 million people, is under an excessive heat warning that runs until Friday night. Wednesday was the fifth straight day that the day time high reached or exceeded 46 C, a record.[3]

The news sends a chill down our spine and alerts us to the exigencies of global warming. We remember the observation of Kate Rigby in "Discoursing on Disaster: The Hermeneutics of Environmental Catastrophe": "Within our own grim historical hour, the most salient fact is that climate change is no longer on the horizon, and possibly preventable: it is already upon us" (24). The report that we have quoted earlier shows in unambiguous terms that climate change is "already upon us'. The questions that we would like to raise are: what is the role of literature and cultural media in the context of this climate crisis? Can literature alert us to the "grim historical hour"? Can literature save the planet from the impending doom? Pramod K. Nayar asserts:

> Literature's role – and more broadly, narrative's – has been central to not only the documentation of disasters but also the *imagining* of future disasters, alternative worlds and an environmental ethics that promises justice to human and the more-than-human.
>
> The literary imagination not only constitutes the cultural response to climate crisis but also determines the constructions of Nature, natural disaster, human adaptability and nonhuman encounters among other themes.
>
> (Nayar, *Vulnerable Earth* 2)

Nayar thus foregrounds the role of "literary imagination" in the context of representing natural disasters. This representation deals with the present and future forms of disaster, evokes the image of "alternative worlds," and gives a direction to "an environmental ethics" which includes among other things the entanglement of humans with the nonhuman other. This representation also speaks of "human adaptability" and human resilience in the face of disaster. Nayar here particularly refers to the role of narratives. Astrid Bracke in *Climate Crisis and the 21st-Century British Novel* (2019) treads a similar line of argument when she observes that "in a time of global climate crisis novels function as experimental spaces in which actual and imagined circumstances are played out, in which ethical and moral dilemmas are considered and in which the world can be understood" (7). These "experimental spaces" are the privileged spaces of novelists engaged in the exploration of climate crisis and the interpretation of the surrounding world.

In the chapter titled "The Novel" in *The Cambridge Companion to Literature in the Anthropocene* (2021), Bracke gives a separate space to "Flood Fictions" and explores how these works of fiction "constitute a broad cultural response to climate crisis in which literal and figurative floods are used to imagine the Anthropocene" (90). The argument that Bracke puts forth is clear. By the word "literal" she refers to "a familiar world submerged by extensive flooding" particularly where the readers can identify the connection between floods and "anthropogenic climate change" (90). Thus, one may be reminded of novels like Maggie Gee's *The Flood* (2004), Amitav Ghosh's *The Hungry Tide* (2004) and Clare Morrall's *When the Floods Came* (2015), among others. In Bracke's opinion, flood narratives are also concerned with the figurative depiction of flood "as a watershed moment of irreversible change and crisis." Examples of this kind of fiction, according to Bracke, include Margaret Atwood's dystopian trilogy *MaddAddam*, Emily St. John Mandel's *Station to Eleven* (2014) and Ali Shaw's *The Trees* (2016). After cataloguing these flood narratives under different categories, Bracke makes a significant statement that although some of them can be described as cli-fi, the climate crisis is not made explicit in most of the novels. This suggests that in the Anthropocene, "climate crisis is no longer exceptional but part of everyday life" (91).

In *From Apocalypse to Way of Life: Environmental Crisis in the American Century* (2004) Frederick Buell makes a similar point:

> No longer an apocalypse ahead, critical environmental problems and constraints help construct society's sense of daily normality. Far from going away, environmental crisis has become a regular part of the uncertainty in which people nowadays dwell.
>
> (xiv)

Buell's observation has a special significance in the context of environmental crisis in South Asia and the South Pacific. At one level of understanding, the people who always "dwell" in environmental "uncertainty" in these parts of the globe are the ecosystem people. These "people depend on the natural environments of their own locality to meet most of their material needs" (Gadgil and Guha 3). They belong to either the coastal regions or the forest areas or some other places which are always ecologically vulnerable. So, when a disaster happens, they undergo the worst kind of suffering. But in most cases, their plight remains scantily represented in narratives. In "Towards a Postcolonial Disaster Studies" (2015), Anthony Carrigan observes that "there has been scant consideration of disaster narratives and imaginative depictions more broadly in mainstream disaster research," and he lays emphasis on "exploring how postcolonial texts challenge, reject, or reconfigure key disaster studies concepts such as resilience, risk, adaption, and vulnerability" (123). Carrigan refers to Kenneth Hewitt in this context and shows how Hewitt emphasises the need to listen to "the plight and stories of distinctly more vulnerable members of society" and develop a discourse that speaks "of missing persons or unheard voices; of 'hidden damage' and 'shadow risks' and, more severely, of 'silent' or 'quiet violence'" (qtd. in Carrigan 123). We would like to briefly illustrate the ideas and arguments of Carrigan and Hewitt with reference to Kim Scott's novel *Benang: From the Heart* (1999) at this point.

Let us first briefly speak about the representation of the landscape in *Benang*. The land is barren and "riddled with . . . death" (55). In Thomas Hardy's *Tess of the d'Urbervilles*, we get the picture of "a starve-acre place" – Flintcomb Ash – where Tess and her companions work as migrant labourers in winter.[4] There is a suggestion of cruelty in the description of the weather. However, there is no idea of "death" anywhere. But in *Benang* one may find "Murder everywhere" in the portrayal of nature (484). While illustrating from the novel, Justyna Poray-Wybranowska admires the "word choice" of the author and states that it establishes "the connection between the systematic violence done to Indigenous bodies and the slow violence enacted on the landscape" (92). The novel indeed starkly portrays how the indigenous people are exploited by the ideology of the white settlers and their land and waterbodies defiled by the latter's crafty strategy of environmental racism. The bushfire that takes place towards the end of the novel is inevitable and perhaps that is why it is not portrayed in a spectacular fashion. Poray-Wybranowska gives an interesting viewpoint when she argues: "In a way, the fire is downplayed. It does not become the single most traumatic event the Mason/Coolman/Scat family has ever experienced" (95). But at the same time, it is true that the aborigines – the "distinctly more vulnerable members of society" (Hewitt 120) – only directly experience this disaster. There is a mention of a man who is left on the island when it is put on fire: "He feels the fire racing at his back, feels its heat" (470). Most probably he is trapped by fire, but the novel does not ascertain it. In this open text, the man thus represents those "missing persons or unheard voices" (Hewitt) who often remain neglected in the discourse of catastrophe. *Benang*, therefore, reconfigures the concepts of "risk" and "vulnerability" and adds a new edge to the framework of postcolonial disaster studies in literature.

Amitav Ghosh's *Gun Island* (2019) is not a postcolonial text in the sense Scott's *Benang* is. But Ghosh's novel also redefines various disaster studies concepts like "resilience," "adaption," and "vulnerability." It unfolds that the cyclone Aila did not only shatter the environmental harmony in the Sundarbans spreading over several parts of India and Bangladesh but the harmony in the familial space of the disaster-hit people too. The "Tipu" chapter of the novel deserves a special mention here. It shows the resilience of the youths who went to "faraway cities or even abroad" for employment opportunities. Secondly, it highlights the adaption of the aged who "resorted to begging" when there was no alternative left to them for survival. Thirdly, it sheds light on the idea of vulnerability as the women became prey to the traffickers and were placed in "distant brothels" (48–49). Amitav Ghosh's *Gun Island*, therefore, contains stories of micro-level catastrophes that happened to people who already belonged to the socio-economically disadvantaged groups. Aila was the centre of "catastrophic convergence" in their lives.

Christian Parenti, in *Tropic of Chaos: Climate Change and the New Geography of Violence* (2011), theorises the concept of "catastrophic convergence." Parenti defines this concept as follows:

> Climate change arrives in a world primed for crisis. The current and impending dislocations of climate change intersect with the already-existing crises of poverty and violence. . . . By catastrophic convergence, I do not merely mean that several disasters happen simultaneously, one problem atop another. Rather, I argue that problems compound and amplify each other, one expressing itself through another.
>
> (7)

The last sentence of the quotation hits at the very idea of "catastrophic convergence" where different "problems" including environmental, economic, social and political among others not only "amplify each other" but express each other in complex terms. One may remember a poignant poem by Nnimmo Bassey titled "Bottled Tears or Water Poem" from the anthology *I Will Not Dance to Your Beat* (2011) in this context. Here are a few relevant lines:

> Last night kids danced in acid rain
> Doubly warmed by infernal dragon tongues
> From gas flares
> Empty shells, lifeless sockets, death everywhere
> Forlorn men, backs broken, homes long gone
> Sit on benches of plastic bamboos
> Whistling for fishes from acid lakes
> Craving for mudskippers from tar ponds[5]

The quoted passage opens with an awful picture of environmental degradation where "acid rain" and "gas flares" are part of the everyday reality in the Niger Delta of West Africa. Innocent kids are thus doomed. The images of "infernal dragon tongues," "empty shells," and "lifeless sockets" suggest how death has spread its message "everywhere." The men are exhausted and weary. Their spines are "broken." The word "forlorn" is perhaps suggestive of the fact that they have been thrown out of jobs. They have been dislocated from their "homes" as well. The image of "benches of plastic bamboos" speaks of a sense of unhomeliness. As usual they whistle for fishes from lakes and crave for mudskippers from ponds knowing not that the waterbody has turned toxic. The lines of the poem build up a powerful discourse on the convergence of crises where ecological hazards are intertwined not only with health hazards but with the concerns of migration, dislocation and the problems of livelihood as well.

In the "Petrofiction" of the Niger Delta, too, the aforementioned crises are often represented in such a way that they "compound" and "amplify" each other. Helon Habila's *Oil on Water* (2010), Chimeka Garricks's *Tomorrow Died Yesterday* (2010) and Imbolo Mbue's *How Beautiful We Were* (2021) are leading examples in this regard. Environmental degradation, which unfolds through the invisible processes of "slow violence," fearfully impacts both the human and the nonhuman world and is obviously at the centre of the discourse here. One may be reminded of that section of *Oil on Water* where the novelist describes the ailing condition of Zaq, a veteran reporter:

> His shirt was wet under the arms and at the back. He was still fighting the sudden fever that had dogged him since we left Port Harcourt, and the more his health had deteriorated the more he had

> taken to philosophizing over almost anything: a bat flying overhead, a dead fish on the oil-polluted water, a gathering of rain clouds in the clear sky.
>
> (5)

The fever from which the reporter suffers and which ultimately results in his death is caused by the deadly pollution prevailing in the land. The water is polluted too. The reference to "a dead fish on the oil-polluted water" is unmistakable here. Habila thus portrays the human and the non-human entanglement in a precarious environment.

But the petrofiction of the Nigerian delta also portrays how the limitless exploitation of the poor host communities of the region by the multinational companies results in multiple vices like oil theft, kidnapping and killing of innocents among others. The environmental crimes (including "environmental racism") committed by the foreign companies on the one hand and the insensitivity and callousness of the Federal Government of Nigeria to the suffering of the masses on the other give birth to the emergence of the militants who show their courage in confronting the military force. At the end of *How Beautiful We Were* the dispossessed people organise a resistance movement against the company of Pexton at the cost of their lives. But the ending of the novel is far from being positive as the whole village of Kosawa (the fictional site in the text) became "too contaminated for human presence" and the government "ordered Kosawa burned" (355).

In her Introduction to *Postcolonial Disaster: Narrating Catastrophe in the Twenty-First Century* (2020), Pallavi Rastogi defines "Disaster" as "a cataclysmic event or process, either a singular occurrence or a concatenation of linked catastrophes in which the totalizing scale of its effects is important" (15). For this definition of disaster, Rastogi takes a cue from the anthropologist Anthony Oliver Smith, who suggests that "disaster is a collectivity of intersecting processes and events, social, environmental, cultural, political, economic, physical, technological, transpiring over varying lengths of time" (qtd. in Rastogi). The earthquake that hit several parts of Syria and Turkey in 2023 and claimed the lives of more than 28,000 people may be cited as an illustration of this point. The report published in *The Telegraph* dated 13 February 2023 (six days after the disaster) reveals that this cataclysmic episode (described as "the world's seventh deadliest natural disaster") involved "intersecting" events.[6] In the first place, the event of disaster immediately developed a "political" edge. We read in the report: "In Syria, the disaster hit hardest in the rebel-held north-west, leaving many homeless for a second time after they were displaced by a decade-old civil war, though the region has received little aid compared to government-held areas." However, the "European Union's envoy to Syria urged Damascus not to politicize issues of humanitarian aid." But apart from political issues, the disaster also involved ghastly socio-economic consequences. We read in the report: "On Saturday, Gizem, a rescue worker from the south-eastern province of

Sanliurfa, said she had seen looters in the city of Antakya." This speaks of the disruption of social order and human values in the aftermath of disaster. Another crucial dimension of this disaster was medical emergency and the problem of hygiene. The report reads: "With basic infrastructure in ruins, survivors feared disease. 'If people don't die here under the rubble, they'll die from injuries. If not, they will die from infection,' said Gizem. 'There is no toilet here. It is a big problem.'" So, the earthquake that shook various parts of Syria and Turkey was scarcely a single event; it combined events that transpired over varying lengths of time; it disrupted and even destroyed the normal way of living for those who were directly or indirectly affected by the disaster.

Haruki Murakami in the story titled "UFO in Kushiro" included in the volume *After the Quake* (2003) shows how the normal rhythm of a family was devastated by the occurrence of an earthquake. Right at the beginning of the narrative, we see the impact of the catastrophe on the psyche of a woman:

> Five straight days she spent in front of the television, staring at crumbled banks and hospitals, whole blocks of stores in flames, severed rail lines and expressways. She never said a word. Sunk deep in the cushions of the sofa, her mouth clamped shut, she wouldn't answer when Komura spoke to her. She wouldn't shake her head or nod. Komura could not be sure the sound of his voice was even getting through to her.
>
> (2)

The story alludes to the Great Hanshin Earthquake that occurred on 17 January 1995. It took a toll of more than 6000 people, mostly from Kobe, and damaged numerous buildings, elevated roads and rail bridges. As we read the story, we find that the woman (Komura's wife) "had no relatives who could have been hurt in Kobe" (2). Yet the impact of the disaster on her was terrible. She left her husband soon without a notice and never came back. Komura tried to handle the crisis in his own way. But the disaster made an ontological change in his life. He did not find peace in any corner. He tried to have sex with Shimao whom he recently met. The lady advised him "to lighten up and learn to enjoy life a little more" but his attempts failed. Towards the close of the narrative, he was haunted by the images of the earthquake:

> Images of it had come to him one after another, as if in a slide show, flashing on the screen and fading away. Highways, flames, smoke, piles of rubble, cracks in streets. He couldn't break the chain of silent images.
>
> (17)

The "chain of silent images" speaks of the circle of trauma in which the individual (the protagonist) was trapped.

In Rastogi's opinion, "Disaster terminates normalcy for the individual and for the social collective, often creating a new-normal condition for existence" (15). This observation, in our opinion, best describes the literature on the COVID pandemic with the reference to which we began this essay. Sarah Moss's *The Fell* (2021) is a fine example. The novel is set against the backdrop of the second phase of the lockdown in Britain when people were confined indoors day after day, as in other parts of the world, in order to ensure safety against the pandemic. The novel unfolds how this abnormal situation aggravates the psychological crisis of a woman in her personal domain. She cries out: "When did we become a species whose default state is shut up indoors?" (57). In order to get rid of this claustrophobic situation, she escapes one night into the world of nature. She achieves some sort of communion with nature's wilderness but meets a deadly accident as well. She barely survives. The novel, however, seems to contain a positive note where the protagonist makes a life-affirming utterance: "Life, then, to be lived, somehow" (180). This statement attests to "the fact of continuity of life" (Kucala 47).

During the lockdown-induced isolation, people accepted the continuity of life but sometimes they succumbed to a state of despair. This is poignantly reflected in Grace Cavalieri's "Isolation." The opening lines of the poem may be quoted here:

> At some point you will have to forget about being in all
> those houses
> different but the same with the neighbours singing songs all
> around you
> you always knew old friends were alive but now are only in
> memory
> the thought of memory
> the people who knew something about you
> you would not recognize them now and they may not recognize you.[7]

These lines shed light on the "new-normal condition for existence" during the pandemic. The poetic persona can't wholly reconcile herself to the altered state of reality. The houses are the "same" but at the same time "different" where the neighbours sing songs perhaps in order to relieve themselves of depression. The friends are "old" but in a period of continued isolation they are now confined to the domain of "memory." So, there is every possibility that the old friends won't "recognize" each other after a long period of time. The crisis that is represented in this pandemic poem is scarcely an imagined one; it was a lived reality for many of us across the globe when we were cocooned in our own private spaces, far away from family and friends, fighting against own experiences of trauma in 2020 and 2021.

Disaster literature cutting across genres is largely based on a lived experience, and it often speaks of a vivid sense of immediacy. Obviously, imagination has a crucial role to play here. In other words, imagination has a meaningful dialogue with the reality in disaster literature (particularly dealing with natural calamity), one terribly impacting the other. One may remember Kiran Bhat's poem "30th Century Babel" at this point. The opening lines of the poem are as follows:

> The oceans are curdling in flame.
>
> And I am at the bottom of it.
>
> So are the landscapes.
> So are the temples.
> So are the buildings.
>
> That was our world.
>
> We have been flooded over,
> And if there is not water on top of us,
> There is sand hotter than melting coal,
> Winds bouncing it around at a whiplash.[8]

The lines evoke a frightening picture of a natural disaster. Underwater volcanic eruptions, we know, can displace the ocean and cause tsunami waves in the coastal region which is close by. The opening line with a reference to "flame' is an indicator that the poetic persona imagines that he is near the centre where the eruption takes place. He is at the "bottom" of the waves along with the "landscapes" and the human constructions including "temples" and "buildings." The line "That was our world" reflects that the tsunami waves have washed away all the constructions on the surface of the earth. When the flood water recedes, there is no sense of relief because the winds then cause "a whiplash" and the sand "feels hotter than melting coal." The lines of the poem thus build up a terrible sense of immediacy. When one reads the poem, one may easily visualise what happens either at the depth of the water or on the seashore.

This sense of immediacy is also unmistakable in Romesh Guneshekera's *Reef* (1994) where the author portrays the picture of an impending disaster:

> When I looked up again I would glimpse the sea between the trees bathed in a mulled gold light. The colour of it, the roar of it, was overwhelming. It was like living inside a conch: the endless pounding. Numinous. You couldn't get away from it. . . . During those two nights we spent on tour I felt the sea getting closer; each wave just a grain of sand closer to washing the life out of us.
>
> (60)

In *The Great Derangement: Climate Change and the Unthinkable* (2016), Amitav Ghosh observes that climate change is largely absent from serious fiction of contemporary times. "It is as though in the literary imagination," he argues, "climate change were somehow akin to extraterrestrials or interplanetary travel" (9–10). In other words, climate catastrophes are often too "improbable" ("unthinkable"?) to be portrayed in stories of day-to-day life. It can be described as "a crisis of culture, and thus of the imagination" (12). But when we read between the lines from the novel quoted earlier, we feel that climate disorder can be captured by literary imagination, and, moreover, climate change is not always "akin to extraterrestrials or interplanetary travel." In the passage, the representation of the "roar" of the sea is made with the help of a powerful simile: "It was like living inside a conch." Immediately the "overwhelming" nature of the sea's roar is brought home to us. The image of the approaching sea: "each wave just a grain of sand closer to washing the life out of us" is thoroughly uncanny and even apocalyptic. It may induce in us "eco-trauma," which "draws attention to the *contingency* of life on earth" (Nayar, *Ecoprecarity* 3).

In his Introduction to *Ecoprecarity: Vulnerable Lives in Literature and Culture* (2019), Pramod K. Nayar acknowledges his debt to Anil Naraine for the use of the term "eco-trauma" (which Narine employed in connection with "Eco-trauma Cinema") and elaborates the idea: "It shows the human race as potentially open to destruction, as *species death* or loss of *species identity*" (3). In ecodisaster movies, the representation of eco-trauma is not a rare feature. One may think of films like *Take Shelter* (2011) and *The Road* (2012) which show that even predictions of a "climate apocalypse" or of the collapse of human civilisation due to climate-related catastrophes may lead to a kind of trauma in the victims.

A recent short film titled *Wade* (2020) hints at the destruction of the human race in a powerful audio-visual medium. The film is located in the largely drowned city of Kolkata where humans struggle with the predators from the Sundarbans for survival. The elements of horror and the uncanny are dominant in the film, and the message is clear. The planet we inhabit is fragile and life forms on the Earth are vulnerable. And this is largely due to the eco-hazardous activities in the Anthropocene. This message to a fair extent aligns *Wade* with central themes in the genre of "Ecocinema."

In "Shifting Paradigms: From Environmentalist Films to Ecocinema" Paula Willoquet-Maricondi writes:

> An important distinction . . . between "environmentalist" films and ecocinema is the latter's consciousness-raising and activist intentions, as well as responsibility to heighten awareness about contemporary issues and practices affecting planetary health.
>
> (45)

We thus get a couple of insightful points about "Ecocinema." But before we discuss these issues, we have to clarify what is meant by "environmentalist" films. In the Preface to the edited book titled *Framing the World: Explorations in Ecocriticism and Film* (2010), we read:

> While environmentalist films can have an impact on audiences' environmental values, the environmental themes in these films function primarily as backdrops to plot development, not as a call to action.
>
> (xi)

These delicate points of difference between environmentalist films and ecocinema to which Willoquet-Maricondi draw our attention are really interesting and instructive as well. Post-apocalyptic films like *The Day After Tomorrow* (2004) and *Flood* (2007) may therefore come under the description of environmentalist films because with their more or less spectacular effect they create "an impact on audiences' environmental values."

Post-apocalyptic films, however, are not always of the same type. They may create a sense of eeriness in the minds of the audience when they can't wholly recognise their homes or other familiar spaces on the screen. This is theorised as the "architectural uncanny" by Nayar (59). Films like *Book of Eli* (2010) or *I Am Legend* (2007), which Nayar cites as examples in this regard, defamiliarise urban spaces and represent a sense of horrid emptiness in streets and public spaces like "malls," "cinema houses," "parks" and "open but empty homes" (60). These movies are thoroughly dystopic in nature and although they have certain dominant elements of environmentalist films, they cannot be categorised as ecocinema.

Ecocinema promotes "environmental awareness," and it is often charged with "activist intentions." We would like to refer to two contemporary films at this point: Ravi Kumar's *Bhopal: A Prayer for Rain* (2014) and Peter Berg's *Deepwater Horizon* (2016). *Bhopal: A Prayer for Rain* deals with the horridness of an industrial disaster (which is absolutely man-made) and can very well come under the discourse of "the risk, the hazard, and the catastrophe" (Wallace 18–19). The movie contains certain scenes which are based on real-life incidents. The death of an employee of Union Carbide Corporation (UCC) due to a sudden exposure to chemical poisoning is one such incident. But the implication is that this episode can't be considered in isolation in the overarching context of a disaster. The Bhopal activists "repeatedly emphasized that the event is not singular or particular" (68). *Bhopal: A Prayer for Rain* successfully stages Wallace's idea of a "global risk" (64) and sends a message (which was highlighted in the accident-related pamphlet entitled *No Place to Return*) to the audience: "We should all be concerned: Bhopal is a lesson for us all" (qtd. in Wallace 68).

The discourse of "the risk, the hazard, and the catastrophe" is also at the centre of the box-office hit movie *Deepwater Horizon*. The film is based on the disastrous marine oil spill in the Gulf of Mexico in 2010 when the explosion took place in an off-shore semi-submersible drilling rig namely Deepwater Horizon. The disaster claimed human lives and the oil spill terribly impacted the lives of marine species like gulls, turtles, pelicans and herons among others. There is a scene in the film that focuses on the death of an oil-covered pelican. While reviewing the film in *The Guardian*, Benjamin Lee wrote: "It's spectacularly constructed, yet it doesn't forget about the loss of life, ensuring that, despite thin characterisation, the impact is felt".[9] One may very well say that *Deepwater Horizon* is close to the idea of ecocinema as it highlights the loss of life not only of humans but also of the nonhuman other due to an ecological disaster.

The present volume of critical essays covers multiple aspects of eco-environmental disaster (ranging from natural calamity to anthropogenic catastrophe, from climate crisis to nuclear disaster) as they are represented in largely contemporary literary texts and cultural media from different parts of the world like South Asia, East Asia, South Pacific, Western Africa, Eastern and North Western Europe and various regions of America. Ecocatastrophe and the Anthropocene disaster, apocalypse and the dystopian imaginaries, vulnerability, planetary precarity and resilience, imperialist politics, climate injustice and decolonisation, climate anxiety, tierratrauma and mental health are the major themes that come under the theoretical purview of the essays included here. The literary texts cut across various forms and genres such as novels, poems and narrative montages, whereas the cultural media range from movies to cartoons to digital games. The essays have been clubbed together in six sections under different headings that shed light on the thrust areas of the chapters of the book. However, it may be mentioned that the section-divisions are not rigid as the issues and concerns of the essays sometimes overlap.

The essay "'We Have So Little Time Left': Portrayal of Environmental Catastrophe in Selected Poems from *Reckoning*" by Joyjit Ghosh explores that poetry has its own role to play in the crisis of the planet Earth. The poems in the US-based magazine *Reckoning: Creative Writing on Environmental Justice*, Ghosh argues, often conform to the genre of nature writing of the Anthropocene in their marvellous portrayal of various forms of human-induced environmental degradation and disaster. They reflect that in our limitless thirst for prosperity and comfort we have systematically destroyed the ecosystem, and the life on earth is therefore uncertain and vulnerable. Environmental risks are omnipresent and climate imperilment is a lived reality. With a sense of urgency, the poems spread the idea of environmentalism and caution mankind against "the wrong kind of strength" that makes the planet inhabitable for not only humans but all other species of the earth as well.

Mahinur Gözde Kasurka's paper "Unmasking the Risks of Climate Change in Liz Jensen's *The Rapture*" studies the precarious condition of the humans on earth due to an assumed ontological superiority of the humans and their passion for controlling the future with advanced technology in the modern era. The author argues that this anthropocentric notion of modernity had placed mankind before an unpredictably catastrophic situation. Using the German sociologist Ulrich Beck's idea of "risk society," Kasurka seeks to examine Liz Jensen's novel *The Rapture* (2009) and demonstrates how the text sheds light on anthropogenic climate change by reflecting the fragile condition of the human in the face of apocalypse. In its attempt to find a survival strategy in the current scenario of ecological degradation, the essay seeks to establish the relationality between "risk society" and ecohorror by clarifying the possibility of "staying with the trouble" in Donna Haraway's sense of the term.

In her essay "Maja Lunde's *The End of the Ocean*: A Narrative of Climate Change and Environmental Crisis," Saumya Priya reads the Norwegian author Maja Lunde's novel *The End of the Ocean* (2017) from the ecocritical perspective, particularly how it represents the environmental issues like global warming, melting glaciers, scarcity of freshwater and the consequent ecological disruption caused by the anthropogenic climate change. Along with the representation of the disastrous consequences of climate change, the chapter also draws our attention to the multiple emotional and psychological problems experienced by climate refugees as a consequence of catastrophic environmental disruptions. Through a comparative analysis with the other eminent works on climate change novels, the author discusses how the novel sharply deviates from the previous novels of the same genre as the novel adds an ethical dimension to anthropogenic climate change, reflecting on the moral responsibility that the current generations bear towards the coming generations and future of the planet in general.

In "Subverting Anthropocentrism: A Critical Study of J. G. Ballard's *The Wind from Nowhere*," Bapin Mallick and Nikhilesh Dhar explore the historical occurrences and cultural practices that contributed to the growth of an anthropocentric worldview that subverted the notion of equality of all species, especially by creating dualistic and binaristic oppositions between human and non-human. Anthropocentrism, the authors argue, has proved to be the most dominant factor for the contemporary environmental crisis and anthropogenic climate change. The authors focus on J. G. Ballard's *The Wind from Nowhere* and examine how the novel predicts the vision of an eco-catastrophe, in the form of cyclonic winds, caused by human-induced global warming that renders the Earth unfit for humans to survive. The paper also examines the frightening images and implications of the disastrous environment caused primarily by the anthropocentric notion of the superiority of humans over nature.

"Kim Stanley Robinson's *The Ministry for The Future*: A Tale of Transcending Climate 'Catastrophism'" by Tushar Kanti Karmakar is a study of Kim Stanley Robinson's novel *The Ministry for the Future* (2020) in light of its potential for transcending the limitations of tales of climate "catastrophism." Karmakar argues that although Robinson's *The Ministry for the Future* employs some of the conventional tropes of the tales of climate "catastrophism," it ultimately transcends the limitations of this genre, firstly, "by attributing agency to the characters as well as organizations in his novel who try to bring in a change in the existing system in order to fight the climate crisis," and, secondly, "by having a utopian ending to the novel, he challenges the paralyzing apathy of the climate "catastrophism" tales." Thus, Karmakar's reading of the novel makes it evident how the novel challenges the limitations of the popular dystopian climate change narratives and mobilises climate change concerns in the current scenario.

Sk Tarik Ali's essay titled "'Climate Plague', Precarious Lives and Resilience in a Post-apocalyptic World: Vignettes of Vulnerability in Sequoia Nagamatsu's *How High We Go in the Dark*" explores the representation of a pandemic caused by an unknown virus unleashed from the Arctic region. The essay in its introductory part locates Nagamatsu's fiction in the tradition of pandemic literature and then sheds critical light on the portrayal of ecological uncanny in the vignettes of the narrative. How a pandemic outbreak can disrupt the system of biosecurity and make all forms of life on the planet Earth vulnerable is at the centre of Nagamatsu's fiction, and Ali through a theoretically informed reading of the text shows that the author here has blended the elements of pandemic fiction, climate fiction and speculative fiction into an organic whole to present montages of a post-apocalyptic world. The essay also reflects on the optimistic message of Nagamatsu's narrative that collective action and solidarity can help mankind overcome the feelings of loss, bereavement and agony in a world of climate peril and ecological uncertainty.

Elwin Susan John's essay "Precarious Selves, (Dis)abled Bodies and Post-apocalyptic Narratives" analyses the various ways in which disability has become a central trope to define the category of "human" in post-apocalyptic narrative. Through a study of the representation and intersection of the three broad categories in popular visual culture – the human subject, the discourse of disability and the narrativisation of the post-apocalypse – John's analysis unravels the vulnerability of human selves and the challenge in the positioning of the normative healthy body in post-apocalyptic narratives. The post-apocalyptic narratives reject the exclusivist, mainstream definitions of a healthy, able body and re-define them in inclusive ways. The reconfiguration of the discourse of disability, the author argues, is "not only a projection of humanity's imagination, curiosity and fear of the unknown, but also a fashioning of the novel contours of the body."

In the essay "Ecocatastrophe in the Literary Imagination: Confronting the Anthropocene through Narratives of Ecoprecarity from North-East India," Paddaja Roy dwells upon ecocatastrophe in the Anthropocene through a critical reading of selected fictional narratives from North-East India. The essay explores various nuances in the representation of ecocatastrophe in literary imagination. Easterine Kire's *Son of the Thundercloud*, Anuradha Sharma Pujari's *The Forest Wails* and Ankush Saikia's *The Forest Beneath the Mountains* have been closely read to show the prevalence of ecoprecarity and ecoanxiety in the context of ecodisaster and climate crisis. Roy posits that in the context of India's North-East region people have traditionally lived in harmony with nature, a connection that has been disrupted by anthropocentric domination and developmental intervention. Roy also addresses the disparity in the vulnerability to ecocatastrophes, drawing attention to how people in the margins are disproportionately exposed to ecodisaster. The essay also engages with how storying plays a crucial role in fostering hope for sustainability by inspiring people to imagine an optimistic future.

Mir Ahammad Ali's essay "Eco-Anxiety, Trauma and Resilience of the Dongria Kond Tribe of India: Locating the Literary and Cultural Responses of the Niyamgiri Movement in the Global Scenario" explores how the songs and narratives of the indigenous people of the Dongria Kondh tribe living on the hills of Niyamgiri express their voice of protest against mindless industrialisation, their grief and anxiety and also their power of resilience to fight against their own trauma. Ali locates the article in the global context of rapid urbanisation, "Eco-imperialism," particularly corporatisation of the earth's natural resources, chiefly forests, land and water, which inevitably lead towards an environmental apocalypse. The chapter makes an elaborate discussion on the issues of "Ecological grief," "Eco-anxiety" and such other allied terminologies of "Eco Psychology" and explores the interconnections between "Ecology" and "Mental Health" while discussing the psychological impacts of ecological degradation on the indigenous people of Niyamgiri regions.

Shankha Shubhra Mandal's essay titled " 'We Should Have Known Our Land Would Soon Be Dead': Resource Curse, Petro-capital Extractivism and Survival Environmentalism in Imbolo Mbue's *How Beautiful We Were*" reads the novel *How Beautiful We Were* as a petro-fiction to highlight the politics of resource extractivism and oil politics represented in Mbue's text. Mandal's essay looks into the representation of the resource curse, state-backed exploitation of indigenous people and the environmental injustice inflicted on the people of colour in Mbue's novel by reading it through the lens of Petrocriticism. Mandal sees how environmental racism is implicated in the politics of oil extraction in Africa. The essay also focuses on the environmental resistance movement of the local people affected by neocolonial intervention, highlighting how, unlike postmaterialist environmentalism, their concerns for the immediate environment stem from their lived experiences of environmental violence and the destruction of their culture.

Inna Sukhenko's essay "The Literary Dimensions of Pagan Spirituality in Fictionalizing the Nuclear Tierratraumatic Experience" makes a critical enquiry into the Soviet literary tradition, which employed pagan spiritualism as a necessary condition for survival in the aftermath of the Chernobyl nuclear disaster. Pagan spiritualism, stemming from the fundamental belief in the sacredness of nature and the spiritual meanings of natural cycles of birth, growth and death, along with the orthodox Christian beliefs in hope for protection and healing, and the Soviet atheistic policies were employed as an underlying recurrent pattern and essential narrative strategy for the writers in the radiation-contaminated environment and the post-traumatic society.

While the chapter focuses on Andrea White's *Radiant Girl* (2008), it outlines the literary imaginaries of implementing paganism, its beliefs and spiritual practices for physical/spiritual survival in the radiation-contaminated area via envisioning resilience of the nuclear "tierratrauma"-experienced communities, as well as facilitating the debates on the spirituality of radioactivity.

Anna Frieda Kuhn's essay titled "Some Things Are More Equal Than Others: Or, How to Read *On the Beach*" draws our attention to the often-neglected relationship between Cold War politics and anthropogenic climate change in literary Anthropocene scholarship. Focusing on Nevil Shute's post-apocalyptic Cold War novel *On the Beach* (1957), Kuhn's article explores how the clash for domination between the communist and capitalist systems, and scientific ambitions of the powerful nations for nuclear weapons resulted in the unprecedented ecological and geological disruptions, which had their far-reaching impacts on the planet. Kuhn places the article in the critical debate on the problematics of employing the Enlightenment-humanist-based conceptions of the "anthropos" and goes on to explore the limitations of the conventional classification system, associated with hierarchisation and competition, and metaphors used in the treatment of the environmental issues in the contemporary climate change novels.

Asmae Ourkiya's essay titled "Gender, Famine and Masculinities: An Ecofeminist Insight into the Irish Great Hunger" offers an ecofeminist reading of the Irish Great Hunger of 1840s to show the disproportionate impacts of this catastrophic disaster on women, men and children. Ourkiya focuses on the issues of gender and masculinities involved in the Great Irish Famine, arguing that it was not merely a natural calamity but a complex historical event influenced by the political, social and economic dynamics of the colonial era. Ourkiya sees the crisis as one caused by the failure of the colonial state to address the problems of a growing population and overdependence on potatoes as a food source. With references to the trafficking of teenage girls to Australia and the crisis of "breadwinner masculinities" during the famine, the essay engages with broader concerns surrounding the issues of gender roles and expectations dominant in Irish society during the Great Hunger. The essay also examines the gendered approach to relief measures taken during the disaster.

Indrė Žakevičienė focuses on the ecology of reading dystopia and the problem of literature's capacity to evoke a particular state of being, defined by Margaret Freeman as the experience of "felt life" in the article "The Ecology of Reading Lithuanian Dystopia: The Cases of *Dorandobongas* by Jurgis Volandas and *Ėko* by Valdas Papievis." In contrast to the Lithuanian rural narratives that are often associated with a specific pastoral space, events in dystopian Lithuanian narratives, such as *Dorandobongas* (2015) by Jurgis Volandas (1933) and *Ėko* (2021) by Valdas Papievis (1962), typically unfold in urban settings. The author employs the concept of "unimaginable," discussed by Jennifer Wenzel, and the idea of "imagination loop," presented by Tania Zittoun, Vlad Glăveanu and Hana Hawlina, for analysing Volandas' and Papievis' modes of writing. The article concludes with the assumption that specific dystopias like those by Volandas and Papievis have a greater effect on the reader than fantasy novels alluding to dystopian futures.

In "*Don't Look Up*: Political Satire Crashes into the Contemporary Disaster Film," Georgios Dimoglou demonstrates how Adam McKay's film *Don't Look Up* (2021) engages with the convention of eco-disaster film in order to satirise the socio-political lethargy towards climate change. By comparing the film with the previous comet killer narrative of disaster films, such as *Meteor* (1979, directed by Ronald Neame), *Armageddon* (1998, directed by Michael Bay) and *Deep Impact* (1998, directed by Mimi Leder), Dimoglou shows how McKay's film *Don't Look Up* sharply deviates from previous disaster films as it narrates the growing distrust of the people against government and corporations, primarily because they seem to ignore or downplay the existence of climate change and other global-scale threats, and examines the role of internet as a social platform for leaking government-suppressed news, for expressing public opinion in a cyber-democratic environment, and sometimes for spreading fake news and misinformation.

"The City That (Never) Dies: Film Noir Imagines the Urban Disorder, Disease and Disaster" by Dimitris Papacharalampous is an exploration of the genre of film popularly known as the "Film Noir." In the essay, Papacharalampous focuses on R. Polanski's film *Chinatown* (1974) and analyses how the film embodies the quintessential elements of the Noir city, intertwining mythic elements with themes of corruption, moral ambiguity and ecological disruption. The author's analysis of the film makes it evident how the film, while it depicts the urban settings with its symbolism of disorder, disease and disorder, brings to the fore multiple environmental problems and natural hazards, which make the human condition in urban civilisations extremely fragile and precarious. Through a comparative analysis of *Chinatown* with other popular Noir films such as *Notorious* (1946), *The Lady from Shanghai* (1947), *D.O.A.* (1949), *Atomic City* (1952) and *Split Second* (1953), the essay investigates the intricate symbolism of the Noir city and its connections to broader cultural representations, depictions of Nature and existential realities.

Deviating from the dominant mode of interpretation that considers Peter Weir's film *The Last Wave* (1977) as a psychological drama, Will Underland and Matthew Spencer's essay " 'Toward an Otherwise': Decolonising Epistemology and Ecology in *The Last Wave*" reads the film primarily as a narrative of decoloniality that deconstructs the rigid rationalistic forms of Western epistemology. The authors' reading of the film unravels the inherent repressive nature of the Western epistemology that is manifested through the practice of domesticating nature and the repressive attitude of the White citizens towards the aboriginal community in Australia. Christianity and the legal system, the authors argue, are the two major cultural institutions, as the film shows, to embody the rigidly rationalistic worldview. The authors' interpretation of the film makes it evident how Weir consistently depicts the Western rigid rationalistic worldview as "ensuring a return of the repressed in the form of ecological catastrophe and how such a position is consistent with contemporary discourse on both climate catastrophe and decoloniality."

Amit Mandal's essay titled "Scorched Earth and Precarious Existence: Representation of the Anthropocene in Bollywood Films *Kadvi Hawa* and *Jal*" engages with climate disaster and its adverse impacts on the marginalised people in the context of two ecologically vulnerable regions in India. Through a critical reading of two contemporary Bollywood feature films, Mandal shows how climate change-induced disasters disproportionately affect impoverished people, who suffer most from anthropocentric environmental exploitation. Mandal first situates the "Anthropocene" by highlighting India's vulnerability as a Global South country where the impacts of the climate crisis are more immediate and severe than in the developed countries. He then studies Nila Madhab Panda's *Kadvi Hawa* to highlight the plight of the farmers of the Bundelkhand region due to extreme weather events. Girish Malik's movie *Jal* has been analysed to address the problem of water crisis, which is one of the critical problems in India today. The essay points out the threat hydroxide poses to both human and animal life.

Laura Akers' essay "From One World to Another: Immersion in Digital Games and its Relevance for Climate Change" reviews the potential effects of video games on their players, focusing particularly on how digital games, otherwise intended for recreation, serve as a medium for communicating climate change concerns. Akers discusses how digital games engage gamers on emotional and intellectual levels, and examines the potential of climate change games like *Survive the Century*, *The Climate Game*, *Fate of the World* and *The Climate Trail* to foster understanding and awareness of climate change issues. The essay dwells on the role of video games in addressing climate anxiety through a critical study of climate change games like *Kibou – The Guiding Light* and *Garden Story*. Akers also shows that digital games based on the theme of climate crisis contribute to enhancing resilience among gamers to face the challenges of climate peril.

The essay titled "Geopolitics of Climate Change Cartoons: Exploring Everyday Resistance Through Visual Discourse" by Shifana P. A. and Asha Susan Jacob explores the effectiveness of political cartoons in climate change communication, particularly their role in subverting the mainstream representations and geopolitical perspectives on climate change. By employing James C. Scott's concept of "everyday resistance" as a theoretical framework, the essay examines the subversive mechanisms inherent in political cartoons depicting climate change. In geopolitics, the authors argue, the cartoons with their seemingly innocuous visual symbolism "function as potent instruments of resistance, illuminating overlooked issues, exposing hypocrisy, and stimulating dialogue and debate." The political cartoons destabilise established norms, subvert dominant narratives by amplifying marginalised voices and foregrounding alternative viewpoints, and, thus, "can challenge prevailing ideologies, critique authority figures, and spotlight social injustices in a manner accessible to diverse audience."

We hope that the present volume on the representation of ecocatastrophe in literary and cultural texts in the global context will make the readers aware of the climate urgency and at the same time bring home the role of literature and cinema and other cultural media to envision the routes to save the planet Earth from the brink of eco-environmental disaster.

Notes

1 The poem is included in *Singing in the Dark: A Global Anthology of Poetry under Lockdown*, edited by K. Satchidanandan and Nishi Chawla. Vintage: An imprint of Penguin Random House, 2020, p. 34.
2 Nnimmo Bassey, "Reclaiming Our Humanity, Our Memory." *I Will Not Dance to Your Beat.* KRAFTGRIOTS: A Literary Imprint of Kraft Books Limited, 2011, p. 8.
3 Alan Blinder and Isabella Taft, "La Vegas Heat Break Records." *The Telegraph*, Calcutta, 13 July 2024, p. 2.
4 Thomas Hardy, *Tess of the d'Urbervilles.* Edited by Scott Elledge. Norton Critical Edition, 1991, p. 223.
5 The lines are quoted from "Bottled Tears or Water Poem," included in *I Will Not Dance to Your Beat*, p. 55.
6 "Quake Toll Tops 28,000," *The Telegraph*, 13 February 2023, p. 2.
7 The poem is included in *Singing in the Dark*, p. 44.
8 The poem is included in *Count Every Breath: A Climate Anthology.* Edited by Vinita Agrawal, Hawakal, 2023.
9 www.theguardian.com/film/2016/sep/14/deepwater-horzion-review-mark-wahlberg-toronto-film-festival.

Works Cited

Bassey, Nnimmo. "Reclaiming Our Humanity, Our Memory." *I Will Not Dance to Your Beat.* Kraftgriots: A Literary Imprint of Kraft Books Limited, 2011.

Bhat, Kiran. "30th Century Babel." *Count Every Breath: A Climate Anthology*, edited by Vinita Agrawal. Hawakal, 2023.

Bilal, Maaz Bin. "Who Did It?" *Singing in the Dark: A Global Anthology of Poetry under Lockdown*, edited by K. Satchidanandan and Nishi Chawla. Vintage: An Imprint of Penguin Random House, 2020.

Bracke, Astrid. "The Novel." *The Cambridge Companion to Literature and the Anthropocene*, edited by John Parham. Cambridge University Press, 2021.

Buell, Frederick. *From Apocalypse to Way of Life: Environmental Crisis in the American Century.* Routledge, 2004.

Carrigan, Anthony. "Towards a Postcolonial Disaster Studies." *Global Ecologies and the Environmental Humanities: Postcolonial Approaches*, edited by Elizabeth DeLoughrey et al. Routledge, 2015.

Cavalieri, Grace. "Isolation." *Singing in the Dark: A Global Anthology of Poetry under Lockdown*, edited by K. Satchidanandan and Nishi Chawla. Vintage: An imprint of Penguin Random House, 2020.

Chakrabarty, Dipesh. *The Climate of History in a Planetary Age.* Primus Books, 2021.

Clark, Timothy. "Nature, Post Nature." *The Cambridge Companion to Literature and the Environment*, edited by Louise Westling. Cambridge University Press, 2014.

Gadgil, Madhav and Ramachandra Guha. *Ecology and Equity: The Use and Abuse of Nature in Contemporary India.* Penguin Books India, 1995.

Garrick, Chimecka. *Tomorrow Died Yesterday*. Paperworth Books, 2010.
Ghosh, Amitav. *The Great Derangement: Climate Change and the Unthinkable*. Penguin Books, 2016.
———. *Gun Island*. Hamish Hamilton, 2019.
Guneshekera, Romesh. *Reef*. Penguin Books India, 2014.
Habila, Helon. *Oil on Water*. Penguin Books, 2011.
Hardy, Thomas. *Tess of the d'Urbervilles*. Edited by Scott Elledge. Norton Critical Edition, 1991.
Hewitt, Kenneth. "Sustainable Disasters? Perspectives and Powers in the Discourse of Calamity." *Power of Development*, edited by Jonathan Crush. Routledge, 1995.
Kucala, Bożena. "Walking to Stay Alive: Sarah Moss's Lockdown Novel *The Fell*." *American and British Studies Annual*, vol. 15, 2022, pp. 37–48, www.researchgate.net/publication/366483206_Walking_to_Stay_Alive_Sarah_Moss's_Lockdown_Novel_The_Fell.
Mbue, Imbolo. *How Beautiful We Were*. Canongate, 2021.
Moss, Sarah. *The Fell*. Picador, 2021.
Murakami, Haruki. "UFO in Kushiro." *After the Quake*. Vintage, 2003.
Nayar, Pramod K. *Ecoprecarity: Vulnerable Lives in Literature and Culture*. Routledge, 2019.
———. *Vulnerable Earth: The Literature of Climate Crisis*. Cambridge University Press, 2024.
Parenti, Christian. *Tropic of Chaos: Climate Change and the New Geography of Violence*. Nation Books, 2011.
Parham, John. "Introduction." *The Cambridge Companion to Literature and the Anthropocene*, edited by John Parham. Cambridge University Press, 2021.
Poray-Wybranowska, Justyna. *Climate Change, Ecological Catastrophe, and the Contemporary Postcolonial Novel*. Routledge, 2021.
Rastogi, Pallavi. "Introduction." *Postcolonial Disaster: Narrating Catastrophe in the Twenty-First Century*. Northwestern University Press, 2020.
Rigby, Kate. "Discoursing on Disaster: The Hermeneutics of Environmental Catastrophe." *Tamkang Review*, vol. 39, no. 1, December 2008, pp. 19–40.
Scott Kim. *Benang: From the Heart*. Fremantle Arts Centre Press, 1999.
Wallace, Molly. *Risk Criticism: Precautionary Reading in an Age of Environmental Uncertainty*. University of Michigan Press, 2019.
Willoquet-Maricondi, Paula. "Preface." *Framing the World: Explorations in Ecocriticism and Film*, edited by Paula Willoquet-Maricondi. University of Virginia Press, 2010.
———. "Shifting Paradigms: From Environmentalist Films to Ecocinema." *Framing the World: Explorations in Ecocriticism and Film*, edited by Paula Willoquet-Maricondi. University of Virginia Press, 2010.

Part I

Anthropocene, Ecocatastrophe and Apocalypse

1 "We Have So Little Time Left"[1]

Portrayal of Environmental Catastrophe in Selected Poems from *Reckoning*

Joyjit Ghosh

Our engagement with the planet earth has turned toxic. We have defiled and degraded the earth. Thames is no longer "sweet" as it was in Spenser's day, writes T.S. Eliot in the early twentieth century. In the twenty-first century, writers engage themselves with the question of "post nature" (which involves an inescapable feeling of uncertainty and fragility), even with what Bill McKibben describes as "the end of nature."[2] We remember McKibben's ominous words articulated in *Eaarth: Making a Life on a Tough New Planet* (2003): "The planet on which our civilization evolved no longer exists. . . . The earth that we knew – the only earth that we ever knew – is gone" (27). We have systematically destroyed the ecosystem. Our craving for more and more prosperity and comfort, and our obsession with the idea of "development" pose a threat to the Amazon rain forest or to the mountainous region in Uttarakhand or to the Sundarbans, a huge mangrove area, spread over parts of India and Bangladesh, to cite a few examples. During the worldwide forced confinement due to the COVID pandemic, we experienced blue vapourless skies. The song of the earth was heard. Wild birds were seen moving in the streets of big cities. Leading articles show that global CO_2 emissions decreased considerably. But this clean air was "only a footnote in the narrative of environmental catastrophe."[3] Indeed, immeasurable changes have begun, and they impact us directly. Sometimes we have a feeling, "perhaps this might be your last century," and so we have to prepare our "trembling lungs" and our "hummingbird hearts," to echo the words from Ashley Bao's well-known poem "when the coral copies our fashion advice."[4] The question of "Anthropocene" becomes important at this point because "beyond its strictly geological reference it becomes the one name for a slow but massive shift in the tectonic plates of human self-conceptions" (Clark 80). Clark speaks of "a sense of the plurality," "unpredictability" and "compromised condition of the natural world" (80). But along with the "compromised" condition we have to speak of the chaotic condition of the natural world as well. Earthquakes, cyclones, tsunamis, flash floods, forest fires and other forms of environmental disaster have become a part of everyday reality across the globe.

DOI: 10.4324/9781003602750-3

When we engage with the literary/cultural representations of ecodisaster, fiction and films obviously occupy a major part of the discourse and poetry occupies a marginal space here. To probe the reason behind it is beyond the scope of this chapter. In this context, one may remember the words of Dipesh Chakrabarty

> an essential ingredient of the process by which humans make sense of crises in public life – or feel inspired to work towards solutions – is stories: narratives we tell ourselves in order to find our bearings in a new situation. . . . Our success in developing a globally concerted response to the climate crisis, for instance, will depend on the degree to which we can tell stories that we can all agree on.[5]

So, the value of stories/narratives in forming "a globally concerted response" to environmental crisis/catastrophe is inestimable. But one may argue that poetry can also perform its distinctive role in this regard. For example, in its capacity as a ballad, Coleridge's *The Rime of the Ancient Mariner* also tells a "story" – story of a mariner's wilful killing of a bird of good omen and how that killing brings in a disaster when the ship is carried north into the tropic seas where it is becalmed and how all the mariners drop down dead one by one and finally how the ancient mariner achieves redemption with a vision of the creatures of God. The poem at the end contains a moral ("He prayeth well, who loveth well / Both man and bird and beast") with which we can all agree. So, poetry in its own inimitable way can address the topic of climate change and make us aware of ecological disasters. But can poetry save the climate-imperilled earth? This is exactly the question that is raised by John Felstiner in his book *Can Poetry Save the Earth?: A Field Guide to Nature Poems* (2009). In the Introduction to the book, Felstiner quotes Rachel Carson's statement in *Silent Spring* (1962): "The aim of science is to discover and illuminate truth. And that, I take it, is the aim of literature." In Felstiner's view, poetry has a significant role to play in the crisis of the planet earth. "Science, policy, and activism point the way toward solutions," Felstiner argues, "but something deeper must draw us there." To echo Felstiner, "It can be found in poetry's musical lift, attentive imagery, and shaping force, which stem from prehistory and live on in today's magazines, slim volumes, readings, slams, songs, Web sites, blogs" (13–14). The *Reckoning: Creative Writing on Environmental Justice* is such a magazine (https://reckoning.press) that may come to one's mind at this juncture. In the present chapter, I will read selected poems from different issues of this US-based online magazine to explore how they represent environmental catastrophe and give a warning to mankind to protect not only themselves but also the flora and fauna of the earth from extinction.

The title of the chapter contains a precise reference to the title of D. Dina Friedman's poem "We Have So Little Time Left." The opening of the poem speaks of a sense of catastrophe:

> Already, the sunlight is shrinking like an old shirt
> That barely covers the belly, even while it glows

> Gilding the dried up cattails, the snapped branches
> That pierce the cloudless sky like a severed bone. (1–4)

The lines have a tremendous sense of immediacy, and they are amazing for their "attentive imagery," to borrow Felstiner's phrase. The image of dwindling sunlight resembling "an old shirt" that gradually shrinks is marvellous. It gilds the upright cattails that are getting "dried up" day by day, while the "snapped branches" resembling "severed" bones "pierce" the "cloudless sky." We have thus a picture of an arid land where everything withers away at a fast pace. The sense of despair is accelerated as we move through the poem and at a certain point the poem sounds apocalyptic: "The air will sizzle, and some far away bed / of ice will implode, or simply drip." The lines strikingly bring home the effect of climate warming on the sheet of ice in the "far away" region – the Antarctica. The poem ends on a note of gloom:

> But tonight, like every night,
> the sun will set in its predictable pattern, cutting off
> another sliver of our lives. And we might crawl
> into bed with a cup of tea, a fantasy
>
> story while fat squirrels scutter up snags
> of what once upon a time were trees. (19–24)

In *Nature Writing of the Anthropocene* (2017) Christian Hummelsund Voie speaks of "a dystopian possibility of regarding every landscape as a potential landscape of fear" (25). Voie to an extent is indebted to Stacy Alaimo's idea of "landscapes of strange agencies" (*Exposed: Environmental Politics and Pleasures in Posthuman Times* 78) in this context. The landscape that the poet portrays here emerges as a terribly vulnerable one and fills us with a feeling of terror as the squirrels are left with only the "snags" – the dead trees – to "scutter up." These snags are the "strange agencies" that compel us to identify ourselves with the wider environment.

The poems in the *Reckoning* often poignantly represent the slow violence caused to environment. This is evident at the very opening of Chloe Clark's "Sidelong Catastrophe":

> I'm not sure who the sky is
> when it's not the sky
>
> but I think I know this river
> was once a beautiful woman
>
> viewed from above all bodies
> of water look like someone
>
> you once loved . . . (1–7)

This is how the poet obliquely hints at human-induced environmental degradation. There is everywhere a touch of defamiliarization: the sky is unrecognisable and so is the river. The river "was once a beautiful woman" (a telling ecofeminist idea), but the poet no longer finds an affinity with her. The poet can look at the sky, but he will miss the familiar face. The concluding lines of the poem carry an ominous note:

> and I'll miss the sun most
> when the clouds weep the ghosts
>
> of rivers for days on end (21–23)[6]

So, the disaster is imminent. The image of clouds weeping "the ghosts of rivers for days on end" signals to a catastrophe. These lines in their tone and spirit offer a point of stark contrast to those in Shelley's "The Cloud":

> I am the daughter of Earth and Water,
> And the nursling of the Sky;
> I pass through the pores of the ocean and shores;
> I change but I cannot die. (73–76)[7]

Shelley's Cloud is an immortal agent of nature. It changes its form, but it is never dead. Shelley's "unification of joy in the living thing and interests in the principle of life is found in here."[8] But when we read Clark's poem, we catch a note of warning that derives its force from the author's awareness of "the human's doubled role in the Anthropocene . . . as both perpetrator of environmental decline, and victim of environmental decline" (Voie 25). Mankind has polluted the land and the water through ages and is now at the mercy of environmental degradation.

This is also reflected in Mohammad Shafiqul Islam's "Rivers Lament." The poet gives a voice to the agony of the rivers:

> O agonized rivers
> You don't shout like human beings absolutely
> Vacuum within! You're calm like trees gifting
> Nature with gentle breeze. Desperate and evil
> Land grabbers ravage banks extending gluttony
> To the rivers, eating out life, destroying beauty
> And disturbing cadence like Grendel. (9–15)[9]

A river does not shout. It remains calm even when it is slowly destroyed. The allusion to Grendel, the monster in *Beowulf*, the Anglo-Saxon epic, is very much apposite here. Grendel ravaged the mead hall of Heorot and killed its inhabitants from time to time before his mortal defeat at the hands of Beowulf, the eponymous hero. The gluttonous land grabbers remind the poet of Grendel. In an interview with Michael J Deluca, the poet clarifies the point at issue: "Those who grab riverbanks, grab land, destroy rivers, devastate nature and torture human beings illogically are modern Grendels."[10] So, Grendel is imagined as an agent of "environmental devastation" in the poem. The "golden days" are over when the "rhythmic sounds" of water "made many a shoal of / Small White fish swim, jump and fall in between / Giant boulders." The water has become toxic now. It spreads diseases "making distance / Between humans and rivers." While discussing Ryan Walsh's poem titled "Wild Perfection," included in *Reckonings* (2019), Scott Slovic shows how poetry emerges as a "mental divining rod" (borrowing Sandra Steingraber's phrase) that may enable "readers to understand the difference between the places we dream of and the places we actually inhabit" (Slovic 159). One may say that Shafiqul Islam's "Rivers Lament" also acts as a mental divining rod that draws the readers' attention to the difference between an almost idyllic world when water was clear and life was rhythmic and the present world when touching water is "to catch incurable diseases."

In the Anthropocene, we disown the Romantic idea of "Return to Nature" as nature does not remain pristine. Nature can no longer be identified with W.B. Yeats's "The Lake Isle of Innisfree," where the city-dweller wants to build "a small cabin" far away from the madding crowd:

> I will arise and go now. For always night and day
> I hear water lapping with low sounds by the shore;
> While I stand on the roadway, or on the pavements grey,
> I hear it in the deep heart's core. (9–12)[11]

The use of the adjective "grey" in the phrase "the pavements grey" is loaded. It suggests the lifelessness of the city as opposed to the "purple glow" that prevails in the domain of pristine nature. But the same adjective "grey" is used in an altogether different sense in Adelia MacWilliam's "Paddling in the Sound" where the cry of a loon is compared to "a soliloquy from the sea's grey throat."[12] The use of the colour "grey" in this context suggests the changed nature of the natural world and hints most probably at ocean acidification. As we go through the poem, we find that there is a lack of "certainty" in the poet's "heart" as she listens to "each note" of the bird. This is truly characteristic of eco-poetry in the Anthropocene as it often reflects a sense of unpredictability and vulnerability.

In *Hyperobjects: Philosophy and Ecology after the End of the World* (2013) Timothy Morton describes the present age as the Age of Asymmetry (2). In Morton's view, we now experience a turn of events when the normal certainties are inverted, or even dissolved. The "Hyperobjects" are entities "massively distributed in time and space" that confound conventional notions of what a thing is in the first place. Black hole, for example, is a hyperobject. These objects involve profoundly different temporalties than the human scale ones we are used to. This obviously necessitates a new thinking/writing style

(5). In "Earthspun" Krista Hoeppner Leahy gives a powerful expression to the idea of inversion of normal certainties in the natural phenomenon:

> A curvy-topsy tornado funnelled up, not down
> uprooting the sky-trees from the cumulus-sea-above
>
> dumping them on the ground where they are all
> dying from lack of clouds and too much media attention. (1–4)

Tornados appear in different shapes and sizes. They are often visible in the form of a condensation funnel originating from the base of a cumulonimbus cloud, or in rare cases, the base of a cumulus cloud. The tornado that the poet portrays here uproots "the sky-trees" from an imaginary upper region where the sea and the cloud merge. The uprooted "sky-trees" are then dumped on the ground where they die "from lack of clouds." The representation of an environmental calamity thus adopts a new thinking/writing style. It is asymmetrical and weird. The poet asserts, "There is no app to map a sky-fallen forest." But the informed reader will immediately intervene and state that what the app fails to map, the poet here catches with her extraordinary imagination:

> Several have melted completely into the ground
> leaving behind no trace but an ineffable sense of loss
>
> nearby if you're walking barefoot, with or without an Iphone.
> Some, arboreally brave, linger even as they disintegrate –
>
> branches split, leaves torn, roots weeping past reflection . . . (9–13)[13]

The lines evoke vivid images of disintegration in the realm of nature. The expressions like "branches split" and "leaves torn" will at once catch one's attention. The last line in the quote with its poignant touch of pathetic fallacy ("roots weeping past reflection") appeals to our deeper sensibility.

The poems in *Reckoning* often conform to the genre of nature writing of the Anthropocene in their apocalyptic portrayal of various forms of man-made environmental disorder. Man has disturbed the harmony in the ecosystem both on land and in water for centuries. But at the moment of environmental crisis, the sense of identification with the flora and fauna may only help. The alternative is to suffer from nightmarish stress and trauma. This is what we find in Jennifer Mace's "letters from the ides":

> I try to think of us as coral. (19)
> . . .
> even as polyps retreat
> against the coming storm,

we breathe and breathe and
breathe. creatures, for once,
made flesh; and in that, unified.
. . .
I try to think of us as coral
because the alternative
is to wake in the night
as red blood cells, as marrow (29–32)
. . .
while the skeleton of Seattle
struggles on without us – because
if there is anything to learn
from the rot and rent of centuries,
it is that even bones can crumble,
given time. (35–40)[14]

While speaking about "eco-trauma" Pramod K. Nayar states that it "draws attention to the *contingency* of life on earth" and "shows the human race as potentially open to destruction, as *species death* or loss of *species identity*" (3). The "skeleton of Seattle" evokes an image of eco-horror and sheds light on the vulnerability of human life, the planet earth and species identity. The poet believes that "we are none of us hermetic, none immutable under strain." And this fear exists with us as "a nebulous companion." However, as the poem marches to its close we find that the poet wants to conquer her fear ("I *will* not be my fear"), trusts her "courage" and "patience" and attaches importance to "a love for one another." This gives a positive dimension to the poem.

But as we read Adam Stemple's "The World Ended in Ice," we find that there is no escape route for mankind when the catastrophe takes place. One must confront what is happening around him/her at the moment. The poem may be quoted in full:

The world ended in ice
as scientists predicted
lab coat Noahs
ignored
as they counted the animals
two by two
into their ark
and lifted off

We who are left
shiver
in the cold
in the dark

in our ignorance and greed
forgotten sinners
with no dove
to fly from Ararat (1–16)[15]

Robert Frost's "Fire and Ice" opens with the lines: "Some say the world will end in fire, / Some say in ice." The poem is thus a speculation on the end of the world. The real charm of the poem however lies in its use of epigrams that distil the essence of human wisdom. In Frost's imagination "desire" is the "fire" whereas "hatred" is the "ice." In the imagination of Adam Stemple, the "cold" and the "dark" represent the doom of mankind as he is possessed by both "ignorance" and "greed." The poem is thoroughly bleak in tone: it is poised between an allusion to the past ("The world ended in ice") and a warning regarding the apocalyptic future. The use of the Biblical myth of Genesis is quite radical here. The "lab coat Noahs" scarcely take note of what is happening around them in the world of nature. They commit grave mistakes. The consequences are awful. Mankind has no saving grace. He rots in the dungeon of despair. No "dove" comes to his rescue. He is bound to Ararat – the snow-capped dormant volcano.

Kelly Madden's "Will We Be Good and Kind at the End"[16] is another noteworthy poem in this context. It poignantly addresses the question of human responsibility to fellow creatures at the time of environmental disasters. The poem exhorts mankind to be compassionate to the caged animals like macaques, chimps, baboons, dogs, rabbits and others and release them from the labs when the winds become turbulent and "buildings begin to buckle/at the knees." This obviously evokes an image of the earthquake and the contemplation on it strikes terror in our hearts. The poem also contains a picture of tsunami when "the sea's belly" threateningly swells up, leading humans to "grab hold of anything" for their survival.

The tremendous sense of desperation in the time of disaster is felt when we read the poem. The images used by the poet are, to borrow the ideas of Timothy Clark, not only "bizarre" but also "deeply counterintuitive" because they represent the devastating forces of nature with an "artistic implication" that entails "a deep suspicion of any traditionally realistic aesthetic" (Clark 81).

The poems in *Reckoning* have, however, another prominent feature of nature writing in the Anthropocene, and that is, they reveal what Yi-Fu Tuan describes in *Landscapes of Fear* (1979) as "the paradoxical fear that plants and animals, even rivers and lakes, may die through human abuse." Tuan adds that the "fragility of nature, not its power, now makes us almost constantly anxious" (212). One may remember Seth Garcia's "Displaced Water" in this context:

over 60,000 displaced
in Sri Lanka, two hundred dead and counting.
This is late May, 2017 AD. Flashbacks

of headlines some years back –
that heatwave which claimed some 2,000
in India but can this be more than read, felt
as the steadfast lamprey must feel, mooring
the stones of its own deathbed in the cool
lunar hollows? (7–15)[17]

These lines make an appeal to the reader to confront human impacts on the planet. Along with the poet we also raise our voice that this should be "felt" in the deepest core of our hearts rather than merely "read." The image of the "steadfast lamprey" confined to its "deathbed" in the waterless "hollows" creates in us a "paradoxical fear" that it is mankind that has led the creature to its cruel destiny, and moves us to "radical compassion." The power of the poem lies in its sense of urgency as we head "towards a future where hope / has long gone become obsolete." This sense of urgency is also prominently reflected in Danika Dinsmore's "How Can I Look Up":

Somebody
do
something
do
something
I want to scream
Let go of every device in your hands
And look up are we going to lose
The sky on our way to losing the sea

How can I leave space for us
To breathe (91–101)[18]

Gabriele Dürbeck in "Writing Catastrophes: Interdisciplinary Perspectives on the Semantics of Natural and Anthropogenic Disasters" explores how recent ecocritical research emphasises that we stand in the midst of environmental catastrophe. The argument of Dürbeck is that "risks are omnipresent and can be ignored only at our peril. The challenge is therefore how institutions and practices can be developed to avoid the most dangerous risks and a broader societal crisis" (Dürbeck 2). When we read the poem, we realise that we have to develop a practice of avoiding "every device" in our day-to-day use that cuts us off from our immediate environment. We have to do "something" to save our earth. We must not "lose" the sky and the sea. We have to leave the breathing space free. The poetic persona thus wants to "rekindle passion's poetry without falling into despair."

The root of "despair" is obviously located in environmental decline due to reckless human activities (largely anthropogenic) for the past three centuries like burning of fossil fuel and excessive use of energy causing an alarming increase of "greenhouse" gases, destruction of tropical rainforests impacting species extinction and awful increase in population among other things that push "the Earth into planetary terra incognita" (Steffen et al. 614). In Alicia Cole's poem "Corrupt the World With Drum" the sense of alarm is powerfully reflected:

Where will we go when we don't conceive
ourselves properly? Where will we go
when everything is overrun with the wrong
kind of strength? (32–35)[19]

The phrase "wrong kind of strength" is indeed loaded and seems to indicate the wrong direction in which mankind moves when he exploits natural resources for his own selfish interests and thereby degrades the environment. Stephen Collins, the Canadian poet and activist, describes the present era as "the era of Geophysical capitalism" when "*life itself* – all planetary biological material – is . . . subject to, and the substance of, the extraction of wealth." Collins coins the term "biotariat" to speak of "the enclosed and exploited life of this planet." It would be profitable if we remember F.J. Bergmann's poem titled "Development" at this point. We may quote the opening lines:

First, we made sure the world was devoid of sentient inhabitants.
Its turquoise skies grew greener (we thought) as summer waned.
On the wine-dark sand we drew diagram and planted pennants:
a cultural center here, a spa or hotel there, on an imaginary beach. (1–4)

A close reading of the lines quoted earlier reveals that the surface meaning of the poem serves as a mask of its real meaning. Is it really the development that the poet speaks of? Or is it environmental damage under the mask of development? The opening line of the poem is sarcastic. A sentient being is able to see or feel things through the senses. The people who draw profit-making plans of development on "an imaginary beach" (including the establishment of "cultural centre," "spa" or "hotel") ensure in the first place that there is no sentient inhabitant around them. The "wine-dark sand" thus serves as a "biotariat" within the rigid and dehumanised structures of capitalist exploitation. The poem, as we go through it, contains a note of warning:

At least one of the suns was always rising, and the light perfect
for filming, as if the planet were a vast, floodlight stage. . .
Our shielded clothing and the protective coatings on excavators,
fabricators, and constructors faded quickly under the incessant
illumination. (5–6, 8–10)[20]

The expression "one of the suns was always rising" most probably speaks of the artificial light that is arranged for the project work at night. The planet is defamiliarized and made into "a vast floodlight stage." But there is a potential risk involved in the project, as under the "incessant illumination," the "shielded clothing" and "protective coatings" wither away at a fast pace. This is indeed an awful signal. As we move to the end of the poem, we find that the "off-world concepts" lose their "relevance" and the completion of the project is "indefinitely postponed." The poem thus ends on a note of uncertainty. The title of the poem ("Development") is vastly ironic and alerts us to the emergence of humanity not only as an unpredictable agency but also as a reckless geological force.

In his article "Why Ecopoetry?" John Shoptaw emphasises, "An ecopoem is environmentalist not only thematically, in that it represents environmental damage or risk, but rhetorically: it is urgent, it aims to unsettle" (400–401). This is true of most of the poems included in *Reckoning.* They spread the idea of environmentalism in wonderful ways. One remembers those lines in Brandon O'Brien's "Papa Bois and the Boy," which staunchly critique the onslaught of modernity on the world of nature:

> The *mimosa pudica* closes her doors
> with each tremor of modernity drawing close. (27–28)

The lines beautifully rhyme with each other and evoke a poignant image of the dark side of modern mechanised life that pushes the concern for environment to the margin. The ending of the poem appeals to us without bordering on sentimentalism:

> "The city does
> forget easy. The woods can't."
> I want to live as long as you do,
> hand over hand, be a pleasant memory,
> 'til the city steps on the very last green. (32–36)[21]

The lines don't sound didactic yet delicately warn us against the curse of urbanisation when it goes unchecked and disturbs environmental harmony.

The poems in *Reckoning* often speak of a colossal sense of ecological loss. We echo the voice of Danika Dinsmore that "we are losing every day / pieces of our humanity / of green / of corals and bees / and owls and streams." As human beings, we are not immune to disaster. We are vulnerable like other forms of flora and fauna. The expression "pieces of humanity" is loaded and most probably alludes to the ecosystem people who suffer most due to environmental degradation as they live at the margins and belong to socio-economically disadvantaged groups or "politically less powerful groups."[22] The future, however, is not entirely bleak. In his interview with Erin Hoffman, Michael J DeLuca, the editor of *Reckoning*, states that the aim of this magazine is not merely to promote "new ideas and beautiful art on the subject of environmental justice"; its mission is to bring an "actual change in the world."[23] This "actual change" is possible, Michael J DeLuca believes, when all the authors and artists come together. We certainly agree with Michael and just add that the world is waking up and the concerted efforts of authors, artists and activists can usher in the desired change in the planet we belong to. So, hope has not yet gone wholly obsolete. We will cling to the last leaf. To sum up with the concluding words of Jennifer Mace's poem "letters from the ides":

> we will wait, and watch,
> and see what comes, and tomorrow, perhaps,
> the maple may bud, and perhaps
> we will see it
> with you. (85–88)

Acknowledgements

The author acknowledges gratitude to Michael J DeLuca, the chief editor of *Reckoning: Creative Writing on Environmental Justice* and the poets including Ashley Bao, Chloe Clark, Krista Hoeppner Leahy, Adam Stemple, Seth Garcia, Danika Dinsmore, F.J. Bergmann, Jennifer Mace, Alicia Cole, D. Dina Friedman, Brandon O'Brien and Mohammad Shafiqul Islam for their cordial encouragement and kind permission to quote from their poems that have been published in different issues of *Reckoning.*

Notes

1 This is the title of D. Dina Friedman's poem "We Have So Little Time Left." See the poem in *Reckoning: Creative Writing on Environmental Justice* 5: 11–12.
2 The title of Bill McKibben's book is *The End of Nature: Humanity, Climate Change and the Natural World.* Bloomsbury, 2003.
3 Stephen Waring, letter to *The Observer* (7 June 2020).
4 See the poem ("when the coral copies our fashion advice") by Ashley Bao in *Reckoning* 5: 8–9.
5 Dipesh Chakrabarty. Foreword. *Global Ecologies and the Environmental Humanities: Postcolonial Approaches.* Routledge, 2015, pp. xiii–xiv.
6 See Chloe Clark's "Sidelong Catastrophe" in *Reckoning* 1.
7 See "The Cloud." *Selected Poems: Percy Bysshe Shelley*, edited with an Introduction and Notes by Edmund Blunden. Collins, 1970, pp. 555–558.
8 This is quoted from Edmund Blunden's brief note to "The Cloud," p. 555.
9 See Mohammad Shafiqul Islam's "Rivers Lament" in *Reckoning* 1.
10 Mohammad Shafiqul Islam's Interview is available in Reckoning 1.
11 See "The Lake Isle of Innisfree" in *The Collected Poems of W.B. Yeats*, with an Introduction by Cedric Watts, Wordsworth Poetry Library, 2008, p. 31.
12 See Adelia MacWilliam's "Paddling in the Sound" in *Reckoning* 3.
13 See Krista Hoeppner Leahy, "Earthspun" in *Reckoning* 2.
14 See Jennifer Mace's "letters from the ides" in *Reckoning* 5: 13–16.
15 See Adam Stemple's "The World Ended in Ice" in *Reckoning* 3.
16 See the poem in *Reckoning* 2.
17 See Seth Garcia's "Displaced Water" in *Reckoning* 3.
18 See Danika Dinsmore's "How Can I Look Up" in *Reckoning* 3.

19 See Alicia Cole's poem "Corrupt the World With Drum" in *Reckoning* 3.
20 See F. J. Bergmann's "Development" in *Reckoning* 2.
21 See Brandon O'Brien's "Papa Bois and the Boy" in *Reckoning* 1.
22 This phrase is borrowed from Laura Pulido's "Environmental Justice," *Dictionary of Human Geography*, quoted in *Lawrence Buell's The Future of Environmental Criticism: Environmental Crisis and Literary Imagination*. Blackwell Publishing, 2005, p. 170.
23 See Erin Hoffman's Interview: "Transition" in *Reckoning* 1.

Works Cited

Alaimo, Stacy. *Exposed: Environmental Politics and Pleasures in Posthuman Times*. University of Minnesota Press, 2016.
Chakrabarty, Dipesh. "Foreword." *Global Ecologies and the Environmental Humanities: Postcolonial Approaches*. Routledge, 2015.
Clark, Timothy. "Nature, Post Nature." *The Cambridge Companion to Literature and the Environment*, edited by Louise Westling. Cambridge University Press, 2014.
Coleridge, Samuel T. *The Rime of the Ancient Mariner. Collins Albatross Book of Verse: English and American Poetry from the Thirteenth Century to the Preset Day*, edited by Louis Untermeyer. Collins, 1971.
Collins, Stephen. "Notes towards a Manifesto of the Biotariat." *Beating the Bounds: Notes on Commons, Poetry, and Climate Justice*. 2014, beatingthebounds.com.
Dürbeck, Gabriele. "Writing Catastrophes: Interdisciplinary Perspectives on the Semantics of Natural and Anthropogenic Disasters." *Ecozon@: European Journal of Literature, Culture and Environment*, ecozona.eu, 2012, pp. 1–7.
Felstiner, John. *Can Poetry Save the Earth?: A Field Guide to Nature Poems*. Yale University Press, 2009.
Frost, Robert. "Fire and Ice." *Four Centuries of American Poetry*, edited by Mohan Ramanan. Macmillan, 1996.
McKibben, Bill. *Eaarth: Making a Life on a Tough New Planet*. St. Martin's Griffin, 2010.
Morton, Timothy. *Hyperobjects: Philosophy and Ecology after the End of the World*. University of Minnesota Press, 2013.
Nayar, Pramod K. *Ecoprecarity: Vulnerable Lives in Literature and Culture*. Routledge, 2019.
Shelley, Percy B. "The Cloud." *Selected Poems: Percy Bysshe Shelley*, edited with an Introduction and Notes by Edmund Blunden. Collins, 1970.
Shoptaw, John. "Why Ecopoetry?" *Poetry*, vol. 207, no. 4, 2016, pp. 395–408, www.jstor.org/stable/44015972.
Slovic, Scott. "'Cultivating an Ability to Imagine': Ryan Walsh's *Reckonings* and the Poetics of Anxiety." *Ecozone@: European Journal of Literature, Culture and Environment*, 2020, pp. 154–161, ecozona.eu.
Steffen, Will, Paul J. Crutzen, and John R. McNeill. "The Anthropocene: Are Humans Now Overwhelming the Great Forces of Nature?" *AMBIO*, vol. 38, 2007, pp. 614–621.
Tuan, Yi-Fu. *Landscapes of Fear*. University of Minnesota Press, 1979.
Voie, Christian Hummelsund. *Nature Writing of the Anthropocene*. Thesis for Doctoral degree in English. Printed by Mid Sweden University, Sundsvall, 2017.
Yeats, W. B. *The Collected Poems of W.B. Yeats*. With an Introduction by Cedric Watts. Wordsworth Poetry Library, 2008.

2 Unmasking the Risks of Climate Change in Liz Jensen's *The Rapture*

Mahinur Gözde Kasurka

Introduction

The twenty-first century has given birth to a blooming of apocalyptic and post-apocalyptic literary imaginations due to an increasing fear of environmental devastation and various contagious diseases. Climate change strengthens the fear of a depopulated earth as a result of a possible end-of-the-world scenario. As we live in an age of anthropocentrism by valuing humans more than any other entities, we tend to think of humans as a separate entity freed from the impacts of any sort of environmental devastation. Yet, this understanding of the human as a self-enclosed entity does not function in explaining human's complex relationality with more-than-human worlds. Hence, human-centred agenda lacks the ability to protect the human as a concept from all possible threats. This fragile condition of the human does not align with the highlights of our modernity-driven logic since modernity promises to negate the threats before humanity. Human's vulnerable condition creates a kind of ambivalence together with the risk/fear that we face in our everyday lives. Therefore, it does not come as a surprise that we tend to narrate stories thematised around this risk/fear that undermines assumed dualisms of human-centred ideology.

In accordance with this, Ulrich Beck highlights the collapse of modernity ideal by decoding how society approaches risk/fear all throughout history. In his acknowledgement, he divides society into three main phases such as pre-industrial society, industrial society and *risk society*. In the first phase, risk/fear factor is something humanity cannot foresee as feudal system is based on God's will. Hence, Beck reflects in his book *Ecological Politics in an Age of Risk* that people are born as "lords, peasants or serfs by virtue of birth, and thereby through God's intercession, the social order is natural and therefore good" (15). In this period, natural hazards that are beyond human control might harm humanity and only God has the potential to predict and control these threats before all living species. With the Industrial Revolution, we move towards an industrial society in which humanity is able to implement science, progress and rationality to govern their lives. It is thanks to reason and rationality that humanity is able to provide shelter against

DOI: 10.4324/9781003602750-4

natural disasters. Thus, they no longer constitute a form of risk/fear factor. However, a new form of risk, that is *manufactured risk* is now on the stage. Industrial society promises a safe and secure haven for humanity, but in reality "with the rise of nuclear power and other advanced technology, humanity has once and for all emancipated itself from the bonds of nature" (Sørensen and Christiansen 10). To be more precise, with Fukushima and Chernobyl, we head towards *risk society* in which we have to live together with the uncertainty that modernity vows to prevent. In the post-industrial phase of society, we need to be more alert to the threats as now we have created a series of brand-new risks. What is underlined in *risk society* is that it is not possible to foresee the possible risks today. In what follows Mads Sørensen and Allan Christiansen describe how uncertainty still lingers in the system in their book entitled *An Introduction to the Theory of Second Modernity and the Risk Society*:

> Uncertainty has made its return to the society, and we cannot get rid of it. We simply cannot insure ourselves against the dangers of radioactive emissions, such as those from Chernobyl and Fukushima. Nor can we insure ourselves against the perils of global warming, the holes in the ozone layer or those risks intricately tied to our releasing genetically modified plants and other organisms into nature.
>
> (12–13)

The paradoxical situation mentioned here arises from the fact that modernity-driven logic claims to eradicate probable threats, yet it causes several others that are not possible to be estimated. The paradox here lies in the fact that modernity creates the illusion of eradicating all sorts of risks. In *risk society*, we witness man-made natural disasters in which humans are not able to take full precaution against them. In this respect, anthropogenic global warming and environmental devastation echo the residues of Ulrich Beck's *risk society*. In line with this, Gabe Mythen reflects on Beck's categorisation in *A Critical Introduction to Risk Society* by expressing that the category of Beck "examines the broader interrelationship between humans and the environment, the effects of institutional change on social experience and the changing dynamics of politics" (6). In what follows this chapter will delve into Liz Jensen's novel *The Rapture* as it consolidates a form of *risk society* that is the product of monolithic thinking. The way Jensen's novel challenges modernity showcases how environmental degradation might be problematised by going back to pre-modernity.

The Rapture and Ecological Degradation Emerging as a Risk Factor

Liz Jensen's novel *The Rapture* emerges as an epitome of climate change fiction which is fraught with anthropocentric wrongdoings leading towards the end of life for humanity. The novel beautifully sheds light on how human-induced actions will lead towards annihilation of the whole planet. In doing so, the implied reader witnesses a dramatic change in the climactic conditions and as a result of this situation, the characters need to update their ways of living. The text manifests technological developments and modernity's impact on human enhancement. Yet, we as the readers also see that seemingly positive changes bring with them a kind of threat/risk that humanity is not able to foresee and take precaution to prevent catastrophic results. With an end to decode "risk" as a transformative factor, I will draw attention to how environmental degradation that is intertwined with ecohorror and fear in *The Rapture* redefines hope as an urge to start from where it seems to end. The way the novel reflects nature/culture division leading towards several catastrophes will be elucidated by underlining the failed ideology of separating human and non-human realms from each other. The text portrays fear freed from its pejorative connotations, as it emerges as an affirmative transformative factor which urges characters to take action by heading towards hope. I will also demystify the implied author's "not the end"[1] acknowledgement by presenting hope as a subversive force against the risk. In view of this, I argue that Jensen's novel fictionalises a form of *risk society* in Ulrich Beck's sense of the term by laying bare the ills of modernity, yet she does not totally reflect a nightmarish vision, in contrast hope is indicative of a survival strategy within the ruins of a devastated world.

Before delving into how anthropocentrism transgresses ontological boundaries of humans and non-humans, the following part will reflect a brief overview of the novel. After discussing transformation of environment due to human actions, I will offer a reading of the narrative as a *risk society* that is threatened by a kind of fear that humanity is not able to fight back. I will conclude this chapter by touching on how *ecohorror* instills hope as exemplified in Gabrielle as a character and her unborn baby within the ruins of a devastated world.

The narrator, Gabrielle, is an art therapist who has just been disabled due to a car accident in the beginning part of the novel. She tries to start a new life by leaving behind her past and begins a new position at Oxsmith Adolescent Secure Psychiatric Hospital. The author attaches Gabrielle's personal apocalyptic story to the larger frame by demystifying another apocalyptic narrative. In both situations, we witness demolished worlds and the ones endeavouring to "start anew once again." Gabrielle faces some difficulties in adjusting to her new life and in doing so, she meets her patient Bethany Krall, who has stabbed her mother to death. Bethany claims to have some prophetic visions starting from the beginning of the novel and she is able to predict disasters by giving exact details about their location and time. All throughout the text, Gabrielle and her lover physicist Frazer try to justify Bethany's predictions by linking them with some rational and scientific symptoms. However, they cannot fully explain how Bethany is able to predict these disasters from a scientific perspective. Behind their struggle to rationalise these disasters, the implied reader bears testimony to various ecological crises in various parts of the whole world and these disasters bring the end of the world as we know it.

The narrative is designed in such a manner that the anthropocentric mindset emerges as the guiding mechanism behind environmental crisis. Climate change and the rising temperatures make themselves visible by causing some unrest in society starting from the beginning. As it is not possible to divide life from its material conditions such as anthropogenic global warming, the chaos that persists all around the world linger in the background by preparing the reader for an unsafe world from the very beginning. Gabrielle summarises what she sees on the news:

> Two more suicide bombings in Jerusalem. Abductions, limbs lost, children orphaned. . . . The row about Chinese and American greenhouse gas emissions has ratcheted up a notch, while the heatwave has spread to the whole of Europe felling old people in precise strokes, like an efficient industry.
>
> (Jensen 33)

As it is obvious, there occurs a civil unrest in the society. The increasing impact of global devastation announces the collapse of modernity by making this transformation in environment as the new norm that the whole world is trying to adjust themselves. The narrative gives many instances of these transformations by tying them to the larger frame of reference. For example, we learn from Gabrielle "[f]ive years ago, the British seasons made some kind of sense. Not any more" (Jensen 54). We learn more about the extremity of weather conditions as Gabrielle states:

> June seemed to last for a thousand years. the temperatures were merciless: thirty-eight, thirty-nine, then forty in the shade. It was heat to die in, to go nuts in, or to spawn. Old folk collapsed, dogs were cooked alive in cars, lovers couldn't keep their hands off each other. The sky pressed down like a furnace lid, shrinking the subsoil, cracking concrete, killing shrubs from the roots up. . . . You longed for rain, it didn't come.
>
> (Jensen 3)

It does not come as a surprise to the reader to learn that "the latest projections predict the loss of the Arctic ice cap and a global temperature rise up to six degrees within Bethany's lifetime" (Jensen 23). As a clear outcome of this increase in temperatures, citizens need to adjust their ways of living by struggling to protect themselves because "the heat has become so ferocious that venturing outside is an ordeal, something one must gear up to, armed with drinking water, sunglasses, cream headgear. Items that were once optional accessories are now survival kit" (Jensen 124).

The author does not only make use of global warming to fictionalise a dystopian narrative since Bethany's visions of environmental catastrophic events also become reality one by one. At first, she envisions a tornado in Scotland, then reflects that "there's a thunderstorm coming tomorrow. From the west" (Jensen 58). Bethany also predicts a hurricane in Rio leading to the fall of Christ the Redeemer statue and later an earthquake in İstanbul. Amidst these terrible catastrophes, Gabrielle is shocked, and the world does not make sense for her anymore. She questions herself after watching the fall of the Christ figure:

> I belong to a generation that has seen statues and icons and buildings come tumbling down on TV: Lenin in Russia, the Berlin Wall, Saddam in Baghdad, the Twin Towers. But those topplings meant something to those who caused them. What does this mean? *Who is to blame?* What can one read into a random catastrophe, an out-of-the-blue event, an act of God'? Nothing. In place of an explanation, however grotesque, there is a void.
>
> (Jensen 82, italics mine)

Gabrielle's analysis of the fall of Christ figure is a telling one in terms of the text's preoccupation with themes of risk/fear. As she states, the fall of the oppressive regime icons would mean a liberating act for humanity, or the Twin Towers would give birth to an anti-Muslim stance all over Europe. In all these occasions, the risk factor comes from a source that humanity would be able to take precautions under the guidance of their governmental organisations. Thanks to progress and modernity of their time, people would feel the risk, yet also they would know the fact that it is possible to find a shelter. In the case of latest situation, Gabrielle does not feel in the same way, as she cannot fathom who to blame or how to protect themselves. This inability to spot an exact person to blame bears the residues of climate change narrative.

The fall of the Christ figure appears as a double-coded message in the novel. To be more precise; firstly, it signifies the annihilation of a grand narrative, the erasure of a centre that cannot hold as it is described in the following way: "More than ever, in these terrible times, the Redeemer is a symbol of hope" (Jensen 81). This symbol of hope that unites people around is now gone, and it is not possible to point out who to blame for this. Secondly, this fall also blurs the guilty one to blame. As people are not able to pinpoint the guilty one, it is not also likely to find a shelter for themselves with the fear of forthcoming unpredictable disasters. This inability consolidates the condition of the citizen in *risk society* in Beck's sense of the term.

The narrative is located in a closer time to the implied reader of the twenty-first century in terms of technological advancements, yet these sources do not help in estimating hurricanes, floods and earthquakes as it is only Bethany who can predict the future. Since the sources of fear/risk are natural disasters, they sound more like Beck's reference to agrarian society in which humanity does not have any sort of control over their lives. The problematic situation here is that the author opens up her text within a technologised world, but the paradigm of risk/fear seems to comply with a reference to past. This ambiguity becomes more resolved when the reader is confronted with humanity's solution in the face of oil crisis. As oil becomes a scarce resource, energy companies find a new way of extracting oil. Ned, a scientist, clarifies this new method:

> You can inject hot water beneath the seabed to destabilise the hydrates. Which will force a pressure change and release methane. The gas moves along the cracks and works its way up. Then you can liquidise the hydrate on the ocean floor and pipe it up like oil and gas. Or release frozen chunks of it from the sea floor and trap them at the surface of the ocean in giant tarpaulins. Exploit the hydrate fields safely, and there's no such thing as an energy problem. Methane's cleaner than oil or coal, if you handle it right. You can power anything with it, and it's there in quantities you can't even imagine. But it's highly volatile. Which means it may cost more than anyone's ever paid for anything. Ever.
>
> (Jensen 213)

This new method brings with itself risks/threats that they should not underestimate. As Ned states, these companies are playing "Russian roulette" by fracturing in the deep ocean to find energy sources (Jensen 212). Even though they have the means to predict possible dangers, they do not find it problematic to use such a dangerous method. This attitude is indicative of both anthropocentric arrogance blinding their eyes to reality and also their inability to anticipate a chain of events leading to methane exploitation. The usage of methane hydrates in the novel is an actual method that scientists have been speculating about the risks for a long time. In reality, scientists note that on the condition that we have the technological means to mine hydrates, "methane gas hydrate has the potential to be widely used as a new energy source" (Lee and Holder 186). On the risks of using methane as a new energy source, Eric Otto gives a clarification by stating: "methane warms atmosphere; warmed atmosphere warms ocean; warmed ocean disassociates icy hydrates, liberating more methane; more methane further warms atmosphere, and so on" (113). The risks of methane hydrates are also highlighted by Terry Gifford, who analyses Jensen's novel in line with energy politics. Gifford highlights the high risk of using methane as a new energy source by resonating how progressive rationale falls short in embracing the threats as now, they reside in a form of risk society.

> But a miscalculation could widen the cracks resulting in a pressure build-up which would not only risk the release of huge amounts of volatile methane, but would destabilise the sediment of the seabed itself with a cascading series of submarine avalanches, more cracks, more

> releases, and after the first huge tsunami, a series of others in the now flaming oceans of the world. Since methane is ten times more powerful as a greenhouse gas than CO2, the whole globe will get radically hotter.
> (722)

The novel presents a scenario similar to Gifford's calculation by leading towards a sudden increase in world's temperature leading to other catastrophic events. Even though it has these risks, humanity does not take a step back by ignoring environmental ruination possibilities. In the forthcoming part, I will reveal both anthropocentric arrogance and *risk society*'s incapability to foresee the threats as the implied reader also observes.

Risk Society as a Symptom of Anthropocentric Arrogance

The human-centric discourse tends to think of the human as a separate entity freed from the impacts of any sort of environmental devastation. Yet, this understanding of the human as a self-enclosed entity does not function in explaining human's complex relationality with more-than-human worlds. Considering the human as placed at the top of species hierarchy does not explain the vulnerable condition of the same human in the face of *risk society*. Paul Crutzen coins the term "Anthropocene" by highlighting "an intertwining of geological Earth time and human history" (Parikka 51). Anthropocene marks human impact on the whole planet by transcending assumed borders between human and non-human spheres. To be more precise, observing human impact on earth as a geological force shatters the divided ontological spheres by illustrating humans as standing amidst various networks of agents. Jussi Parikka defines the pathway leading towards Anthropocene as in the following:

> The emergence of agriculture is one crucial threshold that has changed the chemical relations of the Earth. The invention of fire is another. The steam engine gets a frequent mention in narratives of the Anthropocene, as well as modern synthetic chemistry.
> (52)

In line with what Parikka states, *The Rapture* also highlights a meaningful human impact as a geological force bringing the end of civilisation. It is mainly due to humanity's struggle to compensate for a new energy source that they go deep down the ocean. The problematic issue here is that they do not find it problematic to act in such a risky condition. Evidently, they do not find it questionable to exploit oceanic sources for human betterment. I argue that humanity's inability to take precautions for risk is a product of their portrayal of the world as solely existing for their existence. This human-centric motivation prevents their ability to see reality. Liz Jensen also underlines the same egotistical situation in the novel in many instances.

Harish Modak, a fictional environmentalist in *The Rapture*, explicates in an unequivocal manner the position of the human among other species by referring to Anthropocene:

> Harish Modak's view [is that] the Anthropocene era -the reign of man-is hurtling to a close. . . . People who know about the rocks see Earth in a different timescale from the rest of us. To them humans are just another species -a species that happens to be dominant for now but won't be in the future.
>
> (Jensen 88)

In the forthcoming parts of the novel, Modak also states: "the collapse of *Homo sapiens* as a dominant species means the dawn of a new era for a million other life forms. . . . We have been here a mere instant, in geological terms" and he claims that the annihilation of humanity will be a little spot when compared to other species in the whole universe. I argue that the implied author's fabrication of such a character as Harish Modak is a conscious stance. In this way, the text offers a critical engagement by operating as a space of non-binary relations to the other. One of the most powerful messages related to critique of anthropocentrism in the text is enunciated by Modak. I propose that his forthcoming statement manifests our position as a species within the larger network of relations.

> There is colossal arrogance, he maintains, in the assumption that humans will last for ever. If we look at the planet's life across billions of years, rather than in terms of humankind's meagre history as a dominant species, we will see that our presence on Earth has lasted the blink of an eye. 'We are the agents of our own destruction- and when we are gone, extinguished by our own heedless quest for expansion, the planet will not mourn us. Indeed, it will have cause to rejoice. Today, the human species stands at the brink of a new mass extinction which will see if not its disappearance, then its extreme marginalisation.
>
> (Jensen 36)

The road to humanity's extinction lies in humanity's own hands as Modak states. This possibility of extinction is dispersed through the text as a possible scenario. It is an elusive point as characters are cognisant of this possibility, and it is even turned into more than a mere possibility with Bethany's predictions. After each prediction of Bethany, the implied reader is also convinced of the possibility of an approaching doomsday. Yet, the authorities are not able to see the same reality and take action to save humanity. Thus, characters experience a kind of *risk society* that progress-driven rationale does not function in constructing a safe and secure environment for their citizens. If we return to the text's portrayal of *risk society* by humanity's bringing their own doom without predicting a possible methane explosion, we will see residues of a fictional community turning blind to their empirical reality. As Ned explains that oil becomes a rare source by causing a harsh fight among various countries to find a new source, they fall short in decoding risk/threat. It is important to note that Jensen has used "a fair amount of accurate commentary on actually existing environmental issues, or on near future environmental issues" (Otto 110). Moreover, Jensen's fabrication of probable human reaction in the face of doomsday scenarios indicates faith waves as a strategy to fight against the dystopian residues.

Jensen reflects various faith waves to bring an explanation to the natural disasters that the characters experience. When Gabrielle talks to Leonard, Bethany's father who is the leader of one of these faith waves, he explains that this is the Tribulation in Christian understanding. As they observe "plagues, extreme weather, disasters, globalisation, stock markets collapsing, terrorism, atheism," they acknowledge these to be the symptoms of the rapture, during which "true believers will be saved" (Jensen 150). That is to say, the rapture will function as "the beginning of the apocalypse when the faithful will be lifted up to heaven and the remainder punished with natural disasters extinguishing the species" (Gifford 14). Even though there are many followers of this rapture phenomenon, they all die while waiting to be ascended in a stadium together with Leonard. Gabrielle sees that the earth turns into a lifeless place as she flies in a helicopter without knowing where to land. She states:

> Only a few skyscrapers stand proud of the burning waterscape as the land is relentlessly and efficiently erased. The heat is unbearable, as though the sun itself has plunged into the water and is irradiating us from below. It's almost impossible to breathe.
>
> (Jensen 339)

In the following part, I would like to explore the path heading towards the loss of life on earth for humanity, namely the construction of a *risk society*.

Bethany's latest prediction concerning going up a mountain to save themselves refers to a methane explosion. Bethany's research shows that it is "many times more powerful than carbon," and "there are millions of square kilometres of it locked frozen on to the sea floor, all around the world, in the form of a crust" (Jensen 187–188). She also learns that playing with methane is so risky as "it is so volatile that until recently there was no serious discussion about harvesting it for energy purposes. It was too dangerous" (Jensen 188). Together with Gabrielle, we as readers also learn that in the past there were some methane explosions "resulting in the devastation of most of life on earth" (Jensen 188). By bearing in mind the risk they take on, several organisations around the world do not find it problematic to use it as an energy source. Gabrielle's description of the danger is so vivid as she states: "frozen methane is known as the ice that burns," yet it does not prevent these energy companies. It is also meaningful for us to see both the reliance on rationality in terms of methane fracturing and the possibility of miscalculating the impact on earth. Kristin, another scientist working to save the world together with Gabrielle, unmasks this threat of a possible miscalculation. She reflects:

> To free the frozen methane from the ocean floor, they'll have drilled beneath it horizontally and forced a pressure-change. But if they've miscalculated what's down there, and the interference has widened the gap, that's already there, the pressure will build up. When it reaches a critical point, there's a risk that huge amounts of this frozen methane- far more than they ever intended- could be unleashed.
>
> (Jensen 217)

Hence, energy companies' possible mistakes might lead to a very severe disaster though it is not their intention at first. After hearing these possible scenarios, Gabrielle's reaction explains why *risk society* is established on modernity ideal by attaching too much importance to ourselves. She asks: "But, wouldn't the energy company that owns it *know* about it?" and then she continues by giving her rational explanation not to believe in what is happening:

> They are all being absurd. It's science fiction. The fact that there are precedents for such a catastrophe is irrelevant. These things may have happened in prehistory, but they can't happen in the age of man. Nature can't just destroy civilisation. We've come too far. We can cope with things on this scale nowadays. We can prepare for them.
>
> (Jensen 229)

As Ulrich Beck also claims, *risk society* creates the illusion that any form of threat is defeated thanks to the usage of science and technology, as this is the *age of man* in Gabrielle's terms. However, it turns out to be the opposite by leaving humanity vulnerable in the middle of miscalculated dangers. Modern society's acknowledgement of risk can be explained as the urge "to master time, to discipline the future. To provide for the future does not just mean living from day to day and arming oneself against ill fortune but also mathematizing one's commitment" (Ewald 207). Jensen's fabrication of danger resonates with a form of *manufactured risk* "in which the social, political, ecological and individual risks created by the momentum of innovation increasingly allude the control and protective institutions of industrial society" (Beck 27). Even though there is a highly strict control mechanism over the production of industrialism, as Gabe Mythen states: "manufactured risks continue to appear as environmental 'side effects'" (17). In this respect, methane explosion comes together with a promise of the flourishing of energy "by virtue of the interplay of progress with the possibility of annihilation by the ecological, nuclear, chemical and genetic hazards we impose upon ourselves" in Beck's words as he states in *Ecological Risks in an Age of Risk* (83). This form of risk/threat poses a bigger danger for humanity when compared to the ones in pre-industrial or industrial society. Therefore, it evokes a kind of fear that we can define as *ecohorror* in the age of reason.

Ecohorror underlines the fact that assumed boundaries between human/nonhuman, nature/human and nature/culture do not function in clarifying the enmeshed network of agents. In ecohorror narratives such as Jensen's novel, we see that risk/threat does not come from an outside source such as an alien invasion. Instead, what we witness in *risk society* is that ecohorror mainly results from "narratives in which nature enacts violence on humans in response to the damage caused by human behaviours" (Tidwell 538). In a similar manner, Jensen's narrative also underlines risk/threat embedded in nature and it cannot be separated from culture. As Christy Tidwell remarks "it is important, therefore, to note ecohorror narratives that challenge this division of human and nonhuman, internal, external" (539). All throughout the novel, scientists' struggle to save humanity from a possible doomsday resonates with the impossibility of separating human domain from nonhuman. This intricate relationality showcases the fact that human is always intermeshed with more than human world. This idea of enmeshment postulates a new stance in terms of risk/fear representation as now the danger does not come from an external source, or it is not out there in the wilderness. On the contrary, in *The Rapture* the danger comes from human's greed for exploiting earthly others by disregarding this intimate relationality. Christy Tidwell finalises the discussion on ecohorror in *Posthuman Glossary* by stressing the function of ecohorror narratives as in the following:

> Ecohorror reflects our fears about nonhuman nature in a variety of ways. Perhaps animals will attack us, perhaps we will lose our place at the top of the animacy hierarchy, or perhaps we will have to acknowledge our interconnectedness with other beings. In doing so, ecohorror risks reinforcing those fears and the categories they are built upon, but ecohorror also asks us to reconsider some of those fears and to imagine what might happen if we were not to insist so vehemently upon such divisions.
>
> (117)

It is in this point that Jensen's narrative offers a kind of warning about our own actions by also hiding glimpses of hope to lead affirmative transformation in our actions. As characters try to figure out how to warn everybody about the forthcoming dangers, they are not represented as totally pessimistic about the future. Their search for imaginative horizons highlights this same hope embedded within the anthropocentric chaos.

The characterisation of Gabrielle can be taken as an embodiment of hope in this respect, as her personal apocalyptic story intertwines with the larger frame of the novel in several manners. To be more precise, we learn that

Gabrielle becomes paralysed after a car accident with her lover, who is a married man to another woman. She is pregnant when she has the accident, and she loses both her unborn baby and her lover after this accident. Her new life brings with itself many obstacles such as being a paralysed person, and she tries to adjust to these changes by starting to work in a short time after this catastrophic event. Even though it is not easy for Gabrielle to continue living, she strives to start a new life which still echoes residues of hope in the midst of chaos. Her personal chaotic past resembles the current state of the world with the anthropogenic climate change, rising temperatures, natural disasters that herald the end of the world as we know it. Yet, Gabrielle's resolution to continue her life together with its difficulties is symptomatic of her will to act. In this respect, I offer relating the novel with Haraway's concept of *staying with the trouble* as it is defined in her book *Staying with the Trouble* as in the following:

> We- all of us on Terra- live in disturbing times, mixed-up times, troubling and turbid times. The task is to become capable, with each other in all of our bumptious kinds, of response. Mixed-up times are overflowing with both pain and joy- with vastly unjust patterns of pain and joy, with unnecessary killing of ongoingness but with necessary resurgence. . . . Our task is to make trouble, to stir up potent response to devastating events, as well as to settle troubled waters and rebuild quiet places. In urgent times, many of us are tempted to address trouble in terms of making an imagined future safe, of stopping something from happening that looms in the future, of clearing away the present and the past in order to make futures for coming generations. Staying with the trouble does not require such a relationship to times called the future. In fact, staying with the trouble requires learning to be truly present, not as a vanishing pivot between awful or edenic pasts and apocalyptic or salvific futures, but as mortal critters entwined in myriad unfinished configurations of places, times, matters, meanings.
> (Haraway 1)

Haraway is cognisant of both hope and despair we face in the twenty-first century, and what she offers is centred around raising a recognition of our current chaotic situation together with an awareness that grows together with it. In a similar manner to her conceptual map in the midst of this end-of-the-world narrative, we need to realise how *to be truly present*. This requires establishing an awareness of here and now, instead of focusing on future-oriented actions. Gabrielle in the novel tries to be truly in the present as she finds a way to stay with the trouble of living in turbulent times as a paralysed and pregnant woman to Frazer's baby.

Haraway offers to make kin by establishing bonds of intimate relationality with nonhuman animals instead of having children. From this perspective, it can be said that the novel lacks an awareness in laying bare the entanglement of human and nonhuman animals, but it is possible to see the seeds of hope residing in Gabrielle as now she is pregnant with Frazer's child. The chaotic atmosphere that she learns about this pregnancy is also telling as Bethany informs her of this change in the plane that they are trying to escape from the flood. As the world is reaching its end, Gabrielle is somewhere in between hope and despair as now that she will become a mother, she knows that her baby will not find a world that is ready to raise a baby. Gabrielle's predictions resonate with her limbo situation as her inner thoughts reveal that she is both afraid of what is going to happen and resilient about the new world. She articulates her situation as in the following:

> I know already there will be no green fields in Bethanyland, no safe place for a child to play. Nothing but hard burnt rock and blasted earth, a struggle for water, for food, for hope. A place where every day will be marked by the rude, clobbering battle for survival and the permanent endurance of regret, among the ruins of all we have created and invented, the busted remains of the marvels and commonplaces we have dreamed and built, strived for and held dear: food, shelter, myth, beauty, art, knowledge, material comfort, stories, gods, music, ideas, shelter.
>
> (Jensen 341)

As it is clear from this quotation, Gabriel is cognisant of the fact that the new world order will be nothing like the one she knows as her child has to struggle now to find glimpses of hope. Yet, this struggle is not represented as a negative standpoint, which is apparent from the way the novel ends. Jensen prefers ending the novel as "Not The End" by implying a new beginning in which anthropocentricism hopefully will not take over.

Conclusion

To sum up, Liz Jensen's novel *The Rapture* reflects an imaginary apocalyptic scenario that is not constructed so far away from the implied reader's empirical reality. By fabricating a problem that is close to our everyday reality, the text functions as a cautionary tale to warn us to initiate an affirmative transformation in our lives. Together with anthropogenic climate change, the novel also beautifully portrays the grim reality behind usurping a new energy source, methane hydrate disassociation. In this respect, the text illustrates how modernity-driven logic creates the illusion of a world freed from natural threats and dangers. However, rationality-based models also bring their own consequences by leading towards bringing the end of the world with the hope of finding a new energy source.

The novel's portrayal of ecological disasters erases ontological superiority of human-centred ideology by showing the vulnerable condition of the human and the nonhuman to the same degree. This impending risk of the annihilation of life as we know it sheds light on our fragile condition with the total loss of the secure environment of modernity. In this respect, the novel fictionalises a form of *risk society* in Ulrich Beck's sense of the word. Even though the characters are still struggling to find rational explanations for Bethany's visions, they cannot come up with a satisfying solution. Jensen's fabrication of Bethany as the one who predicts all the precise details about the forthcoming disasters is what brings us to premodernity. The implied author ends the novel by reaching towards the apocalyptic part of the narrative and ends it by stating it does not end in this way. Although the implied reader is not presented with what is going to happen after the apocalypse, in pre-apocalypse part of the narrative, the author succeeds in bringing us to premodernity without destroying all the products of modernity-driven logic. Terry Gifford comments on the ending of the novel as "a place of nature without culture" as everything that has been constructed will be ruined together with the disaster (724). Gifford stresses anthropogenic climate change will negate the residues of human-centred ideology in the text. I do not share the same concern with Gifford in this concern, as I think that it is not possible to divide nature from culture, and it would not be correct to think that what remains after disaster would be totally confined to nature. Instead, the remains will be a total enmeshment of both culture and nature, hope and despair. Gabrielle's unborn baby constitutes an amalgamation of several agents by injecting a utopian undertone together with the hope of starting once again by learning from our mistakes.

Note

1 The novel ends on a "not the end" note by leaving the characters in the middle of an apocalyptic event that is presented as the beginning of the end. The way "the end" is presented signals a new beginning not separated from hopeful connotations.

Works Cited

Beck, Ulrich. "The Reinvention of Politics: Towards a Theory of Reflexive Modernization." *Reflexive Modernization: Politics, Tradition and Aesthetics in the Modern Social Order*, edited by Ulrich Beck, Anthony Giddens and Scott Lash, Stanford University Press, 1994, pp. 1–55.

———. *Ecological Politics in an Age of Risk*. Cambridge, Polity Press, 1995.

Ewald, Francois. "Insurance and Risk." *The Foucault Effect: Studies in Governmentality*, edited by Graham Burchell, Colin Gordon and Peter Miller, Harvester Wheatsheaf, 1991, pp. 197–210.

Gifford, Terry. "Biosemiology and Globalism in *The Rapture* by Liz Jensen." *English Studies*, vol. 91, no. 7, 2010, pp. 713–727.

Haraway, Donna. *Staying with the Trouble: Making Kin in the Chthulucene*. Duke University Press, 2016.

Jensen, Liz. *The Rapture*. Bloomsbury Publishing, 2009.

Lee, Sang-Yong & Gerald, D. Holder. "Methane Hydrates Potential as a Future Energy Source." *Fuel Processing Technology*, vol. 71, 2001, pp. 181–186.

Mythen, Gabe. *A Critical Introduction to Risk Society*. Pluto Press, 2004.

Otto, Eric C. "'From a Certain Angle': Ecothriller Reading and Science Fiction Reading *The Swarm* and *The Rapture*." *Ecozone*, vol. 3, no. 2, pp. 106–121.
Parikka, Jussi. "Anthropocene." *Posthuman Glossary*, edited by Rosi Braidotti and Maria Hlavajova. Bloomsbury Publishing, 2018, pp. 51–53.
Sørensen, Mads and Allan Christiansen. *Ulrich Beck: An Introduction to the Theory of Second Modernity and the Risk Society*. Routledge, 2013.
Tidwell, Christy. "Monstrous Natures Within: Posthuman and New Materialist Ecohorror in Mira Grant's *Parasite*." *ISLE: Interdisciplinary Studies in Literature and Environment*, vol. 21, no. 3, 2014, pp. 538–549.

3 Maja Lunde's *The End of the Ocean*

A Narrative of Climate Change and Environmental Crisis

Saumya Priya

Introduction

Climate change has become one of the major issues nowadays. And since "storytelling is an inherently human way of making sense of the world," contemporary writers are using this "fictional space" well to create narratives that help "shape our understanding" of the ongoing climate crisis (Bracke 7). Literature thus has become an increasingly important medium for representing this environmental change. According to Adeline Johns-Putra:

> In the last five years, climate change has emerged as a dominant theme in literature and, correspondingly, in literary studies. Its popularity in fiction has given rise to the term cli-fi, or climate change fiction, and speculation that this constitutes a distinctive literary genre.
>
> (01)

Although cli-fi as a genre is "resolutely contemporary," the climate change novelists have shown dedication, "innovation and creativity" in fabricating new narratives adequate to current conditions (Irr 1).

In the past, climate change novels like Paolo Bacigalupi's *The Wind-Up Girl* and J. G. Ballard's *The Drowned World* have gained considerable public as well as critical attention. However, this paper focuses on a recent and relatively less explored Norwegian novel entitled *The End of the Ocean.* It is a contemporary climate change novel by Maja Lunde, the acclaimed Norwegian author and screenwriter. Her novels are known to have a focus on climate change. Maja Lunde had swept the world off its feet in 2015 with her debut novel *The History of Bees.* Since then, this award-winning novel has been translated into 38 languages. *The End of the Ocean* is the second novel in what is planned by the author to be a quartet of climate-themed books. *The End of the Ocean* was first published in 2017 in the Norwegian language under the title *Blå.* A couple of years later, the novel was translated into English by Diana Otley. The English version came to be known as *The End of the Ocean*, and it was published in the year 2019. This is a gripping and thought-provoking novel revolving around one of the most essential things for human existence – WATER.

DOI: 10.4324/9781003602750-5

The Looming Water Crisis

Water becomes a dominant theme in most cli-fi novels. Either there is too much water with rising sea levels that threaten to engulf coastlines leading to the flooding of cities, or there is a scarcity of water resulting in droughts and even resulting in wars between countries or groups. Similarly, in Maja Lunde's *The End of the Ocean*, water forms the heart and soul of the book. *The End of the Ocean* has two separate but parallel running storylines each of which highlights the impact and extent of climate change and the subsequent water crisis. Interestingly, both the stories are set in different locations – Norway and France respectively – and they take place in divergent timelines. One is set in the recent past in 2017, and the other in the faraway future in 2041. And, it is the looming water crisis that weaves the two stories together.

Each chapter alternately describes the struggles and actions of Signe and David, the two leading characters of the novel. The first story revolves around Signe Haugher, who is a journalist and professional activist. Signe's story is set in the year 2017 in Ringfjorden, the beautiful, glacial village in Norway where she had spent her childhood. As the novel opens, we meet Signe who now is a 67-year-old "aging woman" dressed in "a little shabby and unkempt worn-out parka, with grey locks of hair sticking out from beneath a knitted hat" (Lunde 5). As the story progresses, we learn that Signe had become an environmentalist at a very young age, and since then she has never stopped fighting to save the world. And now, after almost 40 years she has returned to her home village on a serious mission. She has come back to save the receding glacier called Blåfonna. Signe had spent her entire life protesting for environmental causes, but this time, in her fight against ecological disasters, she is enraged upon realising that the damage being done to her homeland is by none other than her former lover, Magnus. Magnus, the businessman and chairman of the Ringfallene Company, is the person responsible for the extraction and export of freshwater from the glacier. Lamenting over Magnus's appalling actions of exploiting the natural resources for his own profit, Signe remarks:

> They are extracting ice from the glacier – pure, white ice from Norway – and marketing it as the most exclusive ingredient, to be put in a drink, a floating mini-iceberg, surrounded by golden liquor, but not for Norwegian customers, no, for those who have really deep pockets; the ice is to be shipped to desert nations, the homes of oil sheiks, and there it will be sold as if it were gold, white gold, to the wealthiest of the wealthy.
>
> (Lunde 8)

While pondering over her biggest fear and concern of environmental destruction and commercialisation by humans for their sheer self-interest in this, Signe recalls another incident from her seventh birthday when her mom gave her a big surprise by covering the backyard completely in snow because Signe had once "wished for a winter birthday" (17). Later, Signe came to know that her mom had driven the snow "all the way down from the mountain" in a truck, with the help of Sønstebø, the farmer and father of Magnus (15). Realising the gravity of the matter, the little Signe had felt that "something was at stake" (15). She also felt guilty for putting forward a wish that became a cause of environmental harm.

Signe is often seen recollecting her life as a kid, and young woman in Norway, in the 1950s–1960s and how she became an activist. She recalls an incident when she was "put in jail" for chaining herself "to a barricade in protest" (36). This gives the readers a clue that the exploitation of natural resources in Norway has been a long ongoing thing. Signe mentions River Breio, Lake Eide and the Sister Falls all being exploited and capitalised upon for the construction of dams and hydroelectric power plants. Signe is unhappy with these developments, and she complains to her Daddy saying,

> The Sister Falls will disappear. Seven hundred and eleven meters of free-falling water obliterated as if it had never been there. Norway has no other waterfalls like those two and now they are going to be diverted through a pipeline.
>
> (220–221)

Signe had strongly opposed this and had also participated in the anti-dam protests as she believed that humans should not intervene and destroy natural resources and landscapes. The readers get to know that she had left no stone unturned to stop all this, but unfortunately, all her efforts went in vain. Yet, she was not giving up fighting for and saving her environment. While speaking her mind to Magnus, Signe says:

> We don't own nature. Just like it doesn't own us. We don't own the water; nobody owns the water. I will continue taking part in protest marches and handing out fliers, as long as I have feet to carry me and hands to distribute them with.
>
> (Lunde 227–228)

Through Signe, Maja Lunde gives the readers a graphic account of how an idyllic setting that once provided a wonderful haven for animals and for the people who lived here is eventually ruined by humans. Commenting on the destructive nature of human beings, Signe says:

> Water has power, a force. I had considered it to be invincible. But not any longer, not when encountering human hands, power shovels, steel pipes, tunnels, not in the face of license revenues, industrialization, and the welfare state.
>
> (Lunde 223)

However, back then, after being disheartened by not receiving support from her mother and her boyfriend Magus, Signe had left Norway. But now in 2017, she is back with a solid plan and is adamant on teaching Magnus and the others a lesson. Referring to Magnu's flourishing ice trading business, Signe says, "I can't destroy everything, but I can destroy this. I can dump the ice in the ocean and disappear" (42). And from here onwards, we follow Signe on her mission. We see how she secretly navigates her way through "the extraction area" where she is horrified to see "four large white bags that are full, ready for pick-up" (09). These as Signe informs the readers will be "loaded" and shipped away, "in small cargo ships" to "other countries, to people who have never seen a glacier, never held snow between their hands, and there it will melt in glasses, in drinks, there it will be destroyed" (40–41). As per her plan, Signe throws off all the "vacuum packed ice" back into the fjord and carries the rest "twelve containers containing ice one thousand years old" into her sailing boat and sets out in search of Magnus (65). Finally, her adventurous voyage comes to an end in France where Magnus now lived. Signe confronts Magus for his actions, and he accepts that all that he has been doing has led to an irreplicable damage to the environment. At the end of the story, as a token of apology, Magus helps Signe in saving water by burying the twelve containers of ice "halfway underground" on "top of a hill" by the canal nearby (383).

Signe's story clearly hints at the changing times for she mentions, "The ice no longer forms on the lakes . . . the lakes don't freeze over any longer and . . . the heat has already arrived" (64). Thus, it is interesting to see how through Signe's story Maja Lunde showcases the destructive process having begun in the near past the consequences of which are sure to be felt and experienced in the future.

Drought and Distress

Have you ever imagined what life would be like in a climate-changed world? Well, through the second story of *The End of the Ocean* Maja Lunde depicts a haunting future of a drought-stricken world where wars are waging, people are dying, and refugee camps are overflowing and falling apart. While a perpetuating drought has ecological, economic and social implications, *The End of the Ocean* depicts the distressing and dehumanising aspects of drought highlighting an erosion of human values, resulting in violence and conflict over the blatant scarcity of resources. The second story in this novel revolves around David – a young dad, and his daughter Lou. David's story is set in 2014, in France of drought-stricken Europe. David is a "climate refugee" who "has no home" for "*Home* was what he had lost" (303). David is seen travelling through the drying-up world, uncertain of where his wife and son are, uncertain of the future, and uncertain of what will happen to him and his daughter. This feeling of uncertainty and lack of understanding is arguably one of the central psychological effects of climate change and is often referred to as eco-anxiety, climate distress or climate anxiety (Clayton et al. 26–27). According to psychologists, people whose homes are transformed or who are forced to leave their homes due to environmental crisis, often develop eco-anxiety, which stems from the disruption of personal identity and attachment to a place. Such people repeatedly grieve about the loss of their property and things and are worried about a greater possible future damage (Wang et al. 28–32). All of this seems to be true in the second story of *The End of the Ocean.*

At the onset of the story, Lunde informs the readers that it has been 24 days since David and his family had to run away from their hometown in Argelès in Bordeaux, France. Unfortunately, David's family of four had been "split up" when they "left the south," but David and his wife Anne had agreed to meet "at a refugee camp in Timbault" (Lunde 29). They were fleeing away towards the "water countries," which referred to "the countries in the north, where the rain didn't just come once in a great while during the cold months, but also in the spring and summer. Where long-term drought didn't exist" (24). Through David's story, Maja Lunde paints a horrifying picture of the environmental crisis in the form of food and water scarcity. It also highlights how climate change can lead to a breakdown of social norms and stability, thus causing conflict and mass migration.

During our first encounter with David, it becomes obvious that he is tired of the five-year-long ongoing drought and is desperately hoping for some rain. While walking in search of a refugee camp David reflects, "above us the sky was blue. Not a single cloud. Blue, always blue. I'd started to hate that color" (21). Maja Lunde presents a vivid picture of David's journey and his life at the refugee camp. First, the readers get a graphic description of David's long and exhausting journey during which he carried his daughter in his arms and "stopped from time to time" at "every single farm they passed" only to be able to find "food in a couple of places, but water in only one place, in an almost empty tank" (314). Throwing light on the grim condition, David remarks, "the houses we passed were abandoned, the fields dry and scorched brown by the sun" (21). Further enlightening the readers on the changed world around him David says:

> During the past year everything was rationed. We stood in lines for a gallon of milk. For a cut of meat. For a bag of apples, every other kind of fruit. The lines for fruits and vegetables were the longest. There were so few bees, so few insects. They had disappeared gradually but when the drought came, the extinction rate accelerated. No insects, no fruits . . . Lou couldn't remember a life without lines.
>
> (Lunde 29)

Finally, David and his daughter, Lou, manage to reach the refugee camp in Timbulant, which was the last one still functioning. Upon arrival, David finds himself standing in a long queue to get registered in order to receive some food and a place to rest because "everything here was rationed, from food to water to even the space for rest" (55). Describing the state of the refugee camp David mentions that it was "a huge run-down factory, divided up into long corridors and small cubicles by old awning cloth that had been hung from the ceiling" (43). Reporting the paucity of water and the resulting miserable condition in the camp David informs the readers that the refugees were allowed to shower only on Tuesdays and to do laundry, they had to seek permission and request for a "little treasure trove of water" (68). As David's story progresses, one can clearly figure out that water was a prized possession for these climate refugees. Like David, the others also tried hard to save every drop of water they received. Through this story, Maja Lunde has been successful in driving home the message that water is one of the elements without which life cannot exist. She makes us realise how precious and valuable even a single drop of water can be for humans in the wake of a crisis. Lunde presents a detailed picture of David trying to take a bath from the three spurts of rationed water allocated to him in the camp:

> The spray hit me on the back of my head, hammering against my skull. I tried to absorb the sensation of each individual drop against my skin, taking pleasure in every single one. Then the spurt of water abruptly came to an end. I turned towards the shower head. A little water was still trickling out, less and less, until there was no more. *One* tenacious drop hung there. Finally, it dripped off the shiny head and fell to the floor. Then nothing more. . . . For a moment I didn't think of anything else, not even of Anne, not of August, just about the soap, the water, about how it was like getting my body back, shedding a layer of skin.
>
> (Lunde 56)

In addition to this, battling thirst and hunger form a major issue in David's story. With the worsening drought, the camp is unable to keep up with the increasing influx of refugees and hence it gradually starts "falling apart" (210). Gradually, David's camp starts facing an abject shortage of water and food due to which even greater restrictions are levied on the already suffering refugees. Reflecting on the deteriorating condition of the camp David thus mentions:

> During the past few days, the halls had filled up. Many people were being pressured to share cubicles with strangers. . . . The food rations had been reduced. I had almost become accustomed to going hungry. To the growling of my stomach. . . . But the worst part was the water. They had cut back the rations even more. We couldn't shower, couldn't wash our clothes. We received exactly as much as we needed to drink.
>
> (Lunde 235–236)

As food rations are slashed, Lunde portrays a moving picture of these climate refugees trying hard to face moral challenges for survival. We find little Lou stealing food from the camp's store house. David was shocked when he first got to know that Lou had stolen a can of corn. But he neither blamed nor scolded her rather the father-daughter relished the little food Lou procured, "taking turns . . . picking one piece of corn at a time out of the can" (241). David feels helpless about the situation and states that Lou "knew that stealing was wrong. . . . But all the same, she had stolen again, because her hunger was controlling her thoughts. Overpowering everything else" (302).

In addition to this water and food scarcity, Maja Lunde has also touched upon the subject of water contamination. There are several references to the availability of mostly "salty, undrinkable water" (368). "An infinite amount of water. And all completely undrinkable. Dead water, not good for anything" (262). However, there is a glaring incident in the novel which highlights this matter. It so happens that one day, tired of hunger and thirst, David and Lou venture out into the neighbourhood where they chanced upon a locked "rainwater tank" (156). After numerous attempts, the father-daughter duo succeeds in opening it to find that it is "only a quarter full" and the content inside "could not exactly be called water. The sour-smelling liquid was light brown in colour and there were flakes of rust swirling through it" (158). David and Lou were disappointed to discover this. Nonetheless, they took out the so-called water and washed and cleaned themselves. David had been mindful and warned Lou to not drink that dirty, contaminated water. But, unfortunately, the little girl was unable to control her thirst, and she took a sip. And the next day Lou gets sick and starts vomiting, which was an indication of stomach infection. Later, David realises that it was "poisoning" due to the "contaminated water" (190). While speaking to some other residents at the camp, he also gets to know that many had lost their children to this sickness. In a desperation to save his daughter, David visits the first-aid counter in the camp. But all in vain for it was closed due to lack of supplies. David's tireless efforts at saving Lou is heart-wrenching. He makes no fuss when he couldn't find any water to clean Lou's puke but only hints at the deplorable condition:

> I mopped up the vomit on the floor with a dry rag I'd found in the broom closet. I hoped I could find some water to wash with in the morning. My fingers smelled. But I grew accustomed to it, became immune.
> (Lunde 185)

Luckily, Lou gets better within a couple of days and then David decides to leave the camp and move on. Unsure of which direction to take "because the war was advancing from the south, and north of us the borders were closed," David decides to head "west (because) that was the only direction open, but there lay the ocean, only the ocean" (258). Surprisingly, this time on their trek, David and Lou come across an old, abandoned boat "on a stand, in a garden, miles away from the ocean" (95). The readers can clearly decipher that this is no other but Signe's boat. David plans to mend and use this boat to travel north to the water countries. Incidentally, David also finds a house nearby which we know was once Magnus's residence. Here, Lou uncovers the twelve "large plastic containers almost hidden by dried vegetation" which we know Signe and Magnus had earlier hidden underneath (377). David and Lou feel elated upon finding water. They then drag the containers one by one, vigilantly placing them in the middle of the living room. These clean, water-filled containers were such valuable objects that they had locked the door to safeguard them. Thus, David comments:

> We owned a treasure, twelve treasure chests. Enough water to enable us to survive for a long time. We would have to find food, but that we would manage. As long as we had water, everything was possible.
>
> (Lunde 386)

David's story ends on a note of hope as there are noticeable signs of rain, which would thus end the long-term drought. The novel closes on David welcoming rain and harbouring a plan to sail west. With the onset of rain, everything around starts coming back to life. Happily noticing this revival David states, "everything has turned green. The water has infused all the grayness with color. . . . Everything is fresh. Everything is fresh. Everything is new" (391).

Critical Reflections

Maja Lunde's novel *The End of the Ocean* makes a significant contribution in the field of climate fiction. It exceptionally focuses on and broadens the general theme of environmental issues presented in climate change novels. The following are some key elements that showcase how this novel distinguishes itself from the others in the genre.

1. **Depiction of Plausible Climate Change Scenarios**
   Unlike some of the best-known climate change novels like Kim Stanley Robinson's *New York 2140* or Paolo Bacigalupi's *The Water Knife*, which portray a relatively far-fetched, kind of hypothetical future, Maja Lunde's *The End of the Ocean* presents a more plausible and realistic scenario of climate change. In this novel, Lunde explores real-world environmental issues like global warming, melting glaciers, scarcity of freshwater and the consequent ecological disruption. The scarcity of water is a major concern in today's world and by focusing on this matter, Lunde makes the narrative more realistic and compelling. Scarcity of water is not a distant threat but a reality in today's world, and Lunde's detail-oriented

approach in presenting the struggle for survival of a community affected by drought and extreme shortage of resources mirrors actual problems faced by communities affected by water shortages today. The portrayal of water storage, rationing and establishment of refugee camps depicted in David's world reflects real-world issues related to water management in areas experiencing severe drought. Thus, the narrative portrays not just a figment of Lunde's imagination but a bleak reality of the current times. This level of realism adds urgency to her narratives, encouraging readers to see climate change as a pressing issue that requires immediate action, not just a distant, speculative possibility. Thus, Lunde's way of highlighting the environmental crisis in the novel not only makes the readers aware of the pressing challenges but also of the complex relationship between humanity and natural resources and survival.

2. **The Cross-Generational Viewpoint**

By using a dual narrative, Maja Lunde not only craftfully weaves the two plots in the novel together but also brings out a realistic presentation of environmental degradation through generations. Lunde highlights how the environmental decisions taken by one generation can profoundly and irreplaceably affect the survival of the next generation. This focus on intergenerational responsibility is a powerful and unique theme in Lunde's work, making her contribution to cli-fi not only an exploration of present-day challenges but also a meditation on the long-term impacts of climate neglect. Lunde's portrayal of how the decisions of today's generation will impact future generations is presented in the novel through Signe's and David's characters. Signe is seen to be trying hard to stop the environmental destruction as much as she can because she is aware of the consequences it will have on the people in future. But she gets frustrated on seeing how irresponsibly people of her generation act instead of trying to mitigate the environmental damage. Moreover, David is often seen to be contemplating on how difficult his life has become just because of the choices people have made long ago in the past. He is often seen wondering how different things in his world would have been only if people of his previous generations would have acted differently. Such thoughts also make David ponder on what he could do to leave a better, habitable world for the coming generations. Reflecting on this serious matter Lunde mentions,

> He (David) knew that the legacy of the past had led to the present crisis, and now it was his responsibility to do what he could to ensure that the next generation faced a world not destroyed by previous mistakes. (Lunde 306)

Thus, Lunde's idea of emphasising this theme acts as a moral reflection on the responsibility that the current generations bear towards the coming generations and future of the planet in general. This is a unique contribution by Lunde as it expands the ethical dimension of cli-fi, reminding readers that environmental decisions are not just about immediate consequences but have long-term, often irreversible, effects.

3. **Emotional and Psychological Response to Climate Change**
Majority of the cli-fi works focus on large-scale, action-driven narratives, glaringly distinct from The *End of the Ocean*, which centres on the emotional and psychological response to changing environment. By focusing on the emotional and psychological aspects of her characters, Lunde humanises the environmental crisis. The personal stories of Signe and David make the abstract issue of water scarcity more tangible and relatable, thereby increasing the emotional resonance of the novel. Lunde introduces a level of intimacy and vulnerability to the genre, portraying how individuals and families are personally affected by environmental crises. Signe's character highlights the deep and often overlooked human-nature relationship. Signe, the environmental activist is seen to be fighting hard to protect the fast-melting glaciers of her hometown. She fights this lonely battle not from a political motive but because of her deep-rooted concern that stems from her memories and her identity associated with natural resources. The emotional weight of seeing the glaciers she once loved to diminish adds a poignant layer to her character's motivations and struggles. On the other hand, David is seen to be struggling, both physically and psychologically while finding adequate resources for his and his daughter's survival. And he gets depressed when he thinks of the kind of world he will leave behind for his daughter, Lou. Thus, in this novel, Lunde creates a narrative that emphasises human resilience, loss and moral responsibility. This emotional depth contributes to the cli-fi genre by shifting the lens from the global consequences of climate change to personal and psychological.

Conclusion

Timothy Morton, in the book *Dark Ecology: For a Logic of Future Coexistence*, mentions that the challenge for a cli-fi novel is to transport the readers into the "felt feeling" of a climate-changed world. It is to create fiction that "walks them through an inner space that is hard to traverse," which requires creation of characters that readers care about, living through challenges that they can imagine (184). In *The End of the Ocean*, Maja Lunde has succeeded in doing this through her slow-paced narrative that invites readers to engage with climate change on a more contemplative level, encouraging reflection on humanity's deep connections to nature and the consequences of losing those connections. The novel's gripping narrative presents two scenarios: Signe's story depicts nature being vulnerable at the hands of humans, whereas in David's story, humans appear vulnerable and miserable trying hard to survive at the mercy of nature. Maja Lunde's contribution to climate fiction is marked by her focus on human-centred, realistic and emotionally rich storytelling. Her work adds significant value to the genre by highlighting specific environmental issues like water scarcity, exploring intergenerational responsibility, and providing a nuanced view of environmental activism. Through her innovative approach and deep engagement with both scientific and emotional aspects of climate change, Lunde enriches cli-fi and enhances its capacity to provoke thought and inspire action. *The End of the Ocean* written originally in Norwegian, a language not widely spoken internationally, makes a substantial contribution to a much-needed emotional, cultural and linguistic diversification of ecocritical research and also opens possibilities for comparative approaches (Henning 3).

Works Cited

Bracke, Astrid. "Introduction: Climate Crisis and the Cultural Imagination." *Climate Crisis and the 21st-Century British Novel.* Bloomsbury Academic, 2018, pp. 1–22. *Bloomsbury Collections*, Web 1 April 2022, https://dx.doi.org/10.5040/9781474271158.ch-001.

Clayton, Susan, Manning Christie, Speiser Meighen, and Hill Alison Nicole. *Mental health and Our Changing Climate: Impacts, Inequities, Responses.* Washington, DC: American Psychological Association and EcoAmerica, 2021, www.apa.org/news/press/releases/mental-health-climate-change.pdf.

Hennig, Reinhard, Jonasson Anne-Kristin, and Degerman Peter. *Nordic Narratives of Nature and the Environment: Ecocritical Approaches to Northern European Literatures and Cultures.* Lexington Books, 2018.

Irr, Caren. "Climate Fiction in English." *Oxford Research Encyclopedia of Literature.* Oxford University Press, 2017, https://doi.org/10.1093/acrefore/9780190201098.013.4. Accessed 6 April 2022.

Johns-Putra, Adeline. "Climate Change in Literature and Literary Studies: From Cli-Fi, Climate Change Theater and Ecopoetry to Ecocriticism and Climate Change Criticism." *Wiley Interdisciplinary Reviews: WIRssES Climate Change*, vol. 7, no. 2, 2016, pp. 266–282, https://doi.org/10.1002/wcc.385.

Lunde, Maja. *The End of the Ocean.* Scribner, 2019.

Morton, Timothy. *Hyperobjects: Philosophy and Ecology after the End of the World.* University of Minnesota Press, 2013.

———. *Dark Ecology: For a Logic of Future Coexistence.* Columbia University Press, 2016.

Wang, Susie, et al. "Emotions Predict Policy Support: Why It Matters How People Feel about Climate Change." *Global Environmental Change*, vol. 50, 2018, pp. 25–40, https://doi.org/10.1016/j.gloenvcha.2018.03.002.

4 Subverting Anthropocentrism

A Critical Study of J. G. Ballard's *The Wind from Nowhere*

Bapin Mallick and Nikhilesh Dhar

The chronological and cultural growth of human beings as the most dominant and privileged species has resulted in a variety of complexity and divisions that have led to the formation of species supremacy. The beginning of the Anthropocene epoch is marked by a profound understanding of the potentially catastrophic effects that could result from climate change caused by human activity. The term "Anthropocene" refers to a geological period that was developed as a concept after considering the extent to which humans have influenced the geological structure of the world. This places humans at the centre of the contemporary environmental structure of Earth and at the highest level of hierarchy since they are the dominant agency. This idea that all things should concentrate on humans' needs gives rise to the philosophical concept known as anthropocentrism. This concept of anthropocentrism, with its multitudes of complicated materialist ontologies, tends to grant authority to humans within the complex ecosystem of humans and non-humans. In her brave queer-feminist investigation titled "Staying with the Trouble," Donna Haraway emphasises that the terra (Earth) must be allowed to exist and recover from the "trouble" that the Anthropocene has generated. She advocates the concept of "Making Kin, Not Babies," in which humans are required to continually connect with non-humans and sustainably reside with them. This interconnection takes place not only with other humans but also with non-human animals. The formation of kinship paves the way for the functioning of intricate interspecies connections without giving up cultural supremacy over the authority of the environment.

The Western world and its practices are often criticised for advocating the central position of humans. Anthropocentrism is a worldview that centres on human experience and perception of the natural world. It places a high value on the human experience and considers it normal. In this context, Timothy Clark argues:

> An 'anthropocentric' view of the natural world thus sees it entirely in relation to the human, . . . so even an aesthetics of landscape appreciation can be anthropocentric. Anthropocentrism is often contrasted with a possible biocentric stance, one attempting to identify with all life or a

DOI: 10.4324/9781003602750-6

> whole ecosystem, without giving such privilege to just one species. . . . Normally, however, 'anthropocentrism' in environmental discourse names the view that human beings and their interests are solely of value and always take priority over those of the non-human.
>
> (3)

It is undeniable that anthropocentrism is responsible for the formation of almost all aspects of human cultures. It has brought order to our lives, but at the same time, it has drawn clear lines regarding what we believe. Humans have a disproportionate sense of self-importance regarding their life histories because they have grown to view themselves as a higher form of life than other species. Even though anthropocentrism has been a method of comprehending the world for a long time, it has led to changes in all areas of human life. Rob Boddice contends that throughout history, humans, and not animals, have almost always been associated with deities.

As a consequence of these, people desire to re-evaluate the part they play in this world. In his post-apocalyptic series of works, novelist J.G. Ballard tries to foreground humans and how they relate to the world around them. Italian philosopher Giorgio Agamben defines the human as:

> Homo sapiens . . . is . . . a machine or device for producing the recognition of the human. . . . It is an optical machine constructed of a series of mirrors in which man, looking at himself, sees his own image always already deformed in the features of an ape. Homo is a constitutively 'anthropomorphous' animal . . . who must recognize himself in a non-man in order to be human.
>
> (26)

Anthropocentrism is a philosophy that is present in all monotheistic worldviews. A renowned professor of philosophy from Bulgaria named Panayot Butchvarov wrote a book titled "Anthropocentrism in Philosophy," in which he argues that all monotheistic religions share the concept that humans were created in the image of God. In this book, Butchvarov asserts:

> with the possible exception of gods, angels, or extra terrestrials – humans alone are "rational," capable of reasoning. We assume also that only humans are capable of moral and aesthetic judgment, and perhaps that only humans enjoy the moral and political status, "dignity," of possessing "rights." The monotheistic religions, which were the home of medieval philosophy and also profoundly influenced modern philosophy, assure us of humans' unique origin and special place in nature.
>
> (6)

James Graham Ballard emerged as a novelist with the publication of *The Wind from Nowhere* (1962). This is his first creative venture into the realm of environmental fiction. Its purpose is to demonstrate the precarious position of humans concerning the natural world. The natural world has become hostile towards human beings due to the damage they have caused to the environment. It is one of the novels written before its time, especially in this day and age when there are constant conversations on environmental issues and terrible disasters.

The narrative recounts the events that take place in the life of Dr Donald Maitland, who makes arrangements to travel from England to Canada but cannot do so due to the rising violence of the wind. Soon after, Maitland and the rest of his team find themselves in a secret underground bunker, and subsequently, they move inside the pyramid-shaped skyscraper that was constructed by the British multimillionaire Hardoon. The story comes to a close with Maitland being evacuated as the underground base begins to crumble.

The winds that play a central role in the story are responsible for the deaths of millions of people and the horrific destruction of everything in their path. In the opening of the text, the reader is given a detailed description of the terrifying strength of the wind:

> Widespread havoc is reported from many parts of the world, particularly in the Far East and the Pacific, where tens of thousands are homeless. Winds of up to hurricane force have flattened entire towns and villages, causing heavy flooding and hampering the efforts of rescue workers. Our correspondent in New Delhi has stated that the Indian government is to introduce a number of relief measures.
>
> (Ballard 19)

The image that Ballard creates at the beginning of the book shows the conflict between humans and nature. This dichotomy of nature and culture is illustrated with a vivid image in which nature plays the role of the dominant force and humans fill the role of the dominated. Humans are currently at the mercy of a gigantic power beyond their control. There's a parallel between humans and little insects desperately searching for food. Ballard shows the condition of humans by setting them against the world of nature. Ballard argues, "Nature herself in revolt, in her purest, most elemental form" (141).

Ballard narrates the story from an anthropocentric perspective to expose its limitations and devastating effects. When it comes to philosophy, anthropocentrism is examined from a moral, ontological and epistemological point of view. This philosophy claims that humans have intrinsic value, while all other non-human organisms have instrumental value, and human activities should, therefore, always take precedence over the activities of non-humans and the environment. As a result, anthropocentrism makes it easier to rationalise the notion of treating nature as if it were nothing more than a tool that rightfully belonged to humans.

The author creates a transparent dichotomy by pitting humanity against nature. Ballard openly acknowledges that man is nature's most formidable adversary, and he continues by emphasising that man does not belong to nature in any way, shape, or form. Hardoon, a wealthy and powerful individual, suffers from an excessive sense of self-importance that spirals downward into despair. Hardoon is a metaphor for the 61 prosperous, capitalist and grasping Western mentalities that attribute the fixed identity of humans by placing nature and non-human organisms on the margin. Stereotypical characters, like Hardoon, frequently forget what it means to be human and start to live in their illusionary world, in which they control the non-human natural world.

In the contemporary world, the amount of greenhouse gases in the atmosphere continues to increase at an alarming rate. The environmental problems have made us more susceptible to natural disasters such as abrupt floods, freak hurricanes, major wildfires, severe droughts and rapid landslides. It was demonstrated once more in the very recent instance of bushfire in New South Wales, which was caused by high temperatures followed by a severe drought. Fire is responsible for the loss of many lives as well as the destruction of a large number of plants and animals. Lynn White, Jr. asserts that Western Christianity has played a significant role in establishing the concept of species supremacy. Since Latin was the only language that could store knowledge at the time, the only people who could access it were members of the Christian clergy. As a result, the ideas and opinions of mediaeval Christian clergy were the only ones considered reliable sources of information. The church held the belief that nature possessed no intrinsic value. It was nothing more than an instrument designed to fulfil a human need. In this context, White says:

> Especially in its Western form, Christianity is the most anthropocentric religion the world has seen. As early as the 2nd century both Tertullian and St. Irenaeus of Lyonswere insisting that when God shaped Adam he was foreshadowing the image of the incarnate Christ, the second Adam. Man shares, in great measure, God's transcendence of nature. Christianity, in absolute contrast to ancient paganism and Asia's religions not only established a dualism of man and nature but also insisted that it is God's will that man exploit nature for his proper ends.
>
> (1205)

There is another worldview known as biocentrism, which stands in opposition to anthropocentrism and plays an important role in environmental studies. The concept of biocentrism can be summed up as "putting life at the centre of everything," and its origins lie in the ancient Greek word bios, which means "life" (Johansson 12). Lawrence Henderson, an American biochemist, is credited for being the first person to introduce the phrase "biocentrism" in 1913. He intended to express the idea that the universe was the one that initially produced life (Jen 42). According to E.K. Campbell, biocentrism can be traced back to Charles Darwin, who proposed that humans and other creatures emerged simultaneously with all other living organisms.

The notion of biocentrism argues that people should be seen as a part of nature rather than as its masters. On the contrary, biocentrists don't think humans are here to save the planet or play any special role in the universe. Consequently, people should accord the same respect to all forms of life (Johansson 13). There are four major tenets of biocentrism, and they are as follows:

> (a) humans and other species are all members of the community of life in the same sense, (b) the community of life is made up of a system of interdependence, (c) each individual living thing has inherent worth, and (d) humans are not superior to other species.
>
> (Jen 43)

Through the use of the metaphor of a wind that always blows from west to east, *The Wind from Nowhere* transports us to a dystopian future in which an increasingly powerful wind is responsible for the collapse of structures all across the world, compelling people to take shelter beneath the Earth for survival. The most remarkable aspect of the book is the shift from blaming anthropogenic actions to an implicit alarm of global warming, which is apparent in the very contentious atmosphere transforming most of the metropolitan area into an area that is unsuitable for the existence of humans. This transition is the most significant aspect of the book. The characters have a hard time getting enough oxygen to sustain their lives. Ballard says, "The air was full of large pieces of grit that flicked and stung at their faces like angry insects" (105).

To emphasise the ferocity of the disaster, Ballard uses evocative images and words, which are indispensable. He highlights that the storm came from nowhere, which means that its source is unidentified, and uses phrases such as "flatten" and "heavy" to show the enormous strength and devastating aspect of nature. As Gregory Stephenson says, "The nature of cataclysm is a global cyclone whose wind velocity ultimately reaches 550 mph, destroying nearly every surface structure on the face of the earth and inflicting untold millions of causalities" (42).

Ballard demonstrates nature in the form of storms that have become unfriendly and oblivious towards human beings. In the name of development, humans have used their technological instruments to abuse the Earth and its resources. The natural world has finally struck back in ferocious cyclonic storms, transforming all of the supposedly evolved cities into hellish ones. He demonstrates "the deliberate act of an outraged Providence, determined to sweep man and his pestilence from the surface of this once green earth" (48).

The world as we know it has been engulfed in the wrath brought about by the strong winds. There is no developed nation that is protected from its grasp. Several people have been killed as a direct result of the wind. All man-made emblems of development, such as skyscrapers in the capital, New York City, Nelson's Column, the Statue of Liberty, or London Bridge, are swept away by the wind. As if to serve as a reprimand to overinflated and overbearing people, the wind is bringing down the human constructions one by one. The eradication of such structures is interpreted as the eradication of the anthropocentric worldview of humans. Ballard intended to imply, through the depiction of the devastation of world-famous structures, that all man has constructed may be demolished in a matter of a few seconds and that this revelation provides a direct indication to man to submit to the ultimate power of nature. There is an underlying suggestion that human intelligence and reason are powerless in the face of the forces of nature.

Ballard uses vibrant and strange imagery to show how Earth, the abode of human civilisation, is transforming into a suffocating planet that will eventually wipe out civilisation. It has become a hostile place for the people who live here due to the claustrophobic environment caused by the strong winds, causing them to perish in a drawn-out manner. This unusual and deleterious environment has severely disrupted the regular activities of the residents. It demonstrates the dangerous situation of humans, with all of their technological advancements, in the face of an inevitable catastrophic event. The so-called scientists cannot correctly identify the factor causing the ever-worsening crisis. Governments worldwide met to try and figure out a solution and failed miserably.

Ballard subtly depicts the psychological consequences of the natural crisis on humankind as they attempt to endure it. Amid a desperate struggle for existence, people turn against one another. The only way for the inhabitants to protect themselves from the crumbling buildings and debris blown around by the wind on the surface is to seek shelter beneath. They are made to live in their specific moles like rats, reflecting their utter pessimism about their situation. Ballard depicts:

> a sub-world of dark labyrinthine tunnels and shafts crowded with countless thousands of almost motionless beings, huddled together on the unlit platforms with their drab bundles of possessions, waiting patiently for the wind to subside, like the denizens of some vast gallery of the dead waiting for their resurrection.
>
> (122)

They had to fight not only with the wind but also with an increasingly serious lack of food and water. The situation has evolved to the point that people who formerly ruled over wide swaths of this world are now confined to cramped, constricted spaces below the surface of the Earth. It is demonstrated repeatedly by the scene of how helpless people are to deal with the repercussions of the extreme ferocity of nature. The unusual setting in the story brings about a significant shift in conventional habits of conduct. They become the frequent targets of unusual and terrifying nightmares, similar to those seen in horror movies. They cannot distinguish between the dream and the actual world around them. The primary drivers of these experiences are perceptions of obscurity, scepticism and uncertainty. Their efforts are continually foiled by strong gusts that envelop them like a parachute and prevent them from escaping. They live in a mental limbo between the devastation caused by an artificially created environment and the instability caused by the natural environment. It becomes clear that Dr Maitland is the personification of the most profound psychological crises caused by their surroundings. He imagines that he has been wounded multiple times and is blinded.

Ballard shows his utter distrust of science. When it comes to morality, scientists have been unable to come up with a collective answer to the problem of anthropogenic climate change. Throughout their careers, scientists have been so tangled up in abstract ideas and wild conjectures that they have been unable to make any real impact on the depletion of the natural world. Ballard argues:

> He realized that he had been too preoccupied with his abortive attempt to escape from England to more than notice the existence of the wind. At the airport he had regarded it as merely one facet of the weather, waiting, with the typical impatient optimism of every traveler, for it to die down and let him get on with the important business of boarding his aircraft. "What do the weather experts think has caused it?" he asked. "None of them seems to know. It certainly has some unusual features. I don't know whether you've noticed, but it doesn't let up, even momentarily.
>
> (18)

The author continues fantasising about creepy things even though his body is exhausted and his subconscious mind is overwhelmed. Because of the stress in his life, he has fallen into a traumatic event. He puts himself in the setting of a medical facility, complete with other patients. Even after he regained full awareness, he could still hear the faint voices of the medics hidden behind the walls. The characters develop a vicious survival strategy, bringing them dangerously close to eliminating one another. Kroll makes several attempts to murder Symington and Marshall. Everyone is working hard to protect his interests, such as Marshal's plan to flee the scene of the chaos with Deborah while leaving the rest of his crew, which includes Symington, Kroll and Chrighton, behind to perish. In this scenario, more fundamental aspects of humanity, such as modesty, compassion, respect and empathy, are secondary factors.

The conflict between man and nature is depicted in the book in a manner that is brought to life most effectively by Hardoon and the monster wind. The character of Hardoon is used to represent the modern man, who has become arrogant due to the advancement of scientific knowledge; the Hardoon Tower is designed to symbolise humans' efforts to dominate natural phenomena. He brags that he can subjugate everything with the assistance of technology, including mother nature. He does not submit to the authority of the natural world. He mocks the other people as cowards who have given the forces of nature and been confined to bunkers, tunnels, and sewers like insects and worms. Hardoon comes out as extremely scornful, bold and confident in his quest to construct a hierarchy similar to a pyramid to triumph over Mother Earth.

However, the final section of the book shows that even though Hardoon is full of great resources in science, power and cash, he is still unable to win the battle against the wind. The purpose of Ballard's portrayal of the disintegration of the pyramid is to drive home the fact that man will never be able to triumph over nature. The strength and authority of nature cannot be questioned or challenged in any way. The story ends with a vivid image of a massive pyramid being obliterated by massive clouds of dust as it is tossed violently by a strong wind. Therefore, the wind that blows throughout the book illustrates the great force that nature possesses, while the character of Hardoon illustrates the ultimate reality of the limitations placed on man. Gregory Stephenson argues, "in the nature of Hardoon's struggle against the wind, and in his defeat, lies the symbolical meaning of the cataclysm" (43).

In his work, J.G. Ballard powerfully depicts the event of an ecological disaster by employing the metaphor of cyclonic winds to reflect upon the rising environmental crises of today. This novel aims to illustrate the severity of the climate problem, its deadly impacts and the impossibility of the survival of humans. The psychological repercussions of an ecological catastrophe on the protagonists as they are forced to contend with a strange environment and their fruitless efforts to combat the calamity form the central theme of the book. It conveys the most important message: there is little hope of survival in the long run unless and until we learn to respect nature.

Works Cited

Agamben, Giorgio. *The Open: Man and Animal.* Stanford University Press, 2012.

Ballard, J. G. *The Wind from Nowhere.* Berkley Medallion, 1962.

Butchvarov, Panayot. *Anthropocentrism in Philosophy: Realism, Antirealism, Semirealism.* Walter de Gruyter Inc, 2015.

Clark, Timothy. *The Cambridge Introduction to Literature and the Environment.* Cambridge University Press, 2011.

Jen, Hu Y. *Understanding Anthropocentric/Biocentric Orientations Toward Natural Parks.* 2010, shareok.org/bitstream/handle/11244/9678/Hu_okstate_0664M_10676.pdf?sequence=1&isAllowed=y.

Johansson, U. *Towards a Biocentric Attitude in Environmental Education.* Linnaeus University, 2012, www.divaportal.org/smash/get/diva2:537935/FULLTEXT01.pdf.

Stephenson, G. *Out of the Night and Into the Dream: A Thematic Study of the Fiction of J.G. Ballard.* Praeger Publishers Inc, 1991.

White, Lynn. "The Historical Roots of Our Ecologic Crisis." *Science*, vol. 155, no. 3767, 1967, www.science.org/doi/10.1126/science.155.3767.1203.

5 Kim Stanley Robinson's *The Ministry for The Future*

A Tale of Transcending Climate "Catastrophism"

Tushar Kanti Karmakar

Since Amitav Ghosh's query "Where is the fiction about climate change?" in 2016, a large number of novels have appeared which endeavour to address the anthropogenic climate change. Many of these novels have imagined a dystopian future with the fear of apocalypse always lurking behind. Earlier commentators like Lawrence Buell read these apocalyptic tales as "the single most powerful master metaphor that contemporary environmental imagination has at its disposal" (qtd. in Cole 4). However, recent critics have argued that the tales of climate "catastrophism" "depoliticize a range of climate related concerns" and have also proved that they are "paralyzing, not mobilizing" (qtd. in Cole 4). This chapter endeavours to examine Kim Stanley Robinson's 2020 novel *The Ministry for the Future* in the light of the tales of climate "catastrophism." It will argue that although the novel makes use of some of the tropes of the tales of climate "catastrophism," it ultimately transcends the narrow scope of this genre. This chapter will propose that Robinson has deployed twofold strategies to achieve this goal. Firstly, he does this by attributing agency to the characters as well as organisations in his novel who try to bring in a change in the existing system in order to fight the climate crisis. Secondly, by having a utopian ending to the novel, he challenges the paralyzing apathy of the climate "catastrophism" tales.

Ghosh's 2016 book *The Great Derangement* makes a significant contribution to the discourse of climate crisis vis-à-vis literature. His central thesis in this book is that not only contemporary culture but also human imagination has failed to address the climate crisis:

> [I]f certain literary forms are unable to negotiate these (climate crisis) torrents, then they will have failed – and their failures will have to be counted as an aspect of the broader imaginative and cultural failure that lies at the heart of climate crisis.
>
> (93)

He puts it more straightforward when he writes "the Anthropocene presents a challenge not only to arts and humanities, but also to our common-sense understandings and beyond that to contemporary culture in general" (114).

DOI: 10.4324/9781003602750-7

Ghosh also challenges the suitability of the novel form to represent the climate change. Erin James, in her book *Narrative in the Anthropocene*, writes that "for Ghosh, the very phrase "climate change novel" is oxymoronic, as he sees the novel as inherently unable to represent the sudden and violent new realities of global warming" (James 12). However, during the last decades of the twentieth century and the early years of the twenty-first century a number of novels, mostly science fiction, endeavoured to capture the crisis. This, according to Ghosh, inevitably raises the question of whether science fiction is more potent than literary novels to portray the climate crisis. He thought that "This might appear obvious to many" (Ghosh 1045). A new genre of science fiction was born in the early twenty-first century, which came to be known as "climate fiction" or "cli-fi."

Caren Irr in her essay "Climate Fiction in English" has described "cli-fi" as "[o]ne of the most successful new genres of Anglophone fiction" (2). Although there is uncertainty about the origin of the term climate fiction or "cli-fi," it is certain that the journalist Dan Bloom played an instrumental role in popularising the term. Cli-fi endeavours to "imagine the impact of drastic climatological change on human life and perceptions" which "can be set in the past, present, or near future of the planet" (2). Setting plays a very important role in cli-fi: "endangered cities, islands, and remote arctic regions are common locations" (2). Alexa Weik von Mossner, in her essay "Cli-Fi and the Feeling of Risk," has defined American climate fiction as "risk fiction" (129). Cli-fi, according to Mossner, endeavours to gauge "the potential future consequences of anthropogenic climate forcing" (129). Whereas the scientific community is still uncertain about the "exact nature and degree" of the consequences of the crisis, storytellers of the cli-fi try to imagine those risks (129). Mossner further elaborates her point by stating:

> American cultural texts that represent potential future outcomes of currently perceived climate risks – the moment when the perception of risk materializes into actual catastrophe – are often apocalyptic or dystopian in nature. Pawel Frelik even argues in his contribution to this forum that *all* cli-fi is necessarily science fiction, precisely because it imagines the *future* catastrophes that might result from anthropogenic climate forcing.
>
> (130–131)

What is common in both Irr and Mossner is cli-fi's preoccupation with the future in general and future catastrophes in particular. This proclivity of cli-fi to portray "catastrophism" has recently been criticised by a number of authors including Amitav Ghosh who writes,

> cli-fi is made up mostly of disaster stories set in the future, and that, to me, is exactly the rub. The future is but one aspect of the age of

> human-induced global warming: it also includes the recent past, and, most significantly, the present.
>
> (1041)

However, it must be remembered that these critics don't question the scientific veracity of these novels. Their point of criticism lies in the interpretation of catastrophes made by these "catastrophism" novels which fail to "solicit commitment and action" (Cole 4). The focal question they are concerned with is "whether apocalyptic narratives afford 'a deepening of the imaginative engagement' or conversely 'a mode of disengagement'" (qtd. in Cole 4). Yuen in the essay titled "The Politics of Failure Have Failed" argued that

> pairing evidence of ecological collapse with 'woefully inadequate injunctions to green consumption or lobbying of political representatives' relegates agency to the market place and to certain forms of 'existential, expressive, and voluntarist politics,' with 'disempowerment and disengagement' the predictable response.
>
> (Cole 4)

A distinct change can be seen in recent years where there is a clear shift from these cautionary tales of climate "catastrophism." Many commentators, like Weik von Mossner, have shown that these catastrophism tales have "the potentially debilitating effects of negative emotional appeals" (qtd. in Cole 8). In an article titled "'At the Heart of Human Politics': Agency and Responsibility in the Contemporary Novel," Matthew Benjamin Cole has encapsulated this attitude:

> Though apocalyptic narratives need not be categorically rejected, the analysis will show that their utility is limited when they do not illuminate opportunities for collective action, in which case they may reinforce patterns of denial and depoliticizations despite their author's intentions.
>
> (04)

Kim Stanley Robinson is probably the best-known American science fiction writer today. His works, ranging from *Science in the Capital* series and *Mars* trilogy to individual novels like *New York 2140* (2017), have explored, among many things, issues like global warming and the fate of the planet. Robinson's most recent novel *The Ministry for the Future* (2020) is a significant contribution to this genre of novel writing. The novel is a gigantic one that addresses a large number of issues related to climate change like eco-terrorism and geo-engineering, which have hitherto not been explored in as great a scale as Robinson does. Like a true "Master of Disaster," Robinson begins the novel with the description of a lethal heat wave which reaches wet bulb temperature and wreaks havoc in a north Indian town. It is very interesting to note that the heat wave occurs not in the present but in a near future scenario which has led many commentators to describe the novel as a climate "catastrophic" novel. Countering this line of interpretation, my endeavour in this chapter will be to establish that *The Ministry for the Future* succeeds in transcending the limitations of the climate "catastrophism" genre.

In *The Ministry for the Future*, Kim Robinson consciously plays with the trope of climate "catastrophism." The novel begins with a gut-wrenching description of the heat wave in near future north India:

> And then the sun cracked the eastern horizon. It blazed like an atomic bomb, which of course it was. . . . The heat coming from it was palpable, a slap to the face. Solar radiation heating the skin of his face, making him blink. Stinging eyes flowing, he couldn't see much. Everything was tan and beige and brilliant, unbearable white. Ordinary town in Uttar Pradesh, 6 AM. He looked at his phone 38 degrees. . . . Humidity about 35 percent. The combination was the thing.
>
> (Robinson 2)

This heat wave, which reaches wet bulb temperature, claims the lives of millions of Indian humans and animals. In an interview with Aryn Baker titled "You Need to Use Hope Like a Club to Beat Your Opponent": Kim Stanley Robinson on Climate Change and Fiction," Robinson defined wet bulb temperature as "the intersection of humidity and heat at which point the human body is no longer capable of cooling itself even in the shade" (n.p.). At the beginning of *The Ministry for the Future*, we find that Frank May, an American aid worker in India, is desperately trying to save the lives of the Indians from the devastating heat wave. Robinson's description of the impact of heat wave is so accurate that many readers found the novel too disturbing to read:

> Wails of dismay cut the air, coming from the rooftop across the street. Cries of distress, a pair of young women leaning over the wall calling down to the street. Someone on that roof was not waking up. . . . Sirens now cut the air, sounding distant and as if somehow submerged. . . . Sirens like the wails of gods and goddesses, the whole Hindu pantheon in distress.
>
> (2)

At the core of the novel *The Ministry for the Future* is a tale of two characters, Frank May and Mary Murphy, who refuse to be bogged down by "climate fatalism" which is described by Mayer and Smith as "personal beliefs that climate change is unstoppable and related lack of agency to make an impact" (qtd. in Mayer and Smith 10). Although Robinson begins with the description of climate catastrophe, he desists from "framing climate change as an unstoppable, apocalyptic problem" which, he is very much aware, "may hinder a collective social response" (qtd. in Mayer and Smith 1). Rather his two major characters display climate efficacy by trying to fight the climate crisis in their own ways. The novel also depicts two organisations – namely Children of Kali, an eco-extremist group, and the eponymous Ministry for the Future – which also try to fight the climate crisis in their own ways.

In the beginning of the novel, Frank May, a young American working for an NGO in the North Indian state of Uttar Pradesh, tries very hard to save the lives of local residents from the devastating impacts of the heat wave. At first, when the electricity goes off, he brings some local people into an air-conditioned room run by a generator. But, when this generator is snatched away by a group of armed people, May leads the people to a nearby lake in the hope of finding respite from the intense heat. But, on the contrary, the immensely hot water of the lake begins to boil the body of May and these people: "He had been poached, slow-boiled, he was a cooked thing. . . . His limbs were like cooked spaghetti draping his bones, but his bones moved of their own accord" (Robinson 12). May's attempt backfires and everyone whom he leads to the lake dies although he himself is saved by the aid team at the last moment. Although his life is saved, Rebecca Onion remarks,

> May experiences intense, disabling PTSD and survivor's guilt, not least because he suspects he may have gotten through that terrible night because he concealed a stash of water from others around him, sipping from it when people are not looking. This act of resource-hoarding haunts him, and he becomes a catalyzing force in the global fight to decarbonize.
>
> (Onion n.p.)

Frank May's direct experience of the disastrous impact of climate change radicalises him. Returning from Glasgow, where he was being sent to recover his health, to India, he first tries to join the extremist group called Children of Kali in order to fight the climate crisis. But, when he is not allowed to join the group, he goes to Switzerland and kidnaps the secretary of the Ministry for the Future, Mary Murphy, with the intention of galvanising the effort of the ministry towards fighting the climate crisis. Influenced by the ideals of Children of Kali, he also attacks the fossil fuel tycoons whom he considers criminals.

Frank's use of violence in his fight against the climate crisis introduces elements of ecoterrorism in this novel. In an article "Environmental Extremists and the Eco-Terrorism Movement," Chad Nilson and Tod Burke defined "eco-terrorism as any direct or indirect use of force, wilful damage or violence against persons, groups, or property that is used to terminate, prevent, or minimize human alteration to any part of the natural environment or its animal species." The eco-terrorism movement emerged in America in the 1980s. According to Nilson and Burke, it is used as "a tactic to stop companies, institutions, organizations, and governments from damaging or altering the environment." Andreas Malm, in his 2020 book *How to Blow up a Pipeline*, has discussed the importance of using sabotage in contemporary climate movements. He has criticised the position of "strategic pacifism" advocated by Bill McKibben and Extinction rebellion. He is of the opinion that "the ruling classes are constitutionally incapable of responding to the catastrophe in any other way than by expediting it" (Malm 12). He also remarks that "the ruling classes really will not be talked into action," so "the movement must learn to disrupt business-as-usual" (19). He differentiates sabotage or violence perpetrated on property from violence perpetrated on humans and animals by quoting Martin Luther King. Malm ends the book with his endorsement of sabotage in combating climate change: "There has been a time for a Gandhian climate movement; perhaps there might come a time for a Fanonian one. The breaking of fences may one day be seen as a very minor misdemeanour indeed" (104).

It is interesting to note that although Malm in his book has differentiated between two kinds of violence, our protagonist Frank May in the novel fails to make this distinction. Ravaged by the direct impact of climate change in the form of a heat wave, Frank becomes obsessed with the desire to attack those who, according to him, are responsible for creating the crisis. Labelling this kind of person as "climate criminal," he says, "Few war criminals would kill as many people as that man would" (Robinson 86). He goes on to comment that a climate criminal's "life's work would drive thousands of species to extinction and cause millions of people to die" (86). Considering himself as "A Child of Kali, here in the West. A fellow traveller," Frank "still wanted to kill one of the people killing the world" (86). He literally strikes a fossil fuel tycoon on the head as a result of which the later dies. However, he later considers this incident an accident. He realises that "[s]hooting someone was not like punching an asshole. He would have to find a different way" (86).

Frank reaches Zurich where the Ministry is based in order to kidnap its head Mary Murphy with the intention of threatening her to take instant action to combat the climate crisis. He tries to convince the minister of his theory of "climate criminal" as the "mass murderer":

> There are about a hundred people walking this Earth, who if you judge from the angle of the future like you are supposed to do, they are mass murderers. If they started to die, if a number of them were killed, then others might get nervous and change their ways.
>
> (Robinson 98)

He exhorts her to shed bourgeois values because we no longer have time for these. He remarks:

> The stakes are too high for you to hide behind them anymore. They are killing the world. People, animals, everything. We're in a mass extinction event, and there are people trying to do something about

> it. You call them terrorists, but it's the people you work for who are the terrorists.
>
> (97)

The significant point that I would like to emphasise here about the character of Frank is that, instead of plunging into climate despair, he courageously combats the climate crisis, albeit using eco-terrorism. What is more, although he stops short of deliberately murdering "mass murderers," he morally supports those who are using violence to stop the catastrophe. However, we have to remember here the difference in opinion that the author has with the protagonist Frank about the use of violence to fight the climate crisis, which will be discussed later.

The group that fights the climate crisis with success, albeit using violence, is the Children of Kali. This group emerged in India after the lethal heat wave struck north India. It used various measures like drone attacks on passenger planes and ships to reduce the use of fossil fuel. What is significant here is that, instead of wallowing in climate catastrophism, they are taking tangible steps to tackle the climate crisis. However, we must not see it as the only way endorsed by Robinson to fight the crisis. Rather, the author dissociates himself from the violence used by both Frank and the Children of Kali:

> I am in a very privileged position and recommending violence that I would never do myself, I would abhor that. Now, that said, the *Ministry For the Future* is a scary and confusing novel, in that there are people in that book that think they are fully justified in their violence against persons. I wish now that the book made a clearer distinction between murder and sabotage.
>
> ("You Need to Use Hope like a Club" n.p.)

Mary Murphy's fight against the climate crisis is of a different order. She, unlike Frank, is not directly affected by any extreme climate change incident. She fights at the organisational level as the head of the eponymous Ministry for Future. The Ministry, an imaginary organisation established under the Paris Agreement in January of 2025, is "charged with defending all living creatures present and future who cannot speak for themselves by promoting their legal standing and physical protection" (Robinson 16). It was created after the Paris Summit 2024 when the members decided to create an agency which would represent the interests of future generations, especially "the interests of those entities that can never speak for themselves, like animals and watersheds" (98). The ministry consisted of a group of experts from various fields like law, science, economics and politics. At the core of the foundation of the Ministry is the notion of justice, as Mary states: "We're trying to set things up so that in future, over the long haul, something like justice will be created. Some long-term ledger of more good than bad" (98).

One of the strategies deployed by the ministry to combat the climate crisis is the introduction of the carbon coin. Robinson borrowed the idea of a carbon coin from the Australian civil engineer Delton Chen's idea of carbon currency. In an interview, Chen differentiated carbon currency from carbon pricing systems. He conceived of carbon currency in terms of reward which, unlike typical subsidy, "will be managed with monetary policies [controlled by central banks]." In the novel, the Ministry introduces a carbon coin which

> is to be a digital currency, disbursed on proof of carbon sequestration to provide carrot as well as stick, thus enticing the global capital into virtuous actions on carbon burn reduction. Making an effective carrot of this sort would work best if the central banks backed it, or created it. A new influx of fiat money, paid into the world to reward biosphere-sustaining actions.
>
> (Robinson 172)

A carbon coin will be given to any individual or nation provided they will sequester carbon underground for a significantly large period of time. A minimum price will be set for it which will increase with the market but will never go down the minimum set price. This coin will in turn be used in a way that will preserve the health of the biosphere by reducing carbon from the air. This process, as the novel shows, is sometimes, following Chen, known as carbon quantitative easing (CQE). At the level of the story, this entire notion of carbon coin, along with the process of carbon quantitative easing, proposed by the ministry, provides a solution to the problem of increase in carbon which is responsible for the climate crisis. It is noteworthy that the novelist is introducing innovative strategies like carbon coin in the narrative, which enables the novel to dissociate it from the catastrophism genre.

Robinson also introduces geo-engineering, considered by many as the third rail topic in climate change studies, to take charge of planetary management. Many thinkers have expressed their disapproval of the very idea of geo-engineering. Important among these thinkers is Frederic Neyrat who wrote:

> The problem with geo-engineering is that, as its name suggests, it would like to consider the Earth (geo) as its "object". But the Earth is precisely not an object, not even a "hyperobject" . . . as proposed by the American essayist and environmental philosopher Timothy Morton. . . . But the Earth escapes any attempt at objectification. I call it a "transject": a trajectory of time and space which began its formation 4.5 billion years ago, and during which life appeared. When engineers or thinkers imagine a "Gaia 2.0" where the climate would be technologically transformed and therefore controlled, they are seeking to erase with a magic wand the billions of years of research and development (as

> the environmental thinker Barry Commoner put it), of trial and error which evolved with the Earth, enabling it to become what it is. These scientists think that a few calculations made by their computers can replace a history on the scale of the universe.
>
> (n.p.)

Irrespective of the debate on the ethics and practicality of the question of using geo-engineering, we have to admit that Robinson's use of it in his novel is yet another instance of emphasising the importance of agency in challenging the stance of "climate fatalism," which can lead us to a catastrophic future. Recognising the undeniable fact that "humans have become a planetary force," Dipesh Chakrabarty, in an interview, remarked, "We might have to do climate engineering, that is, planetary scale management. Whether that is good thing or bad thing is a different question" (n.p.). However, there is a lot of controversy surrounding the word geo-engineering. Robinson succinctly captured this debate in one of his interviews where he said,

> The word geoengineering has been demonized particularly on the left because it looks like a getaway card for the fossil fuel industry or it looks like we're trying to ignore the problem and just MacGyver our way out of it.
>
> (n.p.)

According to him, "Anything that is done at scale is geoengineering." That's why he considered an issue as divergent as women empowerment to be an example of geo-engineering because it controls the population which leads to the betterment of the biosphere. Robinson captures the crux of geo-engineering debate when he says that a lot of objection to geo-engineering stems from its name because "it covers too many separate activities that have different impacts, and different possibilities" (n.p.). For this reason, he suggests "climate restoration" as an alternate word to geo-engineering, because "Climate restoration speaks to the why of it, that we need to restore the climate or we might cause a massive extinction event" (n.p.).

In *The Ministry for the Future*, India, just after being ravaged by the disastrous heat wave, started doing geo-engineering in the form of "a solar radiation management" defying the protocol set up by the Ministry for the future (Robinson 18). It wanted to avoid the repetition of the heat wave which, as the Indian representative states, was caused by Europe, America and China. India endeavoured to achieve an effect similar to the Pinatubo volcanic eruption of 1991, in which sulphur dioxide was shot into the stratosphere in the ash cloud that significantly lowered temperature for a year or two of the earth by a degree Fahrenheit. India started shooting sulphur dioxide in the air which, reflecting the sunlight back to the space, lowered the temperature and thereby avoided the possibility of the repetition of a heat wave. What is important about the episode is the degree of agency provided by the novelist to its characters who effectively combat the climate crisis. This is all the more evident in the words of the Head of India's delegation to the Paris Agreement: "Everyone knows, but no one acts. So, we are taking matters in to our own hands. We'll lower global temperature for a few years, everyone will benefit. And perhaps we will dodge another massacre like this one" (20).

Another instance of geo-engineering in the novel can be found in Robinson's description of the Arctic and Antarctica where scientists are involved in gigantic affairs of resisting the excessive melting of glaciers. In order to reduce melting of the ice in the Arctic, scientists dye ice sheets yellow which reflects sunrays. The glacier melting in Antarctica poses altogether a different problem. The water melted below the glaciers that create cracks is being thrown away by using ice coring and drilling techniques traditionally used by oil companies. Both these cases highlight the extent to which Robinson goes to endorse his point that geo-engineering is inevitable for tackling our climate-ravaged planet. What is more, he emphasises the importance of taking tangible action in his characters, who play a very crucial role in challenging the status quo in climate change politics and thereby averting catastrophic end to the planet.

Sasha Lilley, in the introduction titled "The Apocalyptic Politics of Collapse and Rebirth" to her edited book *Catastrophism: The Apocalyptic Politics of Collapse and Rebirth*, wrote, "Catastrophism presumes that society is headed for a collapse, whether economic, ecological, social, or spiritual. . . . Catastrophists tend to believe that an ever-intensified rhetoric of disaster will awaken the masses from their long slumber" (17). But these catastrophists failed to understand, Lilley goes on to argue, "the crux of the ecological calamity and misses how fear can hinder, rather than help, its attempts to halt the disaster" (18). Yuen's chapter in this book also suggests that the problem with climate catastrophism is its politicisation. The catastrophists believed that information about the impending catastrophe can have a positive impact on ordinary humans who will take tangible steps to avert the crisis. But this assumption has proved wrong: "Numerous apocalyptic scenarios that never came to pass have heightened the ineffectiveness of environmental rhetorics of catastrophe" (21). In the "Foreword" titled "Dystopia Is for Losers" to the same book, Doug Henwood wrote, "recovering a utopian sensibility is about the most practical thing we could do right now. Dystopia is for losers" (16). By choosing to end the novel on a utopian note, Robinson is exploring an altogether different possibility of a planet where humans will no longer be at the centre. At the end of the novel, Art, a friend of the protagonist, Mary, interprets the statue of Ganymede in Zurich as the representative of humankind who is offering the planet back to the non-humans. Robinson's hope comes alive in the following lines: "That there is no other home for us than here. That we will cope no matter how stupid things get" (Robinson 563). I shall end the essay by quoting Robinson who, in an interview with Bill McKibben, himself has advocated utopian writings for our time:

> That whole dystopian postapocalyptic strain – it doesn't serve as a warning, it doesn't make change your behaviour. I reject all that. I write as a utopian-science fiction writer . . . at the moment we're at right now in world history, I have to set a pretty low bar for "utopia." If we dodge a mass-extinction event in this century, that's utopian writing. That's best we can expect from where we are right now. Having put that story on the table as being possible, it suggests that we ought to be trying for it.
>
> (Wickenden n.p.)

Works Cited

Chakrabarty, Dipesh. "The Planet Is a Political Orphan." *NOEMA*, 3 February 2022, www.noemamag.com/the-planet-is-a-political-orphan/.

Chen, Delton. "Could a 'Carbon Coin' Save the Planet?" *The Wall Street Journal*, 08 March 2022, www.wsj.com/articles/could-a-carbon-coin-save-the-planet-11646749801.

Cole, M. B. "'At the Heart of Human Politics': Agency and Responsibility in the Contemporary Novel." *Environmental Politics*, 26 March 2021, https://doi.org/10.1080/09644016.2021.1902699.

Ghosh, Amitav. *The Great Derangement: Climate Change and the Unthinkable*. Penguin Books, 2016. Kindle Edition.

———. "Where Is the Fiction about Climate Change?" *The Guardian*, 28 October 2016, www.theguardian.com/books/2016/oct/28/amitav-ghosh-where-is-the-fiction-about-climate-change-Web.

Henwood, Doug. "Dystopia Is for Losers." *Catastrophism: The Apocalyptic Politics of Collapse and Rebirth*, edited by Sasha Lilley et al. PM Press, 2012, pp. 10–16.

Irr, Carren. "Climate Fiction in English." *Oxford Research Encyclopedia of Literature*, February 2017, https://doi.org/10.1093/acrefore/9780190201098.013.4

James, Erin. *Narrative in the Anthropocene*. The Ohio State University Press, 2022.

Lilley, Sasha. "The Apocalyptic Politics of Collapse and Rebirth." *Catastrophism: The Apocalyptic Politics of Collapse and Rebirth*, edited by Sasha Lilley et al. PM Press, 2012, pp. 17–30.

Malm, Andreas. *How to Blow up a Pipeline: Learning to Fight in a World on Fire*. Verso, 2021.

Mayer, Adam and E. Keith Smith. "Unstoppable Climate Change? The Influence of Fatalistic Beliefs about Climate Change on Behavioural Change and Willingness to Pay Cross-nationally." *Climate Policy*, vol. 19, no. 4, 511–523, https://doi.org/10.1080/14693062.2018.1532872.

Neyrat, Frederic. "Can We Fix the Climate? Interview with Frederic Neyrat." *Philonomist*, 22 September 2022, www.philonomist.com/en/interview/can-we-fix-climate.

Nilson, Chad and Tod Burke. "Environmental Extremists and the Eco-Terrorism Movement." *ACJS TODAY*, January/February 2002, www.unl.edu/eskridge/ecoterrorism.html.

Onion, Rebecca. "I Can't Stop Thinking About the First Chapter of This Climate Change Novel." *SLATE*, 16 August 2021, slate.com/technology/2021/08/ministry-for-the-future-first-chapter-kim-stanley-robinson-interview.html.

Robinson, Kim Stanley. *The Ministry for the Future*. Orbit, 2020.
———. "'You Need to Use Hope like a Club to Beat Your Opponent': Kim Stanley Robinson on Climate Change and Fiction." *TIME*, 8 September 2021, time.com/6086585/kim-stanley-robinson-climate-change-fiction/.
Weik von Mossner, Alexa. "Cli-Fi and the Feeling of Risk." *Americastudien/American Studies*, vol. 62, no. 1, 2017, pp. 129–138, amst.winter-verlag.de/issue/AMST/2017/1.
Wickenden, Dorothy. "Kim Stanley Robinson on 'Utopian' Science Fiction." *The New Yorker*, 30 August 2021, www.newyorker.com/podcast/political-scene/kim-stanley-robinson-on-utopian-science-fiction.

Part II

Vulnerability, Precarity and Resilience

6 "Climate Plague",[1] Precarious Lives and Resilience in a Post-Apocalyptic World

Vignettes of Vulnerability in Sequoia Nagamatsu's *How High We Go in the Dark*

Sk Tarik Ali

Introduction

The transition years from the second decade to the third decade of the twenty-first century have been the years of pangs, pains and panics when a microscopic yet immensely powerful pathogen brought the entire world to a standstill. The COVID-19 crisis and prevalent pandemic situation put a halt to the wheels of civilisation and posed apocalyptic challenges to human survival on the planet Earth. The public health crisis and pandemic-induced social and economic upheaval once again debunked the illusion of human sovereignty and pointed to the reality of uncertainty, vulnerability and fragility of all forms of life on planet Earth. In today's world of climate peril, the precariousness of all forms of life on earth is complemented by the vulnerability of the environment itself, which has been rendered precarious by the excessive human intervention in the system and process of the environment. Scientists and environmental experts have found a critical interconnection between pandemic disaster and climate breakdown, and they have warned that COVID-19 is not the last pandemic to ravage humanity unless human beings realise the interrelation between invasive pathogens and anthropogenic ecocide. Environmental experts have observed that zoonoses will continue to wreak havoc on humanity with increasing frequency, greater intensity and longer duration until human beings put a check on their reckless exploitation and appropriation of the other-than-human. In an interview with *The Independent* Enric Sala, a marine biologist and ocean conservationist, categorically states: "I'm absolutely sure that there are going to be more diseases like this in future if we continue with our practices of destroying the natural world, deforestation and capturing wild animals as pets or for food and medicine."[2] Scientists and environmental experts have identified environmental factors like global wildlife trade, deforestation, agricultural intensification, habitat loss and unusual weather events behind the transmission of infectious pathogens from animals to humans and their subsequent outbreak. The shifting climatic zones caused by anthropogenic climate change have

DOI: 10.4324/9781003602750-9

resulted in a forced migration of wildlife to new places where they interact with unfamiliar species on an unprecedented scale.

Although in her essay "On Being Ill" Virginia Woolf observes that illness has seldom featured in imaginative literature, literary imagination has regularly engaged with human illness and health concerns through the representation of historical and imagined epidemics/pandemics in literary texts. From Boccaccio's *The Decameron* (1353) to Ali Smith's *Companion Piece* (2022), there is a rich gamut of literary representations of the individual and community responses to the pandemic disaster across time, space and a spectrum of media. Pandemic literature records the health crisis, turbulent socio-economic situations, and human resilience amid the fear and anxiety of death and loss. Again, although critics like Amitav Ghosh have observed that serious fiction has failed to come to terms with the climate crisis, there is an independent genre of climate fiction which features anthropogenic climate change in the form of dystopian imagination. From J.G. Ballard's *The Crystal World* (1988) to Lily Brooks-Dalton's *The Light Pirate* (2022), climate fiction imagines a climate-ravaged world which serves as an eco-cautionary trope in making people see an ominous climate future, a catastrophe of their own making. However, the obvious interconnection between pandemic ravage and climate peril has hardly been represented in literary imagination. In this context, the Japanese-American author Sequoia Nagamatsu's genre-bending work of fiction *How High We Go in the Dark* (2022) engages with the uncanny intersection between pandemic outbreak and environmental collapse.

In impulse a dystopian saga and in mode a morbid speculative fiction, *How High We Go in the Dark* is a genre-transcending montage of individual and community experiences of the invasion of plague in a climate-imperilled post-apocalyptic world. The dark interconnected stories in the novel engage with the multitude of responses across time, space and species when an ancient virus is unleashed from the melting permafrost in the Arctic. The melting tundra of Siberia unleashing a virus, *Black Mirror*-inspired euthanasia park for terminally ill children, genetically modified pigs, lost lives preserved within the plastic carcasses of robot dogs, decaying bodies of the virus victims, funeral hotels, intergalactic travel, intergenerational trauma – all these are refractions of a post-apocalyptic dystopia where grief, loss and death give way to flourishing commerce. However, the montages are as much about prescient warning as they are about the resilience of the human spirit and the infinite human capacity to love and dream even in the face of impending doom. This essay reads Nagamatsu's novel to see how the agency and ascendancy of a pathogen foreground the fragility and contingency of human lives in a climate-ravaged world where all forms of life are uncannily precarious. It looks into the different forms of vulnerabilities human lives are exposed to in a world ravaged by hyperobjects like climate crisis, enduring pandemic and capitalism. Finally, it focuses on how the undercurrent of endurance and resilience in the otherwise gruesome stories postulates that there is still hope in human connections even in the face of appalling disaster.

Pathogen Invasion and the Pandemic Imaginary

In her essay "Illness as Metaphor," Susan Sontag observes that epidemic diseases and contagion are not merely biomedical facts. She finds that in case of an epidemic, "feelings about evil are projected onto a disease. And the disease is projected onto the world" (Sontag 58). Again, René Girard's reading of disease as a metaphor in *Oedipus the King* highlights how epidemics also serve as a metaphor for social tensions and crises. In Nagamatsu's *How High We Go in the Dark* too, the disease is not only a biomedical phenomenon; it serves as a metaphor for social chaos, ineffectiveness of borders and the life under threat. The novel consists of a mosaic of post-apocalyptic pandemic narratives where the author envisions the cultural and planetary effects of a devastating pandemic through the pictures of decimated bodies, empty streets, social chaos and capitalism looming large everywhere. It begins in 2030 when a global plague unfolds after some scientists stumble upon a virus-infected young female Neanderthal corpse frozen in the rapidly thawing permafrost in the Batagaika crater site in the Arctic. What follows in a series of supposedly interconnected vignettes is humanity's response to the global pandemic and how people try to survive and deal with their loss and grief leaving behind a trail of destruction.

The virus initially affects children, and it works by "instructing the host cells to serve other functions, like a chameleon–brain cells in the liver, lung cells in the heart. Eventually, normal organ function shuts down" (Nagamatsu 28)[3]. In the second story "City of Laughter," we are introduced to a euthanasia amusement park where terminally ill children from the Arctic Plague can have last few quality days of life before they are euthanised on the park's rollercoaster ride which is "capable of lulling their passengers into the unconsciousness before stopping their hearts" (31). The virus quickly spreads around, and we meet the first adult to contract the virus in the third story where a group of comatose plague patients gather in the dark and become witnesses of each other's lives leading up to shared infection. A scientist forming a father-son bond with a talking pig in the laboratory after he lost his son to plague, elegy hotels with bereavement coordinators where people can spend time with their loved ones died of the plague, speech of dead loved ones programmed in robot-dogs, a forensic scientist falling in love with a dying victim who has donated his body for the lab after death, woman dying of cancer caused by the plague, experimental treatment complications – Nagamatsu imagines a world pervaded by death, dissonance and dissolution of human sovereignty over the earth. The physical process of death, funerary rites and mourning customs in the stories render the typical sci-fi trappings like talking pig, space travel, robot-dogs and virtual reality world macabre and morbid. The invasion of the animated primordial microbe violates the sovereignty of the human and destroys the social order, and the ontological collapse suggests a cultural anxiety that human beings are not likely to continue as the dominant life form on earth. The microbial invasion and the transmission of pathogenic microbes produce "extended demography" (Nayar 18), which includes invisible microbes and alien lifeforms.

Pandemic narratives usually fall into two different types: the apocalyptic pandemic narratives which are oriented towards the final solution of the "glorious rebirth" of human society after its cleansing through disease and the post-apocalyptic pandemic narratives which portray a pandemic-ravaged society characterised by the failure of government infrastructures, social disintegration, decimation of human population and the relapse of humanity into a condition of "total animality" (Lynteris 142). In his book *Cinematic Prophylaxis: Globalization and Contagion* (2005), Kristen Ostherr observes that in bio-invasion narratives the disintegration of the body is the primary indexical sign of alien invasion. The gradually crumbling body of Fitch in "The City of Laughter" and the decaying body of Laird in "Songs of Your Decay" suggest that the coherence and consciousness that mark the human are lost in the gradual erosion of the infected body. The bodies of the children in the "City of Laughter" as hosts to the pathogen indicate the loss of the human self, and the "pathogenized host body" (Nayar 18) is the human transmuted to become Other. The experiences of the human characters in the stories are pervaded by what Morgan calls "contagion horror." The clean lab consists "of a plastic sheet duct-taped from floor to ceiling" (4), a plastic partition preventing parents from holding the affected child, comatose plague patients sharing their dark memories of the disease, people not eating fish as they believe that the disease is waterborne, forensic scientist seeing his sample-lover gradually succumbing to the disease, people suffering from brooding loneliness in the post-plague time – all these produce an antiquarian and uncanny feeling. The stories are, in fact, "outbreak narratives" (Wald) as they portray the loss of the agentive control of bodies and how the boundaries of nations and race have been subtly punctured by the invisible presence of "non-human actors and non-human violence" (Outka 30). The anxiety, trauma and isolation that the characters experience are the manifestation of the unwarranted intrusions of the foreign lifeform which renders the earth-home an inhospitable unfamiliar space where death, displacement and disorder prevail.

In earlier times, pandemic outbreaks were seen in terms of divine retribution. It was believed that just like other natural disasters, the plague had been sent as a mechanism for fulfilling the divine plan of purgation. It was commonly believed that pandemics were apocalyptic chaos for revelation, redemption and renewal, and people who survived the epidemic/pandemic were believed to be righteous, devout and elect people deserving to live on the planet Earth. However, this myth of divine intervention has become irrelevant in today's world where human beings have transformed themselves from a member of the larger ecological community to the planet's most powerful geological force. In today's world, pandemics are usually seen as the result of ecological damage caused by relentless human intervention in the form of global technological and economic development. The virus unleashed by anthropogenic intervention transcends the fragile border between the human and the non-human, and the essentialist interspecies barrier between the human and the more-than-human world is subtly breached. In her book *Bodily Natures: Science. Environment, and the Material Self*, Stacy Alaimo asserts that one's body is

> never a rigidly enclosed, protected entity, but is vulnerable to the substances and flows of its environments. . . . Imagining human corporeality as transcorporeality, in which the human is always intermeshed with the more-than-human world, underlines the extent to which the substance of the human is ultimately inseparable from the environment.
> (Alaimo 17)

A pandemic outbreak, therefore, dismantles the system of biosecurity and breaks open the vital entanglement of the human and non-human actors and factors in their shared vulnerability. A pandemic of existential magnitude strips humanity of its technological mastery and its consequent self-differentiation from the non-human. It suggests a "temporal reversal" (Lynteris 30) in forcing humans to return to their natural/animal struggle for biological survival. In his book *Human Extinction and the Pandemic Imaginary*, Lynteris uses the phrase "pandemic imaginary" to refer to humans' loss of technological mastery and the failure of medical, scientific and governmental infrastructures in the face of an invisible enemy. Lynteris writes:

> In contemporary technoscientific societies, the pandemic imaginary plays an irreducibly dynamic role. On the one hand, it can enable us to conceptualize a future where human mastery over human/nonhuman relations has ceased to be humanity's condition or destiny. It is, in other words, an imaginary that is potentially institutive of human futures without human mastery. . . . This is a vision of humanity having finally arrived at its proper plane of existence, which can either be maintained or lost but in no way overcome – an anthropogenic *telos*; humanity as Being-in-itself.
> (12–13)

Nagamatsu's novel-in-stories points out all these anguish and uncertainties about human existence and identity through the representation of depopulation, xenotransplantation and the difficulties that the characters undergo in adapting to the new normal. The ubiquity and prevalence of death in the stories through continuous references to funeral services foreground the loss of ontological sovereignty and integrity:

> We were just glorified bellhops for the mountains of Arctic plague victims awaiting cremation, for the families who wanted to curl up in a suite beside the corpses of their loved ones and heal. On any given day,

> the deceased from local hospitals lined the basement halls in biohazard bags, waiting to go through the three-part preservation process – sterilization, embalming, and our antibacterial plasticizing treatment. This bought families time to say goodbye while our crematoriums struggled to keep up with the demand.
>
> (103)

This existential fragility of humans and other life forms is complemented by constant references to the vulnerability of the environment and the climate peril in the stories. The next section of the essay focuses on Nagamatsu's representation of the climate crisis and how the twin vulnerabilities of human lives and the environment (i.e. what is called "ecoprecarity" by Nayar) highlight the horror of ecological uncanny.

Climate Peril, Ecoprecarity and the Uncanny

In *How High We Go in the Dark*, the root of the ravaging pandemic is traced back to thawing permafrost in the Arctic caused by climate change-induced global warming. Through his haunting description of a cataclysmic contagion being unleashed because of climatic breakdown in the very first story in the book, Nagamatsu seems to make the point clear that although the montages present graphic descriptions of the manifold repercussions of the ravishing plague the book is not only a pandemic novel, rather it is also a bleak cautionary tale about climate peril and its devastating impact. Nagamatsu has actually embedded the elements of pandemic fiction, climate fiction and science fiction together to imagine a post-apocalyptic world marked by the vulnerability of all forms of life in a vulnerable environment. The stories are, in fact, replete with references to the issues of feverish environment, climate meltdown and freakish weather events exercising an unprecedented impact on humans and other forms of life in many different ways. The references to "dead soil and dead oceans" (27), "floods and record-breaking hurricanes" (178), "planetary destruction" (234), "drowned world" (255), "typhoon and tsunami seawalls," relocation of "flood zone residents" (270) foreground how Nagamatsu imagines a dystopian future where human population are confronted with the twin challenges of widespread pandemic and vagaries of climate breakdown. In fact, we encountered an uncanny preview of this dual predicament of razing pandemic and savage cyclone together in West Bengal, India, in May 2020 when the deadly cyclone Amphan exacerbated the misery of people already hit hard by the coronavirus pandemic.

What Nagamatsu imagined has actually been experienced by people in parts of India and Bangladesh, and this only substantiates the fact that climate change is no more a distant threat happening in some "imagined other worlds located somewhere apart from our everyday one" (Ghosh 72). It is happening now and here. The human exploitation and appropriation of the natural environment have so far crossed the critical threshold that violent weather events and pandemic outbreaks will become more and more frequent in the near future. In an interview for the Grinnell College Nagamatsu has categorically stated that *How High We Go in the Dark* "could be considered climate fiction, or at least climate fiction adjacent. The virus, it's an other-worldly virus. It's a very strange thing. But the virus is unleashed in part because of climate change. There are increasing wildfires, there's sea level rise, there are efforts at mitigating global warming. So, all of that's present in the novel, and how could it not be? We're living in this world now" (Nagamatsu, interview by Marshall Poe).

The pressing threat of climate peril has been addressed in the very first story when Cliff, the father of the dead climate scientist Clara, writes in her journal:

> *You saw a future of dead soil and dead oceans, all of us fighting for our lives. You had a vision of what life would be like for future generations and acted like the planet had a gun to our head.*
>
> (27) (Italics in original)

Again, the seriousness and urgency of the matter are reflected in Seiji's observation that "we have reached a war zone: us versus the planet" (230) in the story titled "Melancholy Nights in a Tokyo Virtual café." Through this "us versus planet" trope Nagamatsu seems to highlight how the planetary constellations had been thoroughly reset by the oversized impacts of human actors and how due to the indelible mark imprinted on the planet Earth by dominant human actions, Nature as an independent and mysterious force, "the separate and wild province, the world apart from man" (Mckibben 41) has been replaced by an earth that we cannot recognise anymore – "more chaotic, storm tossed, disease ridden" (Mckibben XIV). Despite the scientific consensus on anthropogenic climate change, the capitalism-developmentalism advocates are still sceptical about the urgency of climate change, and they still believe that the climate agenda is a hoax. This danger of denialism which continues to impede our response to climate emergencies has been highlighted by Nagamatsu when in "Songs of your Decay" Aubrey, a researcher who studies how the Arctic plague transforms the human body, observes that "the only (radio) stations we can find are filled with evangelical preachers shouting how climate change is a lie or punishment for our sins" (156). This disinformation and denial campaign is run by the fossil fuel lobby and industry advocates to justify their unprecedented exploitation of the environment for their business boom and profit fanaticism. In Nagamatsu's work, once the health crisis is over the business corporations zeroes in on planetary crisis for profit-making:

> When we arrived at the Centauri system, we received a decades-old message from Earth, informing us that a cure for the plague had been discovered – the comatose woke up and people began to rebuild their

> lives. Funerary corporations expanded to focus on climate projects, building seawalls around coastal cities, sponsoring the solar shade project until the end of the century.
>
> (196)

The replacement of funerary services during the pandemic by climate projects in post-pandemic time highlights how capitalism always capitalises on the vulnerabilities of people, be it a pandemic crisis or climate disaster.

In his book *Ecoprecarity: Vulnerable Lives in Literature and Culture*, Nayar coins the term "ecoprecarity" (2019) which is "at once about the precarious lives *humans* lead in the event of ecological disaster . . . and also about the environment itself which is rendered precarious due to human intervention in the Anthropocene" (7). In *How High We Go in the Dark*, Nagamatsu uses "ecoprecarious imaginary" (Nayar 8) across the stories to highlight the shared vulnerabilities of human lives and the environment in a world perturbed by risks, hazards and catastrophes. As we see in the vignettes, the ecosystem of the earth has been rendered fragile and precarious in the Anthropocene, and it has been represented in terms of the widespread pandemic wreaking havoc on human lives on the earth. Ecoprecarity is also about "the erosion of species boundaries and new forms of species alignment" (Nayar 11). In *How High We Go in the Dark*, the coherence and integrity of the human body is irrecoverably breached because of the bio-invasion. The degeneration of the human into something strangely unfamiliar because of the colonisation of the body by a foreign agent marks the disintegration of species boundaries. The victims' bodies become alien to them due to the presence of the virus in them. These diseased bodies no longer fit into the norms of being human. The eroding and crumbling bodies with mysterious ailments show how the slow violence of the disease results in the loss of the autonomy and integrity of human bodies. The prevalence of death, funerary rites and mourning customs in the stories marks the precarious nature of human existence in a post-apocalyptic world where humans find it extremely difficult to cope with the changing conditions of life. Throughout the novel, Nagamatsu uses the aesthetics of corporeal trauma, silence and strong emotionalism to represent the ontological and social vulnerabilities of people. He, in fact, employs the ecohorror trope in his use of the ravaging pandemic as the manifestation of the enraged nature striking back against human tampering with the earth's natural state.

In *How High We Go in the Dark*, the loss of the ontologically stable body and the decayed environment produces "ecological uncanny" which, according to Carroll, is about "exposing the human in the natural and vice versa" (57). The shared vulnerabilities of the pathogenised eroding bodies and the diseased environment generate a sense of ecological uncanny intensified by the loss of the familiar spaces of home and the sense of being at home. The empty streets, overcrowded hospitals, euthanasia park for giving the doomed children one last wonderful day of life, spectral landscapes, mass corpses – all these signify that in the place of the life-sustaining ecosystem, something alien resides, and the eerie calmness on the surface conceals something horrific. The coexistence of the primitive body forms due to the loss of human sovereignty and the techno-scientific experiment with a focus on singularity and interstellar travel produces the "antiquarian uncanny," which is "the effect of the coexistence of current lifeforms meshed contemporary with advanced techno-science with 'primitive' and 'primaeval' forms" (Nayar 52). The commercialisation of death through euthanasia theme parks, elegy hotels and lost lives preserved within the plastic carcasses of robot dogs presents a dystopian world where the hypercapitalism-biopolitics nexus profits off of the vulnerabilities of humans and the environment.

Community, Empathy and Human Connections

In an interview with the *Literary Hub*, Nagamatsu claims that *How High We Go in the Dark* "is a literary speculative novel that never privileges the virus as the focus so much as facets of humanity – community, grief, resilience" (Nagamatsu, interview by Jane Ciabattari). Although the novel imagines a virus-affected and climate-ravaged post-apocalyptic future, it concentrates more on the importance of empathy, resilience and human connections in the face of a terrible disaster. The stories are no doubt awfully bleak while representing the horror of climate change and enduring violence. However, the novel is "ultimately optimistic," as Nagamatsu himself declares, in foregrounding how it is only meaningful connections, selfless love and community engagement that can withstand the challenges of climate peril, pandemic and capitalism (Nagamatsu, interview by Kendra Fortmeyer). The endless human capacity for love and reinvention is reflected through Cliff's love for his deceased daughter Clara in "30,000 Years Beneath a Eulogy," a cynical employee's love for a desperate mother in the euthanasia park "City of Laughter," the bereaved father's love for a pig son in "Pig Son" and the forensic scientist Aubrey's love for her dying patient Laird "Songs of Your Decay." The interconnected stories represent all types of love – love between parent and child, love between married partners, love between unmarried partners and love between human and other-than-human. The scenes of love and human connection bring a note of assurance to the narratives otherwise dominated by scenes of death, grief and fear. It is important to note here that although Nagamatsu was significantly influenced by the speculative fiction of such contemporary authors as David Mitchell, George Saunders and Kazuo Ishiguro, his stories in *How High We Go in the Dark* differ from the works of these great authors in that there is an undercurrent of positivity and assurance in Nagamatsu's envisioning of a disquieting future.

The portrayal of resilience in *How High We Go in the Dark* extends beyond individual characters to encompass the resilience of communities. The power of solidarity and collective action is also reflected in the representation of family bonding, mutual acceptance and community engagement in the stories. The elements of compassion amongst destruction, hope in moments of utter despair and human fellowship in absolute crisis "stand as a testament to human inventiveness in the face of catastrophe."[4] In a conversation with Wellington Square Bookshop, Nagamatsu categorically states that he "was interested in (a pandemic virus) as a backdrop for seeing how characters, and how professions, how society would respond to something very extreme."[5] The astonishing commitment and capacity for resilience of an impressive variety of characters and communities spanning centuries and generations substantiate how the human spirit endures for survival. In the story titled "The Used-to-be-Party," Dan Paul writes in his letter to the neighbour:

> As you know, I never showed up to anything back then. I was never one to connect. I've been that way my entire life. I went to work, kept my head down, and came home. I let old friendships frizzle. I orbited my family and all of you like a distant planet – there and yet nearly impossible to reach. I know I can't survive alone.
>
> (218)

Again, in "Songs of Your Decay," a sister named Orli got to know his brother truly well only after she lost her brother to the pandemic:

> "My brother and I didn't really spend much time together until our mom died," she says. "This plague has taken so much away from us. It's taken Laird. But in a strange way, I've also gotten to know my brother again because of it."
>
> (164) (quotations in original)

The pandemic finally ended up as a cathartic experience for Orli and many other characters in the stories. These experiences of loss, grief and stoicism of the characters certainly produce a feeling of pity and fear in the readers too. In this context, it is important to note that "among the most important resilience mechanisms inherent in fictional storytelling and artistic creativity is a transformative act of catharsis evoking an empathetic response, facilitating acceptance and overcoming of traumatic events."[6] Nagamatsu uses this mechanism quite successfully in transforming an otherwise dystopian work into a work of human acceptance, endurance and fortitude.

In "Melancholy Nights in a Tokyo Virtual Café," Akira, a homeless man without a job and purpose of life, joins an online chat for individuals contemplating suicide. In the chat room, he finds solace and meaning of life in a single mother Yoshiko, his online comrade. However, the virtual presence of Yoshiko ultimately fails to offer Akira the emotional support he needs, nor does Akira's online presence help Yoshiko to come out of her depression and she commits suicide for lack of human relationship. Kurtzman has rightly observed:

> Nagamatsu uses the relationships of his characters and their fates to insist that, though social media and relationships online are easy and readily accessible to everyone, even those who lack a permanent residence, they do not offer the emotional support and connection of an in-person relationship.
>
> (31)

Akira ultimately finds the purpose and promise of a stable life only after his in-person connection and companionship with Seiji. Seiji literally gives Akira a mode of communication and connection with people by giving him the work of distributing newsletters. Akira distributes the newsletters with the call: "Say no to capitalism and yes to community. Say yes to your fellow human being" (231). Seiji, in fact, sees the pandemic from a radical perspective: "The plagues aren't such a bad thing – a hard reset, a cleanse, a chance to make things right" (229). While Yoshika succumbs to her struggle with loneliness, Akira and Seiji are saved by the companion they find in one another. Seiji emphasises that in a dire situation it is only family and community that help people to face the crisis and overcome it:

> Seiji has given him new pages to print. . . . Instead of something far-reaching like planetary destruction or the altered migration patterns of marine life, these deal with topics on a much smaller scale–family and community. "People have forgotten how to care for each other, for themselves. We can't expect them to care about the world if they don't care about what's in front of them," Seiji explains.
>
> (234)

Nagamatsu shows how community fellowship, family ties and empathy for people are primary concerns before we engage ourselves in planetary responsibilities. His celebration of community fellowship and resilience is most powerfully reflected in the preaching of the priest in the story "Grave Friends" when the priest sermonises that "*in suffering . . . we found our heart. In suffering, we found new traditions, a way forward*" (269) (Italics in original). This celebration of the human spirit of endurance and recuperative strength takes *How High We Go in the Dark* beyond the traditional post-apocalyptic speculative climate fiction to resilience fiction which represents stories of coping, surviving and learning lessons in the face of crisis. The individual stories of suffering, trauma and adaptability make Nagamatsu's work a meaningful narrative because, as Cyrulnik observes, "as soon as we put sadness into a story, we give a meaning to our sufferings and understand, long after the event, how we succeeded in turning our unhappiness into something wonderful" (4). In the novel resilience is not portrayed as a triumph over adversity but rather as a continual struggle to find meaning and purpose in a world fraught with uncertainty.

Conclusion

In *The Great Derangement: Climate Change and the Unthinkable* (2016), Amitav Ghosh finds fault with the representation of climate change as something uncanny, unreal and extraterrestrial in the form of science/speculative fiction. Ghosh observes that the relegation of a climate novel or short story to the genre of science fiction discounts the climate crisis as something implausible and extraordinary, a problem of a distant imagined future. However, Nagamatsu observes that as the gap between present reality and the horrifying future shrinks, writers are being increasingly drawn to the genre of science fiction to represent the horror of climate change. Nagamatsu states:

> I also think science fiction in particular is increasingly becoming a default mode for many writers (even those who might consider themselves realists), seeing that the gap between 'our reality' and what one might consider 'the future' is ever shrinking. We are living in the future. We are living in a dystopia. We are living in a climate crisis.
>
> (Nagamatsu, interview by Tan Gim Ean)

The concurrent onslaught of the pandemic and cyclone in 2020 in the eastern part of India and the coastal areas of Bangladesh as already mentioned bears testimony to Nagamatsu's observation. Climate catastrophe is not a distant threat anymore, and neither is pandemic crisis a once-in-a-lifetime event any longer. In the present world of uncertainty and vulnerability when different forms of disasters loom over humanity, we need more and more stories on the multifold crises facing humanity today because storytelling plays a significant role in forming popular perceptions of these crises. The personal and communal experiences of the climate plague recorded in Nagamatsu's vignettes highlight the vital role storytelling plays in our thinking about the climate crisis and in shaping the climate change imagination. Again, the images of the pandemic crisis in these stories take us beyond figures and statistics to reveal how the disease affects the lives of individuals and communities. Most importantly, these images help us make meaning out of the panic, fear and apparent hopelessness that characterise the crisis time. They make us see the importance of community, empathy and connections in facing the challenges in the adverse situation in life. Despite the abundance of tragic scenes and graphic descriptions of loss and death, *How High We Go in the Dark* offers us a glimpse of how compassion for each other and care for our communities can see us through. Nagamatsu's work, therefore, is not only a speculative fiction imagining an implausible future; rather it is a timely intervention in bringing climate change into popular imagination and in engaging people with the seriousness and urgency of the predicament.

Notes

1 This phrase is used in the brief description of the book *How High We Go in the Dark* on Nagamatsu's website. See www.sequoianagamatsu.com/how-high-we-go-in-the-dark.

2 In an interview with *The Independent*, Enric Sala, a marine ecologist associated with National Geographic, observes that human exploitation of the other-than-human will result in more infectious diseases in times to come. He warns that to stop the birth of zoonotic diseases humans should put a restraint on their life of consumption and affluence. The Interview has been published in part in Louise Boyle's report titled " 'We should start thinking about the next one': Coronavirus is just the first of many pandemics to come, environmentalists warn" published in *The Independent* on 20 March 2020, www.independent.co.uk/environment/coronavirus-uk-pandemics-environmentalists-warning-a9413996.html. Accessed 21 January 2024.

3 All subsequent references to Nagamatsu's *How High We Go in the Dark* have been indicated with the page number in parenthesis.

4 This is how Nagamatsu's *How High We Go in the Dark* has been reviewed in Gabrielle Zevin's Favourite Reads of 2022 in the blog *Waterstones*. See www.waterstones.com/blog/gabrielle-zevins-favourite-reads-of-2022. Accessed 16 March 2024.

5 This conversation is available on YouTube. See www.youtube.com/watch?v=QEm1vLtrFvk. Accessed 23 March 2024.

6 This is taken from the abstract of Marina Grishakova's paper titled "Narrating the Extreme: At the Limits of Resilience" for the Academia Europaea 29th Annual Conference 2017 on "Story-telling and Resilience: The Function of Literature in the Overcoming of Trauma" held in Budapest, Hungary. See www.uantwerpen.be/en/projects/literary-and-theatrical-studies/section-activities/2017/story-telling-resilience/. Accessed 5 April 2024.

Works Cited

Alaimo, Stacy. *Bodily Natures: Science, Environment, and the Material Self*. Indiana University Press, 2010.

Carrol, Siobhan. "The Terror and the Terroir: The Ecological Uncanny in New Weird Exploration Narratives." *Paradoxa*, vol 28, 2017, pp. 67–89.

Cyrulnik, Boris. *Resilience: How Your Inner Strength Can Set You Free from the Past*. Penguin, 2009.

Ghosh, Amitav. *The Great Derangement: Climate Change and the Unthinkable*. Penguin India, 2016.

Kurtzman, Sadie. "Isolation in Excess Communication: Social Media in the Writing of Sequoia Nagamatsu." *First Class: A Journal of First-Year Composition*, vol. 2019, no. 1, pp. 30–34.

Lynteris, Christos. *Human Extinction and the Pandemic Imaginary*. Routledge, 2020.

McKibben, Bill. *The End of Nature*. Random House, 2006.

Nagamatsu, Sequoia. "Interview by Kendra Fortmeyer." *Split Lit Magazine*, 14 December 2021, splitlipthemag.com/interviews/1221/interview-with-sequoia-nagamatsu. Accessed 15 March 2024.

———. *How High We Go in the Dark*. Bloomsbury Publishing, 2022.

———. "Interview by Jane Ciabattari." *Literary Hub*, 25 January 2022, lithub.com/sequoia-nagamatsu-on-writing-the-grief-and-connections-of-an-enduring-pandemic/. Accessed 2 April 2024.

———. "Interview by Marshall Poe." *Grinnell College*, 1 March 2022, www.grinnell.edu/news/sequoia-nagamatsu-04-how-high-we-go-dark. Accessed 28 March 2024.

———. "Interview by Tan Gim Ean." *Options*, 7 April 2022, www.optionstheedge.com/topic/culture/sequoia-nagamatsus-debut-novel%E2%80%98how-high-we-go-dark%E2%80%99-tells-story-deadly-ancient-plague. Accessed 10 April 2024.
Nayar, Pramod K. *Ecoprecarity: Vulnerable Lives in Literature and Culture*. Routledge, 2019.
Ostherr, Kirsten. *Cinematic Prophylaxis: Globalization and Contagion in the Discourse of World Health*. Duke University Press, 2005.
Outka, Elizabeth. *Viral Modernism: The Influenza Pandemic and Interwar Literature*. Columbia University Press, 2020.
Sontag, Susan. *Illness as Metaphor*. Farrar, Straus and Giroux, 1977.
Wald, Priscilla. *Contagious: Cultures, Carriers, and the Outbreak Narrative*. Duke University Press, 2008.
Woolf, Virginia. *On Being Ill*, thenewcriterion1926.files.wordpress.com/2014/12/woolf-on-being-ill.pdf. Accessed 21 March 2024.

7 Precarious Selves, (Dis)abled Bodies and Post-Apocalyptic Narratives

Elwin Susan John

This chapter examines three seemingly distinct, yet interrelated categories – disability, the human and the post-apocalypse – to explore the cultural representation of catastrophe. By situating the discourse of disability within post-apocalyptic narratives, it highlights human vulnerability and underscores the fragile nature of human existence. The subsequent sections will delve into the genre of post-apocalyptic narratives, position the human subject within the framework of disability studies, analyse the representation of disability in select visual texts on post-apocalypse and consider how literary imagination has shaped humanity's navigation through post-apocalyptic scenarios.

Topos of Post-Apocalypse

Post-apocalyptic fiction is identified as a sub-genre of science fiction in literature. Any event in the past that has questioned the adequacy of the human subject has generated post-apocalyptic narratives. Historical events like the two world wars, the holocaust, possibility of a nuclear war, cold war, plague and other pandemics, ecological disasters, terrorist attacks, technological overpowering and AI interferences, which were, and are, tantamount to an apocalypse of the human, both literally and metaphorically, transformed into a post-apocalyptic imaginary. This genre of writing has historically grappled with the idea of posthumanism. In other words, every time humans were concerned about the demise of human species because of a certain cataclysmic event, cultural narratives surmised the posthuman. Therefore, the idea of posthuman has been with us for a very long time. It was probably incidental that speculations of the post-apocalypse are mostly accommodated in science fiction.

The popular representation of the post-apocalypse proposes a narrative circuit which includes certain features, patterns and tropes unique to post-apocalyptic narratives. Temporally, the narrative introduces a unique distant future as the time frame or the premise of the event. A year in the future (after the cataclysmic event) is mentioned at the beginning itself and that generates a distinct relationship between the audience and time.

DOI: 10.4324/9781003602750-10

For example, a prelude that says that the year is 3045 and that humanity has been wiped from the face of Earth and the very few people who have survived with massive health problems have been relocated to Spaceship X or something of this kind. It lets the audience imagine a different world and a world that has changed its fundamentals. At times, the narrative begins with Day-1, Day-2, so on and so forth, dated after the event which would also bring in the framework of a new world where the old rules do not apply. While spatially, the terrain and atmosphere will represent the aftermath of an apocalyptic event. The location of the narrative could be within the debris on Planet Earth or a spaceship or even another planet from which, survivors will come to reclaim Earth. Post-apocalyptic narratives may or may not depict the reasons for its current state of affairs, but the narrative illustrates that the familiar forms of violence are replaced by new rules of control and authority. In short, our familiar definitions of time and space rupture here.

Very often a scientific or a religious rationale will be inserted into the narrative in order to justify the events that are to follow. The most likely event of the catastrophe will be either ratified and reiterated by characters playing the role of scientists or through cultural and mythical metaphors like the Doomsday Clock, Gates of Valhalla, Paradise, Aka Manah etc.

The creation of social awareness becomes one of the agendas of these narratives to warn humanity about an impending catastrophic future so that, if possible, such an event can be delayed or postponed. At times, post-apocalyptic narratives turn to become a cautionary tale. The threat of extinction is usually depicted as an external intervention. However, anthropogenic climate changes would place the human back at the epicentre and transform the human versus external threat into a human versus human threat. As the narrative progresses, the few survivors of the catastrophic event will have to fight and win over the causative agent, which could be natural or human-made as in aliens, the irrevocable collapse of technology, fury of nature, pandemic, nuclear disaster, medical mishap, eschatological or depletion of resources.

However, in most narratives, even when the cataclysmic event has occurred due to the mistakes of humanity, humans always play the part of victimhood. Most of these narratives are depicted as a struggle for survival or a long walk/journey to a safe abode. In this struggle for life, the characters will have to face the alpha-enemy and other fellow contenders of survival. Civilisational instinct is replaced by a survival instinct. Population will be drastically reduced, and the future of the planet will be in the hands of the select few remaining individuals. Thus, these narratives usually identify the alpha pair in the beginning itself, as the responsibility to repopulate the world would lie in their hands.

Climate will be exceptionally difficult for the sustenance of life forms. It might be either too cold or too hot or raining or misty or radiation, making survival a life-or-death phenomenon. It will also be unpredictable and therefore prepared-ness for changes is quite impossible. Nevertheless, none of these narratives represents the apocalyptic event as the final or as the end of times event. Rather, it stands for a new beginning. There is always a "post"-apocalypse which is held by hope and anticipates a different future to be built from scratch. With regard to law and order, these speculative narratives will challenge existing hierarchies and an ideological transformation among the characters would be visible.

Interestingly, there have been no boundaries to the different genres through which post-apocalyptic narratives have speculated the future. There are narratives in the form of novels, short stories, poetry, comics, films, TV series, video games, music, anime, manga, paintings etc. Each of the tropes and patterns discussed can be found in different combinations across all post-apocalyptic narratives.

(Dis)ability and the Human Project

The case of disability came to the limelight as a structured academic topic of discussion more recently with the establishment of disability studies in the 1990s. Ever since, the discourse of disability has been analysed through different intersecting power structures like gender, caste, ethnicity, disease and body and through multiple socio-cultural exigencies. As advocated by medical and psycho-social experts, disability is a broad term which is used to indicate various absences of abilities like physical, fiscal, legal, sensory, intellectual etc. The dimensions that define and categorise disability emerge from disciplines like medicine, ethics, law, humanities, and sports, and thus the naming of this category is very often debated. During the nineteenth century, the term "handicap" was replaced by "disabled" as the former term was used in the context of unequal contestants in horse races then. Later on, the terms "differently-abled," "able-bodied," "physical minority," "Persons with Disability" [PwDs], etc., were suggested for political correctness. Nevertheless, contemporary scholarship suggests that while referring to individuals with disabilities, it is preferred to use a language that focuses on their abilities rather than their disabilities. Hence, the naming of this category is in constant flux, searching for a term that better expresses inclusiveness and agency. The shifting terminology debate within disability studies is an indication of the attempts made to de-stigmatise the disabled body.

The discourse of disability is tethered with what is projected as the healthy, normal body. The norms of being normal and disabled are in the same loop, with each one of them complementing and critiquing the other all the time. In other words, the existence of normal is legitimised by the presence of the disabled. According to Lennard Davis,

> the very concept of normalcy by which most people (by definition) shape their existence is in fact tied inexorably to the concept of disability or rather the concept of disability is a function of a concept of normalcy. Normalcy and disability are part of the same system.
>
> (Davis 2)

This validates that disability is a biologically and socially driven concept. It could be understood as the functional consequence of impairment (biological dimension) and also as the norm that is used to define an ableist community (social dimension). In function, ableism is the discrimination and prejudice faced by disabled bodies when they are defined in conjunction with non-disabled bodies.

The category called "human" has been mostly defined and historically scrutinised in terms of a normal-bodied human being. This essay identifies these historical references as the human project inspired by the ongoing interdisciplinary projects to understand the connections between human and non-human life forms, their behaviours and environment. The premise of this current human project can be analysed in detail through the changing patterns in which the human category has been defined through narratives. Initially, by deprecating humans as demonic, monstrous and ignorant, civilisations turned to unseen gods. Then, humans started rationalising their identity and the self, probably with renaissance ideologies and later on with the thought processes of enlightenment. Civilisations equipped themselves for anthropocentric narratives where they celebrated their identities, pluralities and multiplicities. Humanity was seemingly flourishing with a certain "normal" regularised rhythm of life or probably an "old normal" within the vocabulary of Covid-19 as this is a watershed in the recent pandemic history of humanity.

In all these historical constructions of the human, there was always a normal or an able-bodied category, which was juxtaposed with the disabled category. Distinct examples from varied disciplines and epochs like the metaphor of the ship of fools, the practice of witch-hunting or in the model of hospital construction based on isolation, larger socio-cultural apparatuses were engaged in a process of filtering the normal and "abnormal" humans. Firstly, these examples identified and defined the disabled body as the "other" of a certain normal body. Secondly, it signified that the disabled body needs regulation and discipline and, thirdly, if it does not come under the control of a concerned authority, the disabled body will be factored in as the non-productive unit of society and therefore the expendables.

These definitions rely upon how human value is decided by the productivity of their bodies. According to David Mitchell's *Bioethics of Disability*,

> disability was first coined in the mid-1800s to designate those incapable of work due to injury. This grouping identified disabled veterans of the Civil War as eligible for various governmental supports- disability pension payments, state provisions of prosthetics, life skills training, and so on.
> (Mitchell 212)

When the stereotypical representations of disabled bodies are in the form of helpless bodies in need of charity or bodies looking for special treatment or as bodies looking for compassion, the ruling narratives around this discourse will continue to be stigma, social exclusion and rehabilitation. However, scholarship within disability studies critiques the stereotypes. Sunaura Taylor writes,

> historically, disability rights advocates have declared that we do not want to be cared for, we want rights, justice and an accessible society that does not limit or make impossible our involvement and contributions.
> (Taylor 110)

"Disabled People Destroy Science Fiction!" an issue in the science fiction magazine, *Uncanny* clarifies that disabled individuals are,

> fully actualized individuals, living at the intersections and axes of identities. Queer, nonbinary, Jewish, black, PoC, straight. We are all of these things and we are disabled. Disability itself means different things to different people. We are not a monolith. No *one* experience of disability is *the* disability experience.
> (Sjunneson and Parisien n.p.)

Post-apocalyptic narratives have a different take and tale on the depiction of healthy normal bodies. The select visual texts are not necessarily about disabled characters playing parts. Rather, it interrogates the concept of a healthy, able body.

(Dis)ability and Post-Apocalyptic Narratives

Contrary to the general tendency where disabled characters are killed or discarded in films, contemporary scholars on science fiction like Katherine Duckett and Dominik Parisien suggest that disabled individuals find it easier to fit in a speculative narrative. They like to imagine a future where there is space for their expressions, concerns, contributions and anxieties:

> By turning our attention to the future, we are able to explore concerns and realities in the present and amplify them, correct them, highlight the ways they might become better or worse if allowed to continue on their present course. Through science fiction, marginalized people are able to say, *We are here, now, and we will be there later, too.*
> (Sjunneson and Parisien n.p.)

The visual texts analysed here are *Birdbox*, *A Quite Place*, *After Earth* and *Mad Max: Fury Road*, and this chapter proposes that these narratives challenge the normative able body or what is medically perceived as a healthy body.

In Susanne Bier's *Birdbox*, humanity across the globe undergoes unexplained mass suicides. Due to the distorted whispering from a formless blackness, people started exhibiting psychotic behaviour. Once affected, the individual will prompt others also to do the same by just looking at them in their eyes. The domino effect wiped humanity and *Birdbox* is premised five years after the cataclysmic event and the narrative progresses through flashbacks. This narrative has characters explaining the event with religious and mythical references. In order to survive, one has to wear a blindfold, as the movie's tag line says, "Stay Alert, Stay Blindfolded" or "Don't Open Your Eyes."

In *Birdbox*, seeing by touching is endorsed. In this new world of the narrative, vision is a disability costing the characters their lives. The concept of vision is also redefined here because what is perceived as an ability (which is your sense of sight) is in fact killing the characters. An interesting contrast is brought in the narrative when it is mentioned that people at the psychiatric hospital did not use blindfolds because they were happy and wanted to see the whispering apparition as opposed to the other characters in the narrative. This underlines two problems.

- The authenticity of our object of vision: what we see may not be what is seen by everyone else. Here, the people from the psychiatric hospital could see and appreciate something which the others in the plotline were completely missing out on.
- The efficacy of our medium of vision: the movie contends whether our eyes can actually see. In the movie, the eyes are very evidently portrayed to be powerless in front of the apparition; being unable to see it and at the same time causing humanity's annihilation.

The movie has its own loose ends, but it leaves several morsels of information throughout the narrative about humanity's lack of abilities particularly in contrast with other species. The movie is called birdbox as birds could see and identify the presence of the apparition and thereby warn humans.

The tag line of John Krasinski's *A Quiet Place* goes as follows: "If They Hear You, They Hunt You," directing the survivors to not speak, not cry and not make the faintest sound of any kind if they want to survive for one more day. The movie has blind alien creatures that are hypersensitive to sound, and they get agitated at the bleakest possible sound. Just like *Birdbox,* this movie also progresses with a compelling parallel narrative of families trying to survive in the post-apocalyptic timeline. At the same time, *A Quiet Place* also questions human abilities that were celebrated, the sense of hearing and the ability of speech. The alien creature is evidently more sensitive to sounds than humans, to the extent that it gets distressed. The ability to make speech sounds is of no use and to survive one has to unlearn that particular ability. The characters in the movie communicate by signing and lip reading.

In Manoj Night Shyamalan's *After Earth*, after the post-apocalyptic event, humans were evacuated to an inter-spatial land mass named Nova Prime. A group of leaders called Rangers will help humankind to build their new home there. They are under the constant threat of alien monsters called Ursa which were designed and bred, to kill humans. What is interesting here is the fact that Ursa is blind, but it reacts to the pheromones secreted by humans when they fear something or get scared.

Sociologist Barry Glassner's *The Culture of Fear* suggests that fear mongering is a manipulative strategy used by all civilisations. It is suggestive that a collective submission to an unpopular decision can be achieved by instilling fear among the contenders. With a cultural climate of fear and uncertainty, authorities can bring everyone on board by probably saying that a drastic law is the only way out to face an impending disaster. Alternatively, one can say that a culture of fear can be manipulated, and this movie reiterates the need to leave absolutely, no traces of fear in order to stay alive. Therefore, in this imaginative "after earth" any indication of fear denotes weakness and thereby a disability. As the tag line of the movie goes, "danger is real, fear is a choice."

In George Miller's *Mad Max: Fury Road*, the disjunction between the perceived normal and natural is artfully crafted into the narrative. *Mad Max* caricatures an "abnormal normal." In this post-apocalyptic movie, humanity had gone rogue, terrorising themselves through a thermo-nuclear skirmish; there were oil wars, water waters; their bones were poisoned, and quoting from the movie, "the earth is sour and we have become half-lives." There is an authoritarian leader named Immortan Joe, who is the sole custodian of water resources and at the same time a leader with a very unhealthy body. He has oozing lesions all over his body, which is covered with an ornamented plastic chest shield, and he wears a skeleton-smile mask in order to cover his breathing apparatus. Most of the characters have iron deficiency, and the army of Immortan Joe who are called the war boys have access to blood banks or humans who were caught and caged for this purpose.

In that world, the characters who were born into the post-apocalyptic world believed that it is normal to give birth to babies with health problems, it is normal not to find water and it is normal not to find anyone without disabilities. Therefore, differently abled characters are presented in positive and empowering roles. The character of imperator Furiosa is limited by a missing natural limb but augmented by an artificial arm, whereas the survivors of the apocalypse believed that, all of that is not normal, as they have lived in another world which was more natural to them. Therefore, the character of Max, who is a survivor of the apocalypse and also a universal blood donor to the iron-deficient war boys, mourns that he exists in this wasteland running away from both the living and dead, haunted by the ghosts of whom he could not protect and hunted by human scavengers. *Mad Max* imprints that having an able body need not mean the normal as normal does not imply natural. Moreover, to have that standard healthy body as that of Max is a disability in this narrative as he was treated as fodder in Immortan Joe's "abnormal normal" world.

Post-Apocalyptic Humanity

The process of reimagining the discourse of disability begins with the recognition and identification of perceived human abilities (which is called the normal) as not the natural. Normative is a constructed reality and, hence, natural and normal are not the same. The movies that are discussed go to the extent of saying that the perceived human ability is in fact a human disability within the context of an imagined post-apocalypse. Therefore, there is a process of unlearning and relearning, which is involved in liberating humanity from the previously recognised standards and at the same time equipping humanity to survive an apocalypse. As most of these movies have addressed, the old rules do not apply anymore; they are useless in the face of an apocalypse.

Literature has been processing the different ways in which humanity can be steered through apocalypse. It has been offering an alternative imaginary for post-apocalyptic humanism. In all these new forms of humanism, the category of disability has been re-imagined and re-defined. Human abilities have been re-interpreted to create a new normal. In the movie *After Earth*, the survivors are taught to master a technique called "ghosting," which is to leave all traces of fear and therefore to become invisible in the face of the alien monster, Ursa. While in *A Quiet Place*, children were taught to learn sign language and to follow light or utilise the sense of sight instead of speaking and making sounds. *Birdbox* gives a wonderful example of how humanity will have to depend on other species for survival as their own abilities are of no use in the new normal world. And *Mad Max: Fury Road* represents the circularity of human birth and decay. Instead of a linear narrative where the survival journey of the characters culminates in what they were searching for in a distant place, they return to the same locale from where they started their journey. The fury road was a search for a certain "Green Place of Many Mothers" from a bygone era and when they realise that such a place does not exist anymore, they go back to the Citadel from where they started their journey and to start afresh with new rules.

As projected through these examples, these narratives worked on the possibility of a future, which is spatially and temporally located after the apocalyptic event. None of them ended with the cataclysmic event. They can be considered as different ways to refresh humanity's existence and continuance on the same planet. The new definitions of humanism also mean a reassessment of human abilities and disabilities. These different postulations of the future coincide with what is perceived as posthumanism in the following frameworks:

- A "De-centering Project" envisages a future where the human is no more the centre of knowledge production. Anthropocentrism is to be replaced by a new order of things probably by AI, technology, data, environmental changes or even by other species like a virus. There is a clear void that is created by the displacement of the human centre and at the same time reinstated by another category.

- In the "Re-centering Project," the human category is definitely displaced from the centre, but there is an attempt to re-position and re-centre the human by improving their condition. Many scholars address that this improvement of the human condition is technology driven (genetic engineering, CRISPR gene editing, designer babies, etc.), and it is called transhumanism. The German professor of philosophy Peter Sloterdijk calls it "anthropo-technics" (found in his English translation, *You Must Change Your Life* published in 2013). He deliberates the idea that the human will be fused with technological prosthesis through information and data. In *After Earth*, humanism is merged with technology. They regain humanity by controlling one of the emotions that makes you a human; that is, by truncating the prime emotion of fear coupled with technology. The human-machine interface includes motion sensor suits that will help them identify the presence of alien monsters, the suits can also camouflage, if need be, highly efficient wrist communicators, and specially designed fluid capsules that will enhance oxygen intake if they end up in an inhospitable terrain and the list goes on.
- The final "Harmony Project" addresses the relationship between human, environment and multiple species on earth. It speaks about a co-existence and therefore a harmony project where living and non-living forms have an agency. The movie *Birdbox* begins with a message heard on the radio, filled with a lot of hope and confidence: "we have a community here, cross the river and follow the sound of the birds, we are safe." Interestingly, technology fails in this narrative, and it could not resurrect humanity from annihilation. In one of the attempts at survival, it was shown that digital signals could not track and identify the presence of the viral infection. In fact, it permeated through the camera lenses, infecting and killing the observer. And when the survivors finally reach the safe haven by the river, it is revealed that most of the community dwellers were blind. They were living peacefully in sync with nature and with the birds warning the sighted people. In *A Quiet Place*, the survivors figure out a way to live with the alien monsters.

Afterthought

In Marc Foster's *World War-Z*, a zombie apocalyptic movie, to have a diseased body or an unhealthy, disabled body was the antidote to the zombie plague. The following quote from the movie can be a prelude to the precarious selves and vulnerable bodies discussed in this essay.

> Mother Nature is a serial killer. No one is better. More creative. Like all serial killers, she can't help but the urge to get caught. What good are all those brilliant crimes if nobody takes the credit? So she leaves crumbs. Now, the hard part is, and why you spend decades in school, is seeing the crumbs for the clues they are. Sometimes the thing you thought to be the most brutal aspect of the virus, actually turns out to be the chink in its armor. She loves disguising her weaknesses as strengths.

Works Cited

Bier, Susanne, director. *Bird Box*. Bluegrass Films, 2018.
Davis, Lennard J. *Enforcing Normalcy: Disability, Deafness, and the Body*. Verso Books, 1995.
Forster, Marc, director. *World War Z*. Skydance Productions, 2013.
Glassner, Barry. *The Culture of Fear: Why Americans Are Afraid of the Wrong Things*. Basic Books, 2018.
Krasinski, John, director. *A Quiet Place*. Platinum Dunes, 2018.
Miller, George, director. *Mad Max: Fury Road*. Village Roadshow Pictures, 2015.
Mitchell, David T. and Sharon L. Snyder. *The Biopolitics of Disability: Neoliberalism, Ablenationalism, and Peripheral Embodiment*. University of Michigan Press, 2015.
Shyamalan, M. Night, director. *After Earth*. Columbia Pictures, 2013.
Sjunneson, Elsa and Dominik Parisien. "The Disabled People Destroy Science Fiction Manifesto." *Uncanny: A Magazine of Science Fiction and Fantasy*, 2018, uncannymagazine.com/article/the-disabled-people-destroy-science-fiction-manifesto/.
Sloterdijk, Peter. *You Must Change Your Life*. John Wiley & Sons, 2014.
Taylor, Sunaura. "Interdependent Animals: A Feminist Disability Ethic-of-care." *Ecofeminism: Feminist Intersections with Other Animals and the Earth*, edited by Carol J. Adams and Lori Gruen. Bloomsbury Publishing, 2014.

8 Ecocatastrophe in the Literary Imagination

Confronting the Anthropocene through Narratives of Ecoprecarity from North-East India

Paddaja Roy

Introduction: Situating Ecocatastrophe

The Anthropocene epoch has brought about a condition of precarity which engulfs all human and nonhuman entities. With frequent occurrences of epidemics, pandemics, environmental crises, wars, etc., life and survival have been rendered precarious in the Anthropocene epoch. Precarity, which Anna Tsing defines as life without the promise of stability (Tsing 2015), mirrors the contemporary environmental cataclysm rendering the living conditions as vulnerable. The universally accepted parlance of Anthropocene highlights the fact that the earth has been irrevocably altered by human agency (Attebery 16), where humans have turned into geological force from biological beings (Chakrabarty 2021). The precariousness of the living conditions stems from the discernible results of the Anthropocene, such as extreme and erratic temperatures and climatic conditions, rampant pollution, desertification and environmental derangement, which further results in contagious outbreaks and diseases, droughts, heat waves, cyclones and floods, as such (Duffy 2024). Thus, while the term "catastrophe" encapsulates damage and disaster, ecocatastrophe takes into account the actions of humans, a "set of affects than a biogeochemical process" when its contours can be evaluated or narrated (Merola 3). With the emergence of various genres dealing with the environment and the Anthropocene, a number of writings from India have also encapsulated and envisioned human-induced-catastrophe; for instance, *Gun Island* (2019) by Amitav Ghosh portrays the environmental derangement brought about by human agency while at the same time also highlighting the precarity which pervades the living conditions of both humans and the nonhumans. Many writers from the Northeast of India such as Easterine Kire, Ankush Saikia and Anuradha Sharma Pujari have portrayed how human actions have aggravated the derangement of environment-ecology. Literary narratives such as *Son of the Thundercloud* (2016) by Easterine Kire, *The Forest Beneath the Mountains* (2021) by Ankush Saikia, *The Forest Wails* (2021) by Anuradha Sharma Pujari, *Legends of Pensam* (2006) by Mamang Dai have not only dealt with how human actions have resulted in

DOI: 10.4324/9781003602750-11

critical climatic and ecological derangements (Ghosh 2016) but have also reflected on the dire consequences of the same. Such fictional representations of environmental derangement and degradation often portray the interconnectedness as well as the precarity rendered by ecocatastrophes and the Anthropocene (Nayar 2019). Such fiction, therefore, invites us to broach upon the narratives that thematise the derangement, exploitation and destruction of the environment and the ecology at the hands of humans thereby raising concerns and debates that are ongoing in the environmental humanities and Anthropocene writings.

Thus, this chapter brings out an analysis of the literary representations of environmental catastrophe and degradation in select fiction from Northeast India, as they intersect with the contemporary scholarship of environmental derangement and ecoprecarity. This study would draw extensively on Amitav Ghosh's delineation of environmental derangement and the ideas of precarity as theorised by Dipesh Chakrabarty and Pramod K. Nayar to characterise the frame of ecocatastrophe. The literary narratives under study not only broach upon the impending dangers of climate change and environmental degradation but also uphold the traditional and indigenous value systems ingrained in certain traditional and cultural practices. Thus, such narratives dealing with ecocatastrophe, ecoprecarity and the Anthropocene are seen as people's response to the contemporary ecological crisis, which inherently aggravate the socio-political as well as the cultural and emotional reception of the people, often experienced disproportionately thereby rendering certain beings as "planetary subaltern."

Thus, the aim of this chapter is twofold: it will rely extensively on ecoanxiety and ecoprecarity to broach the contemporary environmental cataclysm and will consider how narratives explore balance and hope in the indigenous past and cultural traditions. Taking up an intersectional approach, I will refer to other fictional as well as non-fictional texts to substantiate the assertions made in the course of study.

Ecocatastrophe in the Literary Imagination

The story of the Great Flood, recurrent in a number of myths across cultures, commemorates one of the greatest catastrophes where, aghast at the corruption of man, God decided to flood the earth to restore the goodness of his creation. In various myths and beliefs, the occurrence of disasters, catastrophes and natural calamities such as earthquakes, volcanic eruptions, tsunamis, thunder and lightning, epidemics and plagues is associated with the wrath of god, as an expression of divine providence. However, the epoch of Anthropocene has brought up new perspectives of examining and studying narratives and texts, putting the prime focus on the exploitative and destructive human actions endangering the environment.

The fact that the environmental and ecological derangements were not frequently and sufficiently addressed in literary narratives was voiced by Amitav

Ghosh in *The Great Derangement: Climate Change and the Unthinkable* (2016), where he also postulates how climate change has not been treated as a significant political issue in the neighbouring countries of India, but the impacts of climate change are felt throughout the Indian subcontinent. The impacts of climate change are apparent in the increasing number of large-scale disasters as well as in the form of a slow calamity gradually destroying livelihoods while at the same time aggravating social and political conflicts (Ghosh 2016). What Ghosh postulates in *The Great Derangement: Climate Change and the Unthinkable* gains representation in the literary narrative of his subsequent novel *Gun Island* (2019). In *Gun Island*, Ghosh highlights how human agency has disrupted the environment and resulted in ecological disasters and catastrophes, while at the same time highlighting the precarity which pervades the living conditions of both humans and nonhumans. The novel is dotted with instances of wildfire, tornado, hailstorms, mass suicide of dolphins and migration of animals such as snakes and spiders – the instances pointing to the re-emergence of the Little Ice Age when disorder and precarity prevailed on the planet. In this context, it can be asserted that the knowledge and occurrences of catastrophes and disasters were not foreign or unprecedented in the geological history of the planet; even then, humans exploiting and destroying the environment were rampant. To quote Chakrabarty,

> the knowledge (of catastrophes) did not affect our quotidian sense of an innate assurance that the earth provides a stable ground on which we project our political purposes. The Anthropocene disturbs that certainty by bringing the geological into the everyday.
>
> (Chakrabarty 180)

In *Dancing with Disaster: Environmental Histories, Narratives and Ethics for Perilous Times* (2015), Kate Rigby charts the evolution of the nomenclature of the term "catastrophe" from Aristotle's usage of the term as an element of tragic drama to its apocalyptic dimensions to the contemporary postcolonial ecocriticism and precarity. Catastrophic events such as mining and construction as such are not generally considered under the purview of ecological catastrophes (Rigby 2015). However, Poray-Wybranowska argues that ecological catastrophe is not a series of unfortunate occurrences but a series of conditions that were consciously and systematically produced over time in the aegis of colonialism and the contemporary socio-political conditions (Poray-Wybranowska 204). Again, with the shifting focus upon human agency as a geological force (Nayar 2019) humans have taken the role of the foremost determinant of the environment of the planet (Chakrabarty, "Anthropocene Time" 209). Thus, Anthropocene has rendered a situation where such catastrophes and disasters cannot be isolated from human activity because there is nothing about our current ecological and climactic patterns that humans have not already irrevocably altered (Attebery 2022;

Poray-Wybranowska 2021). At the same time, an ecocatastrophe has multiple impacts ranging from physical and environmental to socio-political as well as cultural effects on a very large as well as global scale thereby implying a major threat to the natural environment as its inhabitants (Thiltges 93). Such discourses of climate change, mass extinction and migration and various ecocatastrophes consequently generate a sense of ecological anxiety or ecoanxiety (Pihkala 19) as well as the looming sense of ecoprecarity (Nayar 2019) suggesting the vulnerability of the human and nonhuman existence. Wallace further voices the ecoprecarity and ecocritical anxiety in ecocatastrophic instances of the hazards of pollution, industrialisation and toxic wastes and products in the age "global risk" (Wallace 2016), thereby further aggravating the stature of human agency as the geological force in the Anthropocene epoch.

A number of Indian fiction, such as *The Hungry Tide*, *The Butterfly Effect* and *Valli*, have portrayed ecological catastrophes and disasters. In a similar context, a number of writers from Northeast India have encapsulated a nuanced representation of ecological and environmental catastrophes. While *Son of the Thundercloud* depicts a terrible famine with mythical connotations, *Legends of Pensam*, *The Forest Wails* and *The Forest Beneath the Mountains* herald the catastrophes and disasters brought about by human agency culminating in environmental derangement.

Storying Ecocatastrophe: Literary Responses from the Northeast India

The movie *Wall-E* (2008) is an American sci-fi produced by Pixar Animations which presents us with an imagined picture of the earth in the twenty-ninth century: the earth has turned into a wasteland due to the catastrophe of ecocide caused by unchecked human exploitation, capitalism and destruction of the environment. The only way of restoring the earth to its previous stature was through the single sapling of a tree left on the face of the earth. The movie, while showing the destruction and devastation that humans are capable of, also underscores the essentiality of planting trees, giving the message of protecting the planet and the environment. Narratives from Northeast India are replete with nuanced depictions of the environment and ecology vis-à-vis the destructive and exploitative activities of humans. *Son of the Thundercloud* (2016) by Easterine Kire, *The Forest Wails* (2021) by Anuradha Sharma Pujari, *Legends of Pensam* (2006) by Mamang Dai and *The Forest Beneath the Mountains* (2021) by Ankush Saikia subtly bring out the confrontations of humans and nonhumans/nature, with ecological disasters and ecocatastrophes brought about by the vagaries of human action and exploitation.

The novel *Son of the Thundercloud* (2016) deals with the oral literatures, folklores and beliefs of the tribes of Nagaland. The novel begins with the arrival of a famine and delves into the nuances of the catastrophe of the famine. Kire records how people began to forget their customs and "taboos," which were the very foundations of their tradition and culture – "Soon they grew careless about the taboo that said that every villager must keep aside some grain after the harvest as seed-grain" (13). And one day, a swarm of mice and insects wiped out the village of its food stock resulting in famine-like conditions. The people of the region shared a close-knit relationship with nature and therefore not maintaining a custom or a taboo resulted in grave consequences, such as complete abandonment of a village, resulting in famine and scarcity. Gradually, the famine spread to the neighbouring villages including the village where the protagonist, Pele, lived. At the death of his wife and children, Pele set off to distant lands where he could see the devastating effects of the famine: "The earth was so dry that the soil no longer looked like soil" (17) and that the soil was the colour of "death-grey" (17), and it didn't seem fit enough to sustain any form of life. On his way, however, Pele came across two mysterious mythical creatures/women who told Pele about the prophecy of the eponymous Son of the Thundercloud – that the birth of the Son of the Thundercloud would bring "rain and mist that softens the soil, and the earth will sprout grain and grass again. There will be food and life" (19). Very soon, the prophecy was fulfilled, the rains came and Mesanuo, the sister of the two mythical women, got impregnated by a single drop of rain and gave birth to a son named Rhalie. It was after the birth of Rahile that famine got over, thus ending a grave catastrophe. Again, the character of Rahile is symbolical as the redeemer or preserver of the ecological balance. The significance of human actions in conserving the ecological balance of the environment is subtly presented through the simple assertions that Rahile puts forth, such as "if we plant more trees there will be no more drought" (67), and that how trees hold the earth and water together, as such, thereby placing human actions at the centre of the geological turn of events.

In this context, it can also be asserted that the novel talks about not only the famine of food and life but also the famine of stories and songs signifying the killing of Hope (48), thereby foregrounding the contemporary capitalist ideology where nature is seen either as a site for exploitation or as mere wilderness abounding in primitivism, a notion propagated with the advent of colonialism. Storytelling was one of the ways through which various customs, taboos, histories and lessons were imparted to the succeeding generations, but with the advent of colonisation, and exploitative and capitalistic ventures of humans, storytelling and the observance of taboos and customs suffered to a great extent – when the storytellers were killed, people grew oblivious of the stories they were told and assuming those as myths, they allowed their minds to accept the darkness:

> So, the drought came as a result of people rejecting the joyful stories and accepting the dark stories.
>
> If the drought was man-made, isn't there a danger that men could create another drought in the future?
>
> There will always be that danger so long as mankind lives on earth.
>
> (64)

Even the relationship between human and nature was altered and, therefore, human encroachment didn't hesitate to trespass or even disregard the boundaries of nature and the nonhuman. The nonhuman entities, in the oral narratives, stories and folklores were seen as powerful agents capable of wreaking havoc and destruction and even harbingers of death. For instance, in *Son of the Thundercloud* one finds a depiction of the significance of the communities to uphold and abide by certain traditions and customs, taboos and superstitions, and how violation of the same could lead to dire catastrophes and consequences such as famines. In another novel by Easterine Kire, *Spirit Nights*, Kire shows how, due to the violation of a terrible taboo, the "tiger had eaten the sun" rendering complete darkness, similar to a catastrophe or epidemic where people were forced to quarantine at their houses (Kire 2022). The catastrophe was, however, spiritual in nature. Again, a similar depiction is also found in Mamang Dai's *The Legends of Pensam* (2006), which explores mythical stories depicting catastrophes and disasters resulting from trespassing the nonhuman – how the sight of a water serpent foretold a death and premonitions of tragedy, how illness was caused and cured by catering to the nonhumans, as such. Thus, *The Legends of Pensam* also situates the carelessness of the human agency in the spectrum of ecocatastrophe resulting in death and disasters: "But sometimes some households forget to observe the rules, and then our men die in the forests" (Dai 36). Such novels foreshadow the careless exploitation humans carry out in the name of "development," which often leads to destruction and displacement of both nonhuman beings and humans. A number of northeast fiction have broached upon the question of displacement and dislocation which often results from "developmental schemes" and the expropriation of all life forms (Chakrabarty 2016; Chakrabarti and Dhar 2009), thus bringing the context of ecoprecarity – the vulnerability that engulfs human and nonhuman existence even at the face of "modernity" and "development," the roots of which are often traced to exploitative colonial practices and subjugation. *The Legends of Pensam* has questioned the idea of development by bringing up the motive of roads: "This one terrible road is all they have managed for us in fifty years! And what does it bring us? Outsiders. Thieves. Disease" (Dai 143). These novels outline how, as subjects living in precarious environments, amidst ecoanxiety, instances of instability and ecocatastrophe have become intertwined with their day-to-day life. Human activities have permeated instances of ecocatastrophes from in everyday life and existence. For instance, construction of roads has quickened the contagiousness of epidemics/pandemics, and the hollowness of capitalistic ventures and exploitations have destroyed ecosystems, thus encroaching disasters and catastrophes in the everyday as well as lived reality of human as well as nonhuman entities.

Eco-Narratives: Fiction and Confrontation

It would, however, not be justified to vest equal responsibility of climate change and ecological derangement upon the entire human entity as a whole. The generalisation of the entire human race as contributing to the "Anthropocene" is an assertion wrongfully made for the lifestyle and consumption of natural resources varies from person to person. While people aiming for capitalistic ventures would have a higher level of carbon footprint in comparison to someone leading a sustainable lifestyle. Thus, there exists disproportionality in both the contribution to ecological catastrophes and disasters as well as experiencing the same. In the conflicts between humans and nature, most of the time, it is the poor and marginalised who are the worst sufferers. Anuradha Sharma Pujari, in the narrative of *The Forest Wails* (2021), which is an English translation of the novel *Iyat Ekhon Aranya Asil*, traces the looming ecocatastrophes heralded by human exploitation and the predicament which the poor and marginalised face at the hands of questionable developmental schemes and capitalism. Set against the backdrop of socio-political and ecological disasters and dangers that loom upon the city of Guwahati and its people, the novel shows the power hierarchy of the various sections of humans and the clash of forests, humans and urbanisation/development.

The novel mirrors the marginalisation and displacement of poor helpless people, as well as the plight of the flora and fauna at the hands of governmental and the so-called "developmental" activities by showing how, according to the information at the Revenue and Forest Department, the area of the hilly region belongs to the Forest Department but the people residing in the hill for a considerable period of time claim to have brought the land legally with their hard-earned money. The narrator encapsulates the clash by observing,

> We take two different stands on many such situations. We become vigilant against the illegal digging of the hill, against the illegal encroachments. But when the Government rises to displace these illegal occupants, we become vigilant again – against the government.
>
> (21)

The Forest Wails outlines the predicament of not only the oppressed section of society but also questions the marginalisation coupled with the displacement of forests, birds and animals in the Anthropocene. There is a pervading sense of precarity in the narrative of *The Forest Wails* which places humans at the epicentre of ecocatastrophes:

> People have died because of landslides during the rains, the irregularities in the rainfall have become increasingly noticeable, natural disasters have increased. But still, people are going on building houses there. And so the desire to get rid of the weeds! It is common to lose some of the worthwhile plants along with the unwanted weeds in the fire meant for discarding the unwanted weeds.
>
> (62)

In the context of ecocatastrophes, one must take into account both human and nonhuman entities, for "the planetary environmental crisis calls on us to extend ideas of politics and justice to the nonhuman, including both the living and the non-living" (Chakrabarty 13). *The Forest Wails* not only holds up the devastations caused by humans but also highlights its repercussions upon the nonhumans, negotiating our understanding of the nonhuman and the relationship which they share with the human. More than often when we talk about the impacts of ecocatastrophe, we generally do not take into account the catastrophes impacting the lives of nonhuman animals, birds and insects so susceptible yet so crucial in maintaining the balance of the ecosystem. Due to the exploitative practices, nonhuman animals and birds are subject to gradual impacts of ecocatastrophes on a daily basis: "There will not be a single bird left in your hill. I have seen the destructions in front of my eyes–the green hill turning almost into a dry desert!" (237). Thus, the fictional responses to ecocatastrophes and ecoprecarity often portray the confrontation between human activities and nature/nonhuman entities:

> Possibly all the seeds of destruction are ingrained in humans. The desire to conquer the oceans, the mountains, the valleys, the rivers, the streams, the air, the sky – have turned humans into the most dangerous monster striding this earth. I can perceive the bleak, hostile future already hanging in the hill near my house-the relentless war between man and nature. Time will say the consequences; the death-soaked city will say.
>
> (Pujari 240)

Even at the face of precarity, even when humans are fully aware of the looming impacts of ecocatastrophes, the exploitative activities of humans continue unfazed, as Pujari notes, "When during the monsoon, the drains of the plains overflow with the muddy water flooding the entire area. The drains in the valley get choked up with the soil eroding from the destructive activities going on in the hill" (15). Pujari further substantiates the assertion by portraying how most of the river beds and river bodies have become shallow due to the erosion of hills dripping downwards like boulders, "There are few trees left to clutch to the soil and to save it from eroding. Flood in the valley is the natural outcome of the destructions done to the forests" (195), destructions carried out by none other than humans. A similar predicament is also delineated in *The Forest Beneath the Mountains* (2021) by Ankush Saikia where he contends, "the more sand and stones were removed from the river, the less flood protection there was in the downstream areas during the monsoons" (93), thereby affecting both humans and nonhumans. Thus, ecocatastrophes like floods and deforestation, as such, pose serious threats to humans as well as nonhumans. One can even postulate that nonhuman animals are one of the worst sufferers of ecocatastrophes. The way certain humans treat animals is seen as a sort of "animal holocaust" – from conducting a number of inhumane scientific experiments upon them to displacement, commodification, slaughter, etc., to name a few of the inhumanities – only to cater to the needs and whims of a section of human beings (Sztybel 2006), which shows the sort of catastrophe that humans directly bring upon the living environments of animals and other nonhuman beings. At the same time, Pujari also shows how humans have turned animals vulnerable and by snatching away their habitat, food and disrupting their entire ecosystem (Pujari 83). In a similar context, Saikia shows how, as a result of overpopulation and large-scale migration leading to disappearance of forests, animals such as the elephants "have no place to stay and no food to eat so they come out into the open, into the paddy fields and villages" (Saikia 31). The focus on the ecoprecarity and the subsequent ecoanxiety affecting the nonhuman entities draws attention to the manner in which contemporary literary narratives from Northeast India negotiate the ecological catastrophes in the wake of climate change which the region is battling with. Focusing on the representation of how human activities have rendered the environment precarious, not only for humans but also for nonhumans, highlights the way ecocatastrophe is experienced on a very personal as well as grassroots level. The focus is analogous to the structural literary representations of the detritus and legacies of colonial environmental exploitation drawing parallels to the precarity and vulnerability under climate change (Poray-Wybranowska 157).

Conclusion: Narrating Hope and Imagining Futures

The recent literary trends in eco-narratives highlight what Ghosh contends in *Gun Island*, "Everybody knows what must be done if the world is to be a liveable place . . . and yet we are powerless" (Ghosh 296). It is the unguarded human destruction and exploitation that has turned this life into a precarious existence, bringing instances of ecocatastrophes into our daily lives. Pujari encapsulates the contemporary conflict between humans and nature by outlining the inherent will to destroy:

> the tussle between man and nature for a little place of shelter, for a little space of protection is ceaselessly going on a razor edge. The result of this war seems to be all enveloping destruction; a hurtful blow to all humanity.
>
> (111)

The representations of the contemporary spatio-temporal realities and environmental catastrophes in cli-fi or disaster writing subscribing to environmental humanities foreground the importance of recognising and broaching the issues of ecosocial disasters and derangements. Such novels outline the premonition of a gradual impending catastrophe, while at the same time raising questions about the mitigation of exploitative activities of humans and imploring for sustainability and recognition of the urgency of climate crisis. *Son of the Thundercloud* subtly addresses the crucial question of why the contemporary ecocatastrophes are taking place through its nuanced narratives of the metaphor of famine. Kire not only focuses on the negligence of humans in adopting sustainable means but also outlines the gradual disappearance of storytelling. Storytelling is an art through which imagination is enhanced, and one of the reasons catastrophes and disasters eventuate is due to the failure of imagination: "If fantasy fiction – if magic – can help humans encounter and 'make kin' of trees, it has a role to play in cultivating an ethic of care" (Duckworth 90). Stories cannot literally grow trees but they can play a major part in growing awareness regarding ecocatastrophes by portraying the pervading ecoprecarity and ecoanxiety. Be it fictional narratives portraying catastrophes as the wrath of an angry god, or climate or disaster narratives placing human action as the epicentre of catastrophes, both cases speculate a global catastrophe focusing on the transgression of nature, and the recognition of nonhuman entities. The roots of the contemporary catastrophes lie in the solipsistic assertion of humans, capitalistic and power-driven motives, and the sin of improper planning, exploitative actions and such. Narratives such as *The Forest Wails* and *The Forest Beneath the Mountains* explicate the contemporary climate crisis and nuances of ecocatastrophes and also acknowledge the heterogeneity of human agency owing responsibility to the Anthropocene epoch.

Thus, this chapter pivots on the nuances of ecocatastrophes by recognising the heterogeneity of human agency while at the same time negotiating the space of human-nonhuman interaction, thereby speculating upon the ecoprecaity which surrounds the contemporary life and existence and the ecoanxiety which can be seen as a natural outcome of ecocatastrophes and precarity. This chapter also recognises further scope of study by situating the fictional narratives referred to in the study as texts encapsulating decolonisation and decentering of the human agency. The texts under study have also highlighted the baseless dichotomies of nature-humans/culture, modern-primitive, by showing how ecocatastrophes and contemporary environmental cataclysm have rendered the entire existence as precarious, and a proposition which can be commemorated in Bruno Latour's words, "No one has ever been modern. Modernity has never begun. There has never been a modern word . . . We have never been modern" (47).

Works Cited

Attebery, Brian. "From the Third Age to the Fifth Season: Confronting the Anthropocene through Fantasy." *Fantasy and Myth in the Anthropocene: Imagining Futures and Dreaming Hope in Literature and Media*, edited by Marek Oziewicz et al. Bloomsbury Publishing, 2022, pp. 16–25.

Chakrabarti, Anjan and Anup Kumar Dhar. *Dislocation and Resettlement in Development: From Third World to the World of the Third*. Routledge, 2009.

Chakrabarty, Dipesh. "Whose Anthropocene? A Response." *Whose Anthropocene? Revisiting Dipesh Chakrabarty's 'Four Theses'*, edited by Robert Emmett and Thomas Lekan. RCC Perspectives, Transformations in Environment and Society,

2016, pp. 103–113, www.environmentandsociety.org/sites/default/files/2016_2_chakrabarty.pdf.
Chakrabarty, Dipesh. "Anthropocene Time." *The Climate of History in a Planetary Age*. University of Chicago Press, 2021, pp. 155–181, https://doi.org/10.7208/chicago/9780226733050.001.0001.
Dai, Mamang. *Legends of Pensam*. E-book, Penguin Books, 2006.
Duckworth, Melanie. "Arboreal Magic and Kinship in the Chthulucene: Margaret Mahy's *Trees*." *Fantasy and Myth in the Anthropocene: Imagining Futures and Dreaming Hope in Literature and Media*, edited by Marek Oziewicz et al. Bloomsbury Publishing, 2022, pp. 89–100.
Duffy, Helena and Katarina Leppanen. *Storying the Ecocatastrophe: Contemporary Narratives about the Environmental Collapse*. Routledge, 2024.
Ghosh, Amitav. *The Great Derangement: Climate Change and the Unthinkable*. Penguin, 2016.
———. *Gun Island*. Penguin Random House, 2019.
Kire, Easterine. *Son of the Thundercloud*. Speaking Tiger, 2018.
———. *Spirit Nights*. Simon & Schuster, 2022.
Latour, Bruno. *We Have Never Been Modern*. Translated by Catherine Porter. Harvard University Press, 1993.
Merola, Nicole M. "Archives of Ecocatastrophe; or, Vulnerable Reading Practices in the Anthropocene." *American Literary History*, vol. 30, no. 4, 2018, pp. 820–835, https://doi.org/10.1093/alh/ajy039.
Nayar, Pramod K. *Ecoprecarity: Vulnerable Lives in Literature and Culture*. Routledge, 2019.
Pihkala, P. "Eco-anxiety." *Situating Sustainability: A Handbook of Contexts and Concepts*, edited by C. P. Krieg and R. Toivanen. Helsinki University Press, 2021, pp. 119–133, https://doi.org/10.33134/HUP-14-9.
Poray-Wybranowska, Justyna. "Reading Catastrophe through Postcolonialism, Ecocriticism, Indigenous Studies, and Animal Studies." *Climate Change, Ecological Catastrophe, and the Contemporary Postcolonial Novel*. Routledge, 2021, pp. 32–57.
Pujari, Anuradha Sarma. *The Forest Wails*. Translated by Bhaswati Parasar. E-book, Walnut Publication, 2021.
Rigby, Kate. *Dancing with Disaster: Environmental Histories, Narratives and Ethics for Perilous Times*. University of Virginia Press, 2015.
Saikia, Ankush. *The Forest Beneath the Mountains*, Speaking Tiger, 2021.
Stanton, Andrew, director. *Wall-E*. Disney Pixar, 2008.
Sztybel, David. "Can the Treatment of Animals Be Compared to the Holocaust?" *Ethics and the Environment*, vol. 11, no. 1, 2006, pp. 97–132. JSTOR, www.jstor.org/stable/40339116.
Thiltges, Sebastian. "The Nuclear Disaster as Metaphor for the Impending Ecocatastrophe in Anticipatory Fiction from Luxemborg." *Storying the Ecocatastrophe: Contemporary Narratives about the Environmental Collapse*, edited by Helena Duffy and Katarina Leppanen. E-book, Routledge, 2024.
Tsing, Anna Lowenhaupt. *The Mushroom at the End of the World: On the Possibility of Life in Capitalist Ruins*. Princeton University Press, 2015.
Wallace, Molly. *Risk Criticism: Precautionary Reading in an Age of Environmental Uncertainty*. University of Michigan Press, 2016.

Part III

Resource Extraction, Eco-Injustice and Resistance

9 Eco-Anxiety, Trauma and Resilience of the Dongria Kond Tribe of India

Locating the Literary and Cultural Responses of the Niyamgiri Movement in the Global Scenario

Mir Ahammad Ali

Introduction

Greta Thunberg, a 21-year-old Swedish environmental activist, expressed her profound concern about climate change in an interview at Davos in January 2019. In a famous speech, widely known as "I want you to panic," Greta emphasised the pressing reality of human-induced climate change and stressed the need for immediate action to be taken to address the crisis. She stated,

> Adults keep saying: "We owe it to the young people to give them hope." But I don't want your hope. I don't want you to be hopeful. I want you to panic. I want you to feel the fear I feel every day. And then I want you to act.
>
> I want you to act as you would in a crisis. I want you to act as if our house is on fire. Because it is.[1]

The "crisis" she refers to is the climate crisis, which has led her to experience panic over the devastating effects of human-induced climate change worldwide. The "fear" or "panic" arising out of the colossal ecological and climate disasters is indeed a significant issue in our contemporary ecological imagination. Greta's environmental activism to resist climate change has attracted global attention, and her sense of fear and urgency resonates globally, not only among environmentalists, climatologists and activists but also among indigenous tribes around the world. The reality is that climate change is indeed one of the biggest challenges of our time, which is why the United Nations Environment Programme has recently laid great emphasis on the concept of "climate emergency." Even the Paris Climate Accord (also known as The Paris Agreement), adopted in 2015, aims to minimise the effects of climate change by controlling greenhouse gas emissions and lowering the global surface temperature. In contemporary global discourse, much emphasis is given to terms like "Ecocatastrophe" and "Eco-resistance" in the context of climate imperilment.

DOI: 10.4324/9781003602750-13

One of the recent trends in the field of eco-environmental studies is to focus on the mental health trauma caused by rapid ecocatastrophes taking place all over the world. Massive environmental degradation and ecological disarrays are some of the major reasons for this trauma. The current geological epoch, known as "The Anthropocene," shows how the earth and its deep ecology have been abused by its chief geological agents, human beings, by means of heavy industrial forces and unimaginable technological advancement. In an era of rapid urbanisation, unmanageable plastic pollution, breakneck industrialisation and resultant global warming, "Eco-imperialism" (particularly corporatisation of the earth's natural resources, chiefly forests, land and water) is leading us towards an environmental apocalypse. Some geologists speculate that due to the colossal avariciousness of human beings and their forced intrusion into Nature, nearly 75% of the world's natural resources and its species will be wiped out from the Earth in a few thousand years.

The potential threats of ecocide can cause "Ecological grief" a particular kind of mourning arising from the loss of the ecosystem, chiefly by human activities. This grief sometimes leads to "Eco-anxiety," a relatively new concept in the emerging field of "Eco-psychology" where victims, including environmental activists, experience bouts of panic attacks and fear psychosis due to ecological degradation. Symptoms like helplessness, chronic depression, anxiety, disorders and paranoia among many others are manifest in the victims. This interface between "Ecology" and "Mental Health" is a space of research that requires further exploration.

One such phenomenon that has recently attracted global attention is the Niyamgiri Movement of Odisha, India, in which a group of 8000 forest dwellers of the Dongria Kondh tribe, living on the hills of Niyamgiri, have been fighting since 2003 to protect the ecological health of the hill forests from the rapacious industrialisation of the aluminium mining industry, Vedanta Resources. The indigenous people of Niyamgiri, who hold their mountains and forests sacred, suffer from "Ecological grief" and "Eco-anxiety." "Eco-anxiety," alternatively known as "Eco-distress" or "Climate anxiety," is a chronic fear primarily arising out of environmental disaster or ecological damage, and it generally affects the mental and behavioural health of the affected. Grief or mourning is also synonymous with anxiety in such a context. "Grief" is a very loaded word in the discourse on ecology and the environment. Generally, grief comes out with the loss of someone or something intensely close or personal to us. When it pertains to the loss of our deep ecosystem and its intrinsic biodiversity, it becomes "Ecological Grief." The well-known thanatologist, Dr Kriss Kevorkian, describes ecological grief as "the grief reaction stemming from the environmental loss of ecosystems by natural and man-made events" (Rosenfeld). The indigenous people of Niyamgiri have been suffering from such anxiety or grief.

Challenging the so-called established myth of industrial development, the Dongria people assert that the return to the indigenous ways of life through sustainable development is the only means of possible survival amidst this impending ecological doom. For them, this peaceful and harmonious cohabitation with nature requires a deliberate discarding of heavy industrial machines and limitless technological advancements. This essay, thus, seeks to explore the psychological effects of relentless industrialisation on the Dongrias of Niyamgiri, with a focus on emerging concepts like "Ecological grief" and "Eco-anxiety" in the field of "Eco-psychology." First, I shall discuss important terminologies like "Ecological Grief" and "Eco-anxiety." Then, I shall explore how the indigenous people of Niyamgiri undergo these distinctive forms of grief and anxiety. This chapter also examines how the songs and narratives of these indigenous people reflect their voice of protest against mindless industrialisation, and their resilience amidst grief and anxiety.

The Dongria Kondh as the Guardians of Niyamgiri

The Dongrias, a community of approximately 8000 people, dwell in the hills of Niyamgiri, situated primarily in the Rayagada and Kalahandi Districts of Odisha, India. The term "Dongria (in Kui language) literally means mountain or forest dwellers who live on the agricultural lands of the hill slopes. The Dongria Kondh, similar to other Kondh communities like Kutia Kondh and Desia Kondh, is a tribal group that primarily speaks the *Kui* or *Ku* language, maintaining a distinct ethnic identity. Since, the word "Kui" literally means "mountain," the term "Kondh" means inhabitants or dwellers of the mountain. As a part of their anthropomorphic belief system, the Dongrias revere the Niyamgiri hills as the abode of their supreme deity, Niyamraja, an all-pervading spirit that sustains and governs the sacred law of the land, streams and mountains. Tatpati et al. mention:

> The Dongria Kondh tribal community has been a part of the landscape for many years, sustaining themselves from the resources of the forests of the Niyamgiri range. . . . They believe in a Supreme deity Niyamraja (the King of Law) who is also their ancestor. The Niyamgiri hills are thus "the hills of Law", the abode of Niyamraja, who rules the hills in accordance to these laws, along with other deities associated closely with nature.
>
> (4–5)

So, the Dongrias of Niyamgiri follow "the hills of Law," an unseen mystical order governed by Niyamraja. This natural law unites the tribes of Niyamgiri.

Sustainable Livelihood and Ecological Stewardship

The Dongrias primarily practise shifting cultivation and agroforestry on agricultural land in the foothills of the mountains and collect minor forest products for their survival. They carry axes as symbols of their traditional ethnic and cultural identity and also as tools to protect their natural environment.

Known alternatively as "Jharnia," meaning protectors of the streams and the rivers, the Dongrias consider themselves to be the rightful royal descendants of Niyamraja (from which the name Niyamgiri is derived). Regarding their sustainable livelihoods and ecological stewardship, they worship Niyamraja and seek his benediction to protect Niyamgiri.

Niyamgiri is famous for its dense forest and deep ecology. It is equally rich for its floral and faunal variation housing more than 300 plant species, including seven rare or endangered species. The Niyamgiri hill range also hosts more than 30 wild animal species, including critically endangered fauna like four-horned antelope, green pit viper, golden gecko, Travancore wolf snake and rare birds like the hill myna and pied hornbill. Along with its extremely rich flora and fauna, the Niyamgiri hill range is also renowned for its variety of streams and rivers. Nearly 100 perennial streams flow from the Niyamgiri hills, with Vamsadhara River being the lifeline of the Kalahandi district.[2] According to a report, published by India's *Centre for Science and Environment* (CSE):

> The Dongaria Kondhs's economy and its major sources of livelihood are directly related with Niyamgiri Forests. They collect various types of forest products and sell it in nearby weekly haats. Around 40 to 50% of their annual income is derived from by selling of forest products like Siali leaves, Myrobalans, Amla and etc. They grow fruit crops like pineapple under the thick forests. Dongaria Kondhs are not at all known for hunting. No literature developed in vernacular or in English suggests that the Dongrias were hunters at any point of time in history.
>
> (19)

So, far from being hunters, the Dongrias are primarily protectors of land and forests with a deep belief in sustainable development and ecological cohabitation. Emphasising this cohabitation with nature and ecological reliance, Jitu Jakeskia, a young Dongria said:

> We get everything from the jungle like the fruits we take to the market. This is like our source of life for our Dongria Kondh people. We are not paying any money to get these fruits, this is free, it is like paradise for us here.
>
> Niyam Raja is our supreme god. His name means Lord of Law, he made all things.
>
> Niyamgiri mountain is the most important place for Dongria Kondh people, it is like Niyam Rajah's temple, that is why our people worship nature, they have to protect nature.
>
> (Grammaticas)

We can comprehend from Jitu Jakeskia's words how closely the Dongrias are connected to their land and god. The mountain that they live in is everything to them and this is why they worship and protect nature. Jonna Eede also emphasises the Dongrias' symbiotic relationship with nature:

> The Dongria Kondh have expert knowledge of the forests, plant species and wildlife of the hills. They believe they are the royal descendants of Niyam Raja Penu, who has conferred to them the right to cultivate Niyamgiri's slopes; they grow over 100 crops and cultivate diverse orchards which produce crops such as oranges, sweet papaya, and the aromatic resin Jhunu. The Dongria also gather foods from the forests, such as wild mango, pineapple, jackfruit and honey.
>
> (Eede)

Their deep connection with the hills, forests, wildlife and plants of Niyamgiri compels them to perform their duty as ecological stewards.

External Threats and Ecological Grief

However, this pristine, prelapsarian kind of living of the Dongrias, rooted in the harmonious cohabitation with nature, did not last long, and it faced a significant threat when a Memorandum of Understand (MoU) was signed between Odisha Mining Corporation Limited (OMC) and Vedanta Resources (Vedanta Alumina Limited), a London Stock Exchange-indexed Indian mining company on 7 June 2003. The agreement between the two parties legitimised the extraction of bauxite, the chief raw material for Aluminium from the Niyamgiri Hills. It was estimated that nearly 73 million tons of bauxite would be extracted over a period of 23 years. Subsequently, a lease agreement was signed between OMC and Vedanta in October 2004, leading to the establishment of an Aluminum Refinery power plant at Lanjigarh, which opened in 2006 with an investment of approximately 4000 crores. This industrial incursion and land acquisition, supported by the State Government, caused acute anxiety and trauma among the Dongrias who mourned the environmental damage and loss of sacred ecology. The entire community has been combating this eco-anxiety, arising out of the massive industrialisation. The panic or the fear of the Niyamgiri people is exactly what Greta Thunberg points out, as mentioned at the beginning of this chapter. This panic can better be understood as "eco-anxiety" and "ecological grief." Johanna Oksala, in the essay titled "Climate Change as an Existential Threat: Environmental Politics in the Shadow of Nihilism," clearly states, "Based on the available climate science, there is no doubt that climate change presents an existential threat in the sense of threatening the continued existence of human civilization" (234). According to Oksala, climate change is primarily a political problem, which may lead to existential, nihilistic or ontological angst. Chronic grief, hopelessness, apathy and anomie are some of the pathological outcomes of "Eco-anxiety" or "Climate-anxiety" (234).

Two other famous environmentalists, Ashlee Cunsolo and N.R. Ellis, state that "Ecological Grief" is "the grief felt in relation to experienced or anticipated ecological losses, including the loss of species, ecosystems, and meaningful landscapes due to acute or chronic environmental change" (275). They argue that this grief primarily affects those "who retain close living, working and cultural relationships to the natural environment, and one that has the potential to be felt more strongly and by a growing number of people as we move deeper into the Anthropocene" (275). The mourning over the environmental loss, massive ecocide and global climate change certainly affects the "emotional and psychological" (275) well-being of the affected individuals. This is poignantly apparent among the Dongrias.

Most of the Dongrias have expressed their concerns about the impending ecological crisis caused by Vedanta. A few young activists like Bari Pidikakka, Lodu Sikaka and Jitu Jakeskia among others time and again asked why they should leave their own land as a result of industrial intrusion. According to a young Dongria activist, Bari Pidikakka, the mountain has given them abundant fruits, vegetables, crops and other essential things for their survival. Their anxiety is obvious as they worry also about the rivers. Bari gets panicked wondering if the water in the rivers from Ambadola to Rayagarda dries up, their crops will fail.

The deep-rooted concern of the Dongrias can be taken as the ecological grief of the people. Bouts of panic attacks and anxiety have followed at both the individual and the collective levels. There is a growing concern for the future generation too. How will they survive in an environment fundamentally altered by human activity? Much like Thunberg, the indigenous people too are terribly worried about their ecological losses. This kind of worry may come under the idea of "Solastalgia." According to Glenn A. Albrecht, a famous environmental philosopher, solastalgia refers to a distressful feeling when one's familiar environment or ecological habitat changes drastically against their wishes. In his words, it refers to the "homesickness you have when you are still at home" (27), that is, reflecting the helplessness felt while witnessing environmental degradation or doom.

Many researchers and scholars working with the Hurricane Katrina evacuees and the Inuit (Eskimo of Northern Canada and parts of Greenland and Alaska) from Nunavut, Canada, have found that many indigenous people suffer from complex forms of ecological grief or anxiety due to the loss of their habitats as a result of the changing global climate. Recently an elderly Inuit from Nunavut stated, "We are people of the sea ice. And if there's no more sea ice, how do we be people of the sea ice?" (Vince). A good number of tribes in Greenland also suffer from eco-anxiety as they mourn the gradual melting of the glaciers due to human-induced global warming. Cunsolo and Ellis rightly point out:

> This grief also emerged from the disruption to sense of place and connection to the land, and was often accompanied by strong emotional

> reactions, such as anger, sadness, frustration, anxiety, distress, hopelessness, depression and despair.
>
> (277)

The same is the case with the Niyamgiri people. Their indigenous rights to protect and preserve the forests and the sacred homeland are poignantly voiced by a few old Dongrias. They complain:

> We used to eat and drink well. We lived on fruits and vegetables from the forest and the farming was good. Then the company came and told us to leave our land and take money in exchange.
>
> Most people said: we'll not leave our village. But the company tried to force them out and harassed them a lot. When people said they wouldn't leave, they came with bulldozers at night and demolished their homes.
>
> (*Survival International*)

From the words of the elderly men, we clearly understand how the mountain people are intimidated by the industrial company. One of the most influential leaders of the Niyamgiri Suraksha Samiti, Lado Sikaka from Lakhpadar village, also expresses his concern over the rapid change that they face. He points out:

> After 10 years or more, I see us as what we are today. We don't want change. Change will mean that everything will be lost – our culture, our language. Some people are stepping out to study, but when they come back they've lost everything. What is a man without an identity?
>
> See what has happened in Lanjigarh. When the company (VAL) was not there, the 'Kui' folk (Kutia and Desia Kondh communities inhabiting the foothills of Niyamgiri) were like us, we lived like brothers. You could identify them as Kandha (Konds). But when the company came, everything changed. Land was lost, culture was lost, and identity was lost. Now, they are labourers. They were kings, owners over their own land before. Now you cannot make out who is pano, who is kandho, everything is mixed. What is the use of that kind of development?
>
> (Tatpati et al. 13)

True to Lado's concern, along with their land and forests, the Dongrias will one day lose their language, culture and indigenous identity if they don't resist the industrial encroachment. According to Lado, the Konds have already been lowered to the level of labourers from their state of glory. Lado asks a very significant question: "What is the use of that kind of development?" The words of Lado Shikha may remind an informed reader of the anguished voice of Sankar Nagesia who represents an ethnic group in the then Bihar (currently Jharkhand) in Mahasweta Devi's novella "Pterodactyl, Puran Sahay, and Pirtha":

> Why did the foreigners come? We were kings. Became subjects. Were subjects, became slaves. Owed nothing, they made us debtors. Alas, they enslaved and bound us. They named us, as bondslaves, Haroahi, Mahidar, and they named us Hali, named us Kamiya, in many tongues. Our land vanished like dust before a storm, our fields, our homes, all disappeared.
>
> (Devi 119)

So, the situation of the Adivasis – original inhabitants of the land – is the same all over India. They were evicted from their own place in the name of development and they became "subjects," even "slaves," at the hands of the intruders whom Sankar described as "foreigners."

Eco-Anxiety and Trauma of the Dongrias

But the agony of the Dongrias has another significant dimension to it. This comes under the idea of what Dr Alpes Panchal, a famous Mumbai-based psychiatrist, describes as "eco-anxiety." In his words:

> While the chronic effect of eco-anxiety is not as visible as acute ecological ramifications of environmental disasters such as earthquakes or floods, they do manifest in the form of fear, anger, frustration and despair. . . . The idea of eco-anxiety brings out the basic fear of dying. It's about existential conundrum which in layman terms can be called mid-life crisis and in younger people existential crisis.
>
> (Shalkie)

In their interviews, the people of the Dongria tribe time and again give voice to their sense of "fear," "anger," "frustration" and "despair," which all contribute to an "existential conundrum." One may refer to the sense of fear as expressed in the words of Dodhi Pusika:

> They will dig our mountain and take away all our wealth. We won't get water and we will perish along with the people in the plains. There are so many people in Niyamgiri – Harijans and Adivasis. There are Adivasis in the plains, farmers and landlords. All will perish.
>
> We have water here because of Niyamgiri and that is why people have lived here for ages. If the water is gone and our corps fail, what will we eat?
>
> (*Survival International*)

Such existential dread over ecological degradation affects the entire Dongria tribe. The Niyamgiri Movement of Odisha bears witness to it. This is, however, not the first environmental movement in Odisha. The state saw multiple resistance movements against large-scale extractive projects in past. Notable movements like Chilika Bachao Andolan (1991–93), Gopalpur Andolan (1995–98), Kashipur Movement (1994–2008) and Anti-POSCO Movement (2006 onwards) against mining and refinery extractions and land acquisition deserve a special mention here.[3] Large-scale resistance as a way out of this eco-grief/ eco-anxiety can be located in their speeches.

> Vedanta Company will give us lots of money and promise jobs but we don't get lured by that. We have to look out for generations and money can be spent in 5 minutes. Will we eat money? Our fortune is our land, our forest. Why should we leave?[4]
>
> (*Survival International*)

One could easily understand the intensity of grief articulated in their words. They mourn over the existential question, "Why should we leave?" The fear of impending doom, caused by massive ecocide makes them traumatised but they try to resist the industrial encroachment through their vehement protests against Vedanta.

Voices of Protest of the Activists

In the face of globalisation and industrialisation, they are aware of their ecological stewardship. This is reflected in the voice of another activist named Dodhi Pusika from the Gorota village:

> We will not allow Vedanta to come here and because we didn't know earlier, they were able to build the refinery in Lanjigarh. But now that we know about them, we won't allow them to come here at any cost. We don't need roads up to Niyamgiri. We don't have any vehicles to use this kind of road.
>
> (*Survival International*)

Dodhi's statement staunchly critiques the so-called notion of development as he points out that they "don't need roads up to Niyamgiri" because they "don't have any vehicles." He again warns:

> Vedanta does not have any right to touch our Niyamgiri mountain. Even if you cut our throats, even if you behead us, we are not going to allow this. We will see how they are going to take over our mountains.
>
> (*Survival International*)

One can understand that they are awfully aware of their indigenous rights. In no circumstances, they will allow the intruders to touch their Niyamgiri mountain. They are ready to sacrifice their lives to protect nature.

The Dongrias have stood long for their ecological rights and challenged the conventional "doom and gloom" narrative of development. The words of Lodu Sikaka, a powerful leader of the movement, are worth quoting at this point:

> It's crazy when these *outsiders* come and teach us development. In our land we don't have to buy water like you, and we can eat anywhere for free.
>
> Our God lives in open space, you keep your God locked up with a key. We won't leave Niyamgiri. If the government and politicians ask for it we will fight. [Emphasis added]
>
> (Eede)

One may ask: who are these "outsiders"? The answer to it is hidden in Lodu Sikaka's words. They are not only the industrialists who invade the ecological space but the "government and politicians" as well who sponsor them in this destructive activity. However, in a larger context these "outsiders" represent the "Anthropoi" as described in Amitav Ghosh's *The Living Mountain* who always believe that they are stronger than the indigenous people and hold their ideas as "universal" against the "false, local beliefs" of the latter:

> They had laughed at our inherited ideas of the Mountain's sacredness: that was all ignorant, pagan superstition, they had said. All mountains were the same, they could all be climbed if only the climbers were strong enough, intelligent enough, resolute enough.
>
> (Ghosh 26)

The Adivasis of the Niyamgiri resist this supremacist attitude of the "outsiders" – the "Anthropoi." They believe in the "sacredness" of Niyamgiri. And they are determined to keep its sacredness intact. They will not leave their land. They will fight till the last.

Song as a Form of Cultural Resistance

This staunch attitude of the people is remarkably evident in their songs. In her conversation with Gayatri Chakravorty Spivak, Mahasweta Devi emphasises the importance of songs in the lives of the tribals: "Every event that the tribals come to know they transfer to song, they do not write. They have retained the memories of their fights, of natural calamities in this way" (Spivak xi).

To a great extent, this is true of the songs of the Dongria Kondh tribe. We select two songs here for discussion. The first one bears the title "Lament of Niyamgiri."[5] The song begins with the assertion that the mainstream people are "unaware" of the story of Niyamraja that he is "the first of the Dongria Kondh." The narrative is: "After making pineapple, mango, jackfruit and grains/Niyamraja said to us 'live on what I have given you.'" The Dongria people believed in their deity and were satisfied with their natural resources. But things took a sharp turn when the outsiders for their own selfish interests invaded the Niyamgiri – their ecological space. Niyamraja sensed the disaster:

> Niyamraja is crying today, the hills will turn into mud
> The rocks will crumble and everyone will die
> Will there be any rivers left if there are no streams?
> Will there be any streams left if there are no hills?

The lines evoke a picture of impending environmental catastrophe when hills will not exist and the existence of the rivers is threatened. The theme of the song thus goes beyond the regional boundaries of Niyamgiri and speaks of the planetary crisis. Psychologists have tried to document numerous cases of climate anxiety around the world. Although not yet indexed in the Diagnostic and Statistical Manual of Mental Disorders (DSM-5), the term, "eco-anxiety," was first introduced in a 2017 report by the American Psychological Association. The report, focusing on "the impacts of climate change on mental health," (Fawbert) defined "eco-anxiety" as "a chronic fear of environmental doom." *Psychology Today* describes it as "a fairly recent psychological disorder afflicting an increasing number of individuals who worry about the environmental crisis" (Fawbert). One such case is Sam Johnston from Manchester, who suffers from eco-anxiety. In one of his sessions, Sam detailed the pathological effects of eco-anxiety:

> When you go to sleep, you start thinking about everything – the state of the planet, really, and the potential future of it – and knowing that there's only so much you can do as one person. I think that's the anxiety – because you just feel a bit powerless in it all.
>
> (Fawbert)

Niyamraja cries because he feels "powerless" in the face of the ecological disaster. We know that the indigenous people are historically the chief protectors of the environment and biodiversity all over the world. This is true of the Dongria Kondh tribe of Niyamgiri as well. But now they face severe eco-anxiety. The narrator in the song thus raises a question: "How will everyone get to know of Niyamraja's miracles?" The miracles most probably speak of the ecological harmony in Niyamgiri maintained by its guardian deity. But that harmony is now at stake. The narrator, through his song, will therefore raise a couple of pertinent questions:

> If Niyamgiri will be not there, I will sing, I will sing
> . . .
> What will the animals do without the big forest?
> What will we do without its plants that save lives?

This is the voice of a man who knows how to spread ecological consciousness through a powerful cultural medium. The repetitive expression "I will sing, I will sing" speaks of the desperation of the narrator to voice the grief of those who represent the ecosystem people in the Niyamgiri region and whose stories remain unrepresented in the mainstream environmental discourse.

The second song is titled "Song for Niyamgiri."[6] At the very opening of the song the note of protest against the state-sponsored mechanism of uprooting the indigenous people of Niyamgiri is struck:

> We will never leave the mountain, stream, forest and our home
> We will not give up the struggle, even if we have to sacrifice our lives
> Dear friends, we will not leave the struggle

The mountain, streams and forest are their "gods." The narrator interprets the words like "devata" and "Bhagaban" from an Adivasi worldview: "devata" is "the one who always gives," and "Bhagaban" is a combination of " 'bha' for *bhoomi* [land], 'ga' for *gagan* [sky], 'ba' for *baayu* [air] and 'na' for *neer* [water]. The narrator thus helps us understand an epistemology that sheds light on an inclusive idea of the environment. The Dondgrias cannot imagine their existence without the hills, trees and rivers. Niyamgiri is their "home." The land with its "greenery, trees, fruits and trees" is a heaven on earth to them. They imagine the water of the river as "the milk of the motherland." They enjoy "freedom" in their land. But they sense a nexus between "the government and the corporates" in suppressing them. They are, however, determined to protect nature and not to be ousted from their land. Their argument is plain and simple: "If the environment is destroyed, all of us will be dead." They give a call for a united protest against the government:

> Hey government! We will not be afraid of your lathis, bullets and jail
> We will sacrifice for our motherland and not give up the struggle!
> Wake up all the tribes and workers
> If we are attacked, we, the children of the soil, will not remain silent
> Our struggle will continue . . .

The lines speak of a "Forest satyagraha." In "Primitive Rebellion and Modern Nationalism: A Note on Forest Satyagraha in the Non-cooperation and Civil Disobedience Movements" Sumit Sarkar describes the "Forest satyagraha" as "the reassertion by poor peasants and tribals of traditional customary rights over forests 'reserved' by the colonial state." Unfortunately, these forest movements do not prominently figure in the "conventional nationalist historiography" (511) and are used only as passing references. Prof. Sarkar shows that in the colonial period, "the centres of forest agitation were all in the hill regions of Maharashtra, Karnataka and the Central provinces" (517). When we speak of the Niyamgiri movement, it is again a hill region where the tribal people organised more than a decade-long protest against the state apparatus to assert their rights over the forests. The nature of the protest is largely non-violent as is reflected in the lines:

> We will not be afraid of your lathis, bullets and jail / We will sacrifice for our motherland and not give up the struggle!" The narrator emphasizes that they are peace-loving but they must fight for their rights. He gives a clarion call to the comrades, "Dear friends, let the corporate government get out!.

Conclusion

These songs have a therapeutic value in the sense that they help in easing out their grief, anxiety and agony of impending ecological catastrophe. By singing their songs collectively and by strengthening their struggle through these songs, the Dongria Kondh people are able to overcome the buried anxiety and trauma and tread the path of resilience. Songs become a mode of cultural revolt with them against all the state-sponsored repressive forces that deny them indigenous rights to land and water and forest. Songs give voice to their activist movements for conservation of forests and creation of environmental consciousness.

Acknowledgement

I am deeply indebted to *Survival International* for granting me the necessary permission to quote a few lines from their YouTube video titled *The Real Avatar: Mine – Story of a Sacred Mountain* (youtu.be/R4tuTFZ3wXQ?si=XSVzpPqwSOy-NuJX), uploaded on 31 March 2009. Their support has been invaluable to the completion of this work.

Notes

1 Greta Thunberg delivered a speech that was later covered by *The Guardian* in an article titled 'Our House Is on Fire': Greta Thunberg, 16, urges leaders to act on climate' on Fri 25 Jan 2019. This is famously known as "'I want you to panic': 16-year-old issues climate warning at Davos," www.theguardian.com/environment/2019/jan/25/our-house-is-on-fire-greta-thunberg16-urges-leaders-to-act-on-climate.

2 The Centre for Science and Environment has published a report titled "A brief report on Ecological and Biodiversity Importance of Niyamgiri Hill and Implications of Bauxite Mining." See www.cseindia.org/report-on-niyamgiri-7654.

3 Interested readers may read Kundan Kumar's article "Confronting Extractive Capital: Social and Environmental Movements in Odisha" published in *Economic and Political Weekly*, vol. 49, no. 14, 5 April 2014.
4 Interested readers may find Dodhi Pusika's full interview that is reported by OTV "Niyamgiri Hill Residents Stage Agitation Protesting Mining In The Hill." Link: www.youtube.com/watch?v=ohcODW8Jhmk&t=193s.
5 The song can be accessed at: www.cultureunplugged.com/documentary/watch-online/play/57/the-lament-of-niyamraja. For the written text interested readers may see p. 12 of *The Niyamgiri Story: Challenging the Idea of Growth without Limits*? by Meenal Tatpati, Ashish Kothari, and Rachna Mishra, published by Kalpavriksh, Pune, Maharashtra, 2016.
6 This song, sung by the poet and activist Rajkishor Sunani at the Niyamgiri festival in 2018, is available at the People's Archive of Rural India (PARI). The song can be accessed in an article titled "The mountain, forest and streams are our gods" written by Purusottam Thakur, and published on 20 April 2018. See. ruralindiaonline.org/en/articles/the-mountain-forest-and-streams-are-our-gods/.

Works Cited

Albrecht, Glenn A. *Earth Emotions: New Words for a New World*. Cornell University Press, 2019.

Centre for Science and Environment Report. "A Brief Report on Ecological and Biodiversity Importance of Niyamgiri Hill and Implications of Bauxite Mining." *Centre for Science and Environment*, www.cseindia.org/report-on-niyamgiri-7654.

Cunsolo, Ashlee and Neville R. Ellis. "Ecological Grief as a Mental Health Response to Climate Change-related Loss." *Nature Climate Change*, vol. 8, no. 4, 2018, pp. 275–281. *Nature*, https://doi.org/10.1038/s41558-018-0092-2.

Devi, Mahasweta. "The Author in Conversation." *Imaginary Maps*, translated by Gayatri Chakravorty Spivak. Thema, 2001.

———. "Pterodactyl, Puran Sahay, and Pirtha." *Imaginary Maps*, translated by Gayatri Chakravorty Spivak. Thema, 2001.

Eede, Joanna. "Dongria Kondh Tribe of India Resist Powerful Mining Company." *Ecologist: The Journal for the Post-Industrial Age*, 2013, theecologist.org/2013/nov/20/dongria-kondh-tribe-india-resist-powerful-mining-company.

Fawbert, Dave. "'Eco-Anxiety': How to Spot It and What to Do about It." *BBC Three*, BBC, 27 March 2019, www.bbc.co.uk/bbcthree. Accessed 24 August 2024.

Ghosh, Amitav. *The Living Mountain: A Fable for Our Times*. Fourth Estate, 2022.

Grammaticas, Damian. "Tribe Takes on Global Mining Firm." *BBC News Orissa*, 17 July 2008, news.bbc.co.uk/2/hi/south_asia/7486252.stm.

Oksala, Johanna. "Climate Change as an Existential Threat: Environmental Politics in the Shadow of Nihilism." *Purple Brains: Feminisms at the Limits of Philosophy*, edited by Annabelle Dufourcq et al. Radboud University Press, 2024, pp. 233–246. *JSTOR*, www.jstor.org/stable/jj.15684204.22. Accessed 16 June 2024.

Rosenfeld, Jordan. "Facing down 'Environmental Grief': Is a Traumatic Sense of Loss Freezing Action against Climate Change?" *Scientific American*, vol. 21, 2016, www.scientificamerican.com/article/facing-down-environmental-grief/.

Sarkar, Sumit. "Primitive Rebellion and Modern Nationalism: A Note on Forest Satyagraha in the Non-cooperation and Civil Disobedience Movements." *Proceedings of the Indian History Congress*, vol. 38, *Indian History Congress*, 1977, pp. 511–523. *JSTOR*, www.jstor.org/stable/44139110.

Shalkie. "Dealing with Eco-Anxiety: Experts Explain How to Tackle the Existential Crises of Climate Change." *Deccan Chronicle*, 4 December 2019, www.deccanchronicle.com/lifestyle/environment/041219/dealing-with-eco-anxiety.html.

Survival International. "The Real Avatar: Mine – Story of a Sacred Mountain" [Video]. *YouTube*, uploaded by *Survival International*, 31 March 2009, youtu.be/R4tuTFZ3wXQ.

Tatpati, Meenal, Ashish Kothari, and Rachna Mishra. *The Niyamgiri Story: Challenging the Idea of Growth without Limits*? Kalpavriksh, 2016.

Vince, Gaia. "How Scientists Are Coping with 'Ecological Grief'." *The Guardian*,12 January 2020, www.theguardian.com/science/2020/jan/12/how-scientists-are-coping-with-environmental-grief.

10 "We Should Have Known Our Land Would Soon Be Dead"

Resource Curse, Petro-Capital Extractivism and Survival Environmentalism in Imbolo Mbue's *How Beautiful We Were*

Shankha Shubhra Mandal

Introduction

In this petro-modern era, which is heavily dependent on the extraction and transportation of crude petroleum, oil spills have been a frequent predicament in all the oil-producing regions around the globe. The Niger Delta region of Nigeria, which is one of the most ecologically vulnerable oil-producing regions of the world, is no exception in respect of this chronic challenge of oil spills which is now an internationally recognised environmental issue. The spillage of petroleum from the oil pipelines destroys the main sources of livelihood, like fishing and farming, of the poor host communities of the Niger Delta, obliterates the natural habitats and brings in disharmony to the ecosystem. Thus, oil spills, which are a curse to the ecosystem because of their ruthless destruction of the flora and fauna and which lead to the pauperisation of the "ecosystem people"[1], find regular media coverage across the world. openDemocracy, an independent international media platform, ventilates it in poignant terms:

> On a sweltering afternoon, Longlife Bob returns empty handed from his search for work at nearby building sites.
>
> Before the oil spill, Bob, 40, was a fisher. On good days, he made an average of 80,000 Nigerian naira (about £85) from selling his catch. With this income, he has taken care of his family for the last ten years.
>
> But things have changed. The spill that contaminated the Oluku river in Aleto, before spreading to Bob's Akpajo community in June, destroyed his fishing nets and boat and polluted the river that had been his livelihood. An alternative livelihood is not forthcoming, at least for now. "We are suffering and life is hard for us," Bob told openDemocracy.[2]
>
> (openDemocracy.net)

Thus, the poisoning of the river due to oil spills gradually leads to the destruction of its ecosystem. The fishes, which were once in huge numbers

DOI: 10.4324/9781003602750-14

there providing food and source of livelihood to the fishermen of the region, disappear forever from those fishing waters. As a result, the local fishermen become helpless having no source of alternative income. Similarly, when we speak of the petrofiction of the Niger Delta, we remember a classic petrofiction by Imbolo Mbue titled *How Beautiful We Were* (2021), which faithfully represents the oil-induced environmental violence which is seen on a large scale in the African continent. The novel engages with the petro-colonial exploitation meted out to an imaginary indigenous African place named Kosawa by an all-powerful American oil company Pexton.[3] It shows how the abundance of oil becomes a curse for the local people of Kosawa who are subject to incurable health hazards, deprivation of livelihood means and socio-political exploitation due to the regular spillage of crude petroleum, blowouts from the oil pipelines and raging fires caused by gas flaring which are the common phenomena in the process of extraction and transportation of crude oil.

These violent catastrophes destroy the flora and fauna of the oil-producing regions and victimise the poor host communities who rely upon their ecosystem for shelter and sustenance. The idea of "resource curse" (Richard Auty) thus becomes very much dominant in this context. The exploitation of the Kosawa people comes full circle in their exclusion from every decision-making apparatus as well as from any share in oil revenue. The indigenous people become victims of "environmental racism"[4] (Robert D. Bullard), which speaks about the unequal distribution of environmental hazards among poor people of colour. However, the final impression of the novel is not completely pessimistic as the villagers succeed in forming a socio-environmental resistance movement in the face of the overwhelming threats of environmental breakdown and the politics of injustice implicated in it. *How Beautiful We Were,* with its explicit criticism of the contemporary petro-capitalist system and the environmental injustices prevalent in the petro-despotic regime of Nigeria is of great critical interest and deserves much scholarly attention. But a careful study of the critical works on the novel brings out the fact that though there are a few articles on the novel that deal with the themes of environmental degradation in the novel, none of them binds together the major tropes of resource extractivism, petro-colonial-capital politics, environmental violence and resistance movement into an organic whole. Thus, the present chapter will critically examine the representation of the politics of resource extraction, petro-colonial-capital violence, state-backed exploitation of the indigenous people and the environmental injustices inflicted on the people of colour in Mbue's novel by reading it from the major perspectives of Petrocriticism. It will also focus on the local people's environmental resistance movement to see how the impoverished people's concerns for their immediate environment come from their lived experiences of environmental crisis. The novel, through the representation of environmental disasters like water and land poisoning due to oil spills, warns us against profit-motivated anthropogenic activities and thus performs the function of an ecocritical narrative by opening our eyes to environmental degradation and encourages us to fight against the developmental intervention of the capitalist or neocolonial powers like the American oil company Pexton.

Resource Curse and the Politics of Extractivism

Amitav Ghosh, the eminent author and one of the pioneering figures writing about climate change and eco-disaster imaginaries, in *The Nutmeg's Curse*, as the title suggests, starkly points out how the abundance of a natural resource or commodity, like the Nutmeg, becomes the cause of extermination of an entire group of islands, the Banda Islands of Indonesia. However, the idea of "Resource Curse" was first propounded by Richard Auty in 1993 in his work *Sustaining Development in Mineral Economies: The Resource Curse Thesis*. According to Richard Auty, "not only may resource-rich countries fail to benefit from a favourable endowment, they may actually perform worse than less well-endowed countries. This counter intuitive outcome is the basis of resource curse thesis" (Auty 1). Thus, "Resource Curse" refers to the paradoxical or ironic situation that countries enriched with natural resources such as petroleum, natural gas and valuable minerals tend to have lower economic growth, much fragile democratic structure and worse environmental situations than countries with less natural resources. In this way, the state, cursed with the abundance of natural resources, suffers from constant political instability, severe economic crises, social unrest and increasing environmental degradation. This theory of Auty has been fairly elaborated by Mitchell Ross in his book *The Oil Curse: How Petroleum Wealth Shapes the Development of Nations* (2012), where he argues that the resource curse is

> more accurately a mineral curse, since these maladies are not caused by other kinds of natural resources, like forests, fresh water, or fertile cropland. Among minerals, petroleum–which accounts for more than 90 percent of the world's minerals trade–produces the largest problems for the greatest number of countries. The resource curse is overwhelmingly an oil curse.
>
> (Ross 1)

However, it would be an oversimplification to say that the abundance of petroleum resources is a curse for each oil-producing nation. On closer inspection, it is found that it depends upon how the vast revenues generated by the oil industry are handled by the government. This is clearly evident when we consider the situation prevailing in the oil-producing Arab countries where we find a petroleum-based utopia as opposed to the African countries, where petroleum endowment is the root cause of human suffering and environmental devastation.

One of the multiple layers of politics associated with the oil industry is the politics of extractivism. Resource extraction is a common phenomenon since the very dawn of modern civilisation. But extractivism, unlike extraction, is not simply a practice; rather it has a philosophy, a politics working behind it. R. Sreejith Varma argues that "extractivism is profit-motivated and insensitive to the harm wrought upon people and the environment" (Varma 657). Thus, in the context of the oil industry, the politics of extractivism becomes instrumental in the functioning of the capitalist, neocolonial, multinational oil companies that are in league with the military or neoliberal governments to gain petro-dollars. The politics of extractivism, which minutely puts the rich and the poor at loggerheads and plays a pivotal role in causing environmental disasters, is conspicuously evident in Imbolo Mbue's *How Beautiful We Were*. The title of the novel itself is suggestive of the irony of the abundance of oil in Kosawa, a fictional African village ravaged by violence caused by crude oil extraction, as it refers to the primordial or pristine state before oil exploration in the region by the American oil company Pexton. The oil village of Kosawa was now on the verge of environmental catastrophe due to the irresponsible activities of the oil company in the region which is expressed in the very opening lines of the novel:

> WE SHOULD HAVE KNOWN THE END WAS NEAR. HOW COULD WE NOT HAVE known? When the sky began to pour acid and rivers began to turn green, we should have known our land would soon be dead. Then again, how could we have known when they didn't want us to know?
>
> (Mbue 3)

After a long period of continuous oil extraction in the region, the villagers realised that the end or the ultimate environmental catastrophe was near. When Pexton first came there after the discovery of crude petroleum in their land, the villagers "began dreaming of how splendid our lives would be" (228) when they would get their "share of the money" (227). But this dream of gaining the petro-dollars by just sitting back and allowing Pexton to do whatever activities they wanted to perform proved to be an illusion in the end. What they experienced instead is a nightmarish tale of exploitation and deprivation. The ominous impacts of the extractivist policies of the American oil company in the region are very much evident from the concentrated harm wrought upon the poor people and their environment:

> We remembered those who had died from diseases with neither names nor cures–our siblings and cousins and friends who had perished from the poison in the water and the poison in the air and the poisoned food growing from the land that lost its purity the day Pexton came drilling.
>
> (Mbue 5)

Regular oil spills, gas flaring, acid rain, etc. gradually destroyed the main sources of livelihood of the villagers who were mainly farmers and fishermen and, thus, the deprived native people were left in utter destitute having no other source of living. The ultimate fate of Kosawa was what might be called a total annihilation of identity and existence. Due to the incremental environmental violence caused by Pexton, the village gradually became uninhabitable and the government ordered it to be burned:

> We were refused one last chance to enter Kosawa and empty our huts. The government decided the land had become too contaminated for human presence. His Excellency ordered Kosawa burned. What once were our huts became ashes. Our mother's kitchen, ashes. Our barns and outhouses, ashes. Our ancestor's pride, ashes. Nothing reminded of Kosawa, except for what we kept in our hearts. Hour after hour, day after day, year after year, winds of every kind came and blew away the ashes of what used to be our home.
>
> (Mbue 355)

Thus, continuous environmental violence gradually turned Kosawa into an uninhabitable foreign land for the villagers which first resulted in, according to Nixon, their "displacement without moving" (Nixon 19) and finally their physical displacement and total annihilation of identity. Thus, the abundance of a natural resource, petroleum, becomes the cause of extermination of an entire indigenous village, Kosawa, and oil endowment proves to be a curse for its "disposable people"[5].

Colonial Intervention and Petro-Capitalist Exploitation

The history of the petroleum industry around the globe shows that it is undeniably exploitative in nature, and it has deep-seated relationship with the history of colonialism as well as the history of capitalism. In this regard, Michael Watts argues that "the entire history of petroleum is replete with criminality, corruption, fabulous wealth, and the worst of rapacious frontier of capitalism. Power politics, ruthless exploitation, and thuggery have always been oil's defining characteristics, and imperialism is its defining moment" (Watts 189–190). Thus, the people of the oil-producing regions have always been at the receiving end of severe exploitation and deprivation. In the colonial period, before the discovery of oil, they were oppressed, tortured and exploited by their colonial masters, mostly the Europeans. In Mbue's novel, the people of Kosawa, for example, were forced to work in the rubber plantations and those who refused were killed brutally. One may remember the words of Yaya in this context: "Nowadays young people talk about the oil as if it's our first misfortune: they forget that, long before oil, the parents of our parents suffered for the sake of rubber" (Mbue 222). The intrusion of the Europeans in the lives of the ancestors of Kosawa is an instance of colonial intervention on the plea of improving the quality of their lives, which is a burden the Europeans must take for the fulfilment of their desired mission.

Similarly, after the discovery of oil, the American oil company, Pexton, gradually takes control of everything in the region, which is evident from the ways it regulates the Government, including the environmental policy making of the state, economic policies, management of oil resources and its revenues. Thus, the American oil company carries forward the legacy of colonialism and functions as a neocolonial force in the region, slowly sucking the vital resources from the imaginary African village and its helpless inhabitants.

The neocolonial exploitation of the oil village of Kosawa by the oil company Pexton can be described as an undeniable case of "petro-colonialism" as the American oil company works in collaboration with the ruling elites of the village to make the oil drilling conditions favourable without caring about the environmental concerns and the ecological disasters that oil industry creates in a petro-capitalist regime. It doesn't matter whether they are Europeans or Americans, the grim shadow of colonial intervention and exploitation hovers over Kosawa as they all believe in the idea that they have the right to do whatever activities they want since they are politically and economically superior: "No matter their pretences, they all arrive here believing they have the power to take from us or give to us whatever will satisfy their endless wants" (103). The colonial intervention of the oil company in the guise of development is clearly articulated in what the representatives of the government, who accompanied the oil explorers, said to the ancestors of the villagers when oil was discovered in the region: "drilling for oil would bring something called 'civilization' to our village. One day, the government representatives said, Kosawa would have a wonderful thing called 'prosperity'" (73). But this long "cherished guest" (73) called prosperity remained elusive forever and what they got instead was death and destruction.

Thus, the exhaustion of the natural resources of the imaginary African space, the incremental violence on the environment and the severe exploitation of the local inhabitants of Kosawa – all originate from a petroleum-based capitalism, which, according to Imre Szeman, is called "oil capitalism." In "System Failure: Oil, Futurity, and the Anticipation of Disaster" Imre Szeman argues that

> Oil and capital are linked inextricably, so much so that the looming demise of the petrochemical economy has come to constitute perhaps the biggest disaster that 'we' collectively face. The success of capital is dependent on continuous expansion, which enables not only profit making but investment in the reproduction of capital that is a necessary condition for its continuation on into the future.
>
> (Szeman 807)

Thus, the basic premise of petrocapitalism is that it must accumulate capital and generate profit based on petroleum and its by-products in order to continue its existence. And it is for this reason that both the environmental and the human sides of petro-violence are scrupulously avoided by the giant multinational oil companies. The petro-capitalist exploitation of the host community as well as the environment of Kosawa by the capitalist American oil company Pexton is very skilfully delineated in *How Beautiful We Were*:

> Paxton's agreement with our government was that Pexton would extract the crude and our government would be responsible for all negative externalities. What this meant, he explained, was that if the case ended up in a trial, Pexton wasn't going to deny that its practices led to spills that famished Kosawa's soil. Its lawyers wouldn't try to dispute that the waste on the big river was from their oil field. They wouldn't even need to argue that the children's deaths had nothing to do with them. All they would need to show was evidence that our government had relieved them of any responsibility to the land and people in exchange for splitting the oil profits.
>
> (Mbue 316)

The government of Kosawa gave the drilling rights to the capitalist oil company and took responsibility for all the negative ramifications of crude extraction, be it oil-induced environmental disasters, like oil spills, or human catastrophes, like child deaths, because the government was heavily dependent on the revenues generated by the oil industry and the share of petro-dollars. This facilitates the capitalist, neocolonial oil companies to indirectly control the government and intervene in almost all the policy makings of the state. Thus, the petrocolonial oil companies' motive is the creation of a petro-capitalist system in which the government remains theoretically independent and legally liable for all the devastating impacts of the oil industry in exchange for a small share of oil profits, but a handsome share of the petro-dollars remains in the hands of the oil companies, and the native communities bear all the brunt of oil extraction and suffer from incurable health hazards and environmental degradation.

Environmental Racism and Environmental Injustice

Throughout the history of human civilisation, environmental disasters or catastrophes have disproportionately affected the humans of different countries, places and societies. But since the beginning of the twentieth century, environmental anomalies have become more and more anthropogenic. Thus, in the Anthropocene, with the increasing planetary devastation by human beings, along with social, political and economic injustices, the concept of "environmental injustice" has gained currency and importance in environmental studies. The term "environmental injustice" basically refers to the activities or policies that negatively impact or harm the environment while victimising specific groups and communities. It is against this environmental injustice that the idea of "environmental justice" becomes crucially important. The term "environmental justice," thus, refers to the conditions where there are equal rights for all people, irrespective of their race, gender, class or colour, in respect of living in a safe, healthy, productive and sustainable environment.

> This term, broadly defined, gives rise to at least two testable proportions: the environmental racism hypothesis, which maintains that unbalanced proportions of environmental hazards may be located in minority communities; and the environmental classism hypothesis, which focuses on whether the same problem affects the poorer communities more so than affluent ones.
>
> (Lester et al. 1)

However, the term "environmental racism" was coined by the African-American civil rights leader Dr Benjamin Chavis in 1982. According to Robert D. Bullard, it refers to

> any policy, practice, or directive that differentially affects or disadvantages (whether intended or unintended) individuals, groups, or communities based on race or color. It also includes exclusionary practices that limit participation by people of color in decision making boards, commissions, and regulatory bodies.
>
> (Bullard 23)

Environmental racism has emerged as a serious threat to minority communities or people of colour around the globe in recent times with the increasing environmental hazards. In fact, it has deep-rooted association with extractivism, which places the environmental burdens of resource extraction or mining, especially oil extraction, upon the native minority communities or indigenous people who are largely people of colour. Thus, it is no surprise to find out that the burgeoning alliance of environmental racism and resource extractivism has become an aggravating factor in the practice of environmental injustices that have been meted out to the people of colour or the host minority communities or the "ecosystem people" who live in the proximity of the oil extraction sites around the globe, especially in the African oil-producing countries. In Mbue's novel *How Beautiful We Were*, the threat of environmental racism constantly hovers over the indigenous community of Kosawa and places them at extreme exposure to man-made eco-disasters or environmental calamities. The imaginary African village, Kosawa, solely consists of indigenous people who are unquestionably the people of colour from the ever-exploited land of Africa. They are "different tribes thrown together with no common dream" (Mbue 335) and now due to the environmental catastrophes, like oil spills, gas flaring and river poisoning, caused by the state-backed insensitive extraction of crude petroleum by Pexton and the discriminatory exploitative practices, they are "crumbling from within" (335). They are not only deprived of their equity of oil wealth but also systematically excluded from every decision-making body or policy-making apparatus. In fact, their land was given to the oil company in a meeting in which there was no representative from the indigenous people of Kosawa:

> Three decades before, in Bezam, on a date we'll never know, at a meeting where none of us was present, our government had given us to Pexton. Handed, on a sheet of paper, our land and waters to them.
>
> (Mbue 11)

The grievous impact of the practice of careless crude extraction by Pexton on the indigenous people is evident from the countless deaths in the village: "We've already had too many deaths, our old man said; we've used up too many burial plots" (6). There are innumerable children who died because of the poison in the water that comes from the all-consuming oil industry:

> It was the poison – the baby was too pure for the filth in the village well's water, the toxins that had seeped into it from Pexton's field.
>
> (9)

Though these poor subsistence people never get impartial treatment both from the oil company officials in Bezam and the government, they have been treated equally by death:

> We are reminded of our parents' lies whenever there was a new death, sometimes in our huts, sometimes in the hut next to ours, sometimes children younger than us, babies and toddlers, children who had barely tasted life, always children we knew. We were young but we knew death to be impartial.
>
> (9)

Thus, the disfranchised people of Kosawa, who are mainly the people of colour, are compelled to bear the environmental hazards for the benefit of the American oil company Pexton, which mainly consists of white people, in conformity to what Bullard argues in "The Threat of Environmental Racism": "Environmental racism combines with public policies and industry practices to provide *benefits* for whites while shifting costs to people of color" (Bullard 23). In the novel we get a contrasting picture of how, on the one hand, the host community of Kosawa suffers from terrible environmental and economic crises and, on the other hand, the elites in Bezam enjoy "lives overflowing with clean necessities and superfluities" (Mbue 10). Thus, environmental injustice combined with environmental racism becomes one of the most recurring motives in the novel as it is evident from the "countless crimes" (343) that Pexton and the government commit in collaboration. However, it would be an oversimplification to say that the indigenous people of Kosawa succumbed to their existential crisis amid the overwhelming threats of environmental injustices because with the audacious act of the madman, Konga, they "declared war on them" (25) and resorted to resist the petro-colonial exploitation of Pexton.

Resistance to Petro-Violence and Survival Environmentalism

In *How Beautiful We Were*, with its multiple narrators, Mbue has vividly presented the saga of obnoxious violence on the environment as well as on the local inhabitants of the oil-producing village by the state-sponsored oil industry. This petroleum-induced violence has been termed "petro-violence" by Michael Watts in his "Petro-Violence: Community, Extraction, and Political Ecology of a Mythic Commodity." According to Watts, petro-violence is a form of violence "that so often attends particular sorts of resource extraction and, quite specifically, the extraction of petroleum–arguably the most important resource of the twentieth century" (Watts 189). Petro-violence has deep-rooted association with the struggles over the control of oil reserves and it refers to "both ecological violence perpetrated upon the biophysical world and social violence–criminality and degeneracy associated with the genesis of petro-wealth" (Watts 189). Thus, petro-violence has multiple levels of signification: on one level, petro-violence refers to the violence or harm that petroleum extraction does to the environment; on another, it refers to the victimisation of human beings by the hydra-headed ramifications of oil industry; and on a third level, transcending the first two, it also implies the violence caused by endless wars between the state-backed military forces and the revolting militants in a petro-state. These violent confrontations between the two struggling forces spring from the conflict over the control of oil reserves, the gaining of petro-dollars and also the resistance to petroleum-induced environmental violence. The unequal distribution of oil wealth and the immense pollution of the environment by the oil industry results in the rise of militancy among the youths, which further leads to the kidnapping of oil expatriates for ransom. What is even more important is that the communities who live in the proximity of the oil extraction sites are mainly the "ecosystem people" who depend upon their immediate environment for existence and survival. In Mbue's *How Beautiful We Were,* the people of Kosawa, especially the ecosystem people of the region, are exposed to tremendous environmental degradation, ecological destruction and various health hazards which make their survival a hell-like experience. Everything in the village, including trees, are "felled all over the valley to make room for the oil field and pipelines and Gardens" (Mbue 73). The essential natural resources like water, soil and air are immensely polluted by the toxic chemical wastes of the oil industry. Oil spills, blowouts and gas flaring destroy their main sources of livelihood like fishing and farming. So, very naturally, the poisoning of fishing waters, farming lands and regular child deaths lead the ecosystem people of Kosawa to unite and wage an endless war against the American oil company Pexton for its countless environmental crimes that made death a ubiquitous phenomenon in the village. The novel shows that the subsistence people of Kosawa formed an environmental resistance movement, led by the protagonist Thula, against the profit-devouring oil company Pexton. Their violent resistance started with the kidnapping of the oil men from Pexton. They knew that starting an unequal battle with Pexton was nothing other than inviting death, but they were bound to do so because it was a necessary survival imperative for them:

> I will take them to see the graves of my sons who died from their poison. I'll show them the graves of every child we've buried, and they'll count the graves so that they'll know the number and never forget it. Then we'll keep them prisoner and I'll guard them until their employer stops killing us.
>
> (Mbue 23)

After Konga's audacious act of taking the car keys of the oil men, the wretched villagers realised that they "can and should fight" (22) against the omnivorous multinational oil company. Their exploitation, deprivation and existential crisis reached such an extent when Lusaka, one of the most peaceful men in Kosawa, firmly declared, "If Pexton doesn't stop killing our children, I'll kill these three of its children with my bare hands" (24). The environmentalism of the minority community of Kosawa is solely what Guha and Martínez-Alier have called the "empty-belly" environmentalism of the poor practised in the Global South. According to them, "the environmentalism of the poor originates as a clash over productive resources" (Guha and Martinez-Alier 18) and this "empty-belly" environmentalism has close connection to the survival and subsistence imperative of the "ecosystem people," who depend upon their immediate environment for their survival needs. Thus, the poor host community of Kosawa stood against the state-sponsored, resource-extractive oil industry for their survival needs and their basic sustenance necessities. However, the poor villagers of Kosawa are not only exposed to the spectacular side of petro-violence like military-militant confrontations, state-backed brutality on native communities and instantaneous destruction of the environment and ecology by the fire of gas flaring but also to "slow violence" which, according to Nixon, is "a violence that occurs gradually and out of sight" (Nixon 2). The presence of both slow violence and spectacular violence has been declared through the mouth of the children, "Some of us cried for fear that death would arrive the next day, others for the illness that might lead to death the next month. We all knew the truth: death was at hand" (Mbue 22). The message is quite clear that if they are not killed by the brutal military forces of the state or Pexton, they would eventually be killed by the nameless diseases or by the poison from Pexton's oil fields that have been infused into Kosawa's soil, air and water. Thus, in the novel, the environmental resistance movement of the poor solely consists of a disproportionate battle between the state-backed, powerful multinational oil company on the one hand and the dispossessed host community on the other hand.

Conclusion

Mbue's *How Beautiful We Were*, thus, provides an uncompromising criticism of the contemporary politics prevalent within the sphere of modern petro-capitalist system and the environmental injustices implied in it. The novel, through the depiction of the environmental plight and violence associated with the state-sponsored oil industry, gives a clarion call for environmental justice. The oil-determined plot of the novel gives voice to the disfranchised and devoiced people who want their story to be told. One may remember the emphatic statement that the author puts on the lips of Sahel: "*This story must be told, it might not feel good to all ears, it gives our mouths no joy to say it, but our story cannot be left untold*" (138). The novel impels us to rethink the crucial role of such narratives within the body of petro-literature in bringing about substantial socio-political change. It delves deep into the politics of denial that tries to suppress the concern for environmental reality about the poisoning of the entire ecosystem of the imaginary indigenous space where the local people live in close connection with their immediate environment. Though the novel, like most of the African petro-narratives, presents an apocalyptic vision through the graphic representation of the total destruction of the indigenous African village, it speaks of lifting "every voice" till mankind may live to see "a glorious morning" when "there'll be no more crying, no more bleeding, no more sickness" (357).

Notes

1 The term "ecosystem people" has been borrowed from Gadgil and Guha, who coined it to refer to the people who "depend on the natural environments of their own locality to meet most of their material needs" like food, shelter and livelihood. See Madhav Gadgil and Ramachandra Guha, *Ecology and Equity: The Use and Abuse of Nature in Contemporary India*. Routledge, 1995, p. 3.

2 The extract is taken from an online news article on recent oil spilling in Niger Delta titled "The Niger Delta community devastated by yet another Shell oil spill" by Saint Ekpalı in openDemocracy, an independent international media platform covering news, dated 12 December 2023, www.opendemocracy.net/en/shell-niger-delta-oil-spills-sabotage-equipment-failure/.

3 In Mbue's novel *How Beautiful We Were*, Pexton is a fictitious American multinational oil company that operates its harmful extractivist activities in the imaginary indigenous African village called Kosawa.

4 "Environmental racism" is a term first used by the African American civil rights leader Dr Benjamin Chavis in 1982 and he defined it as the "racial discrimination in environmental policy making, the enforcement of regulations and laws, the deliberate targeting of communities of color for toxic waste facilities . . . and the history of excluding people of color from leadership of the ecology movements." See en.m.wikipedia.org/wiki/Environmental_racism#:~:text=%22Environmental%20

racism%22%20was%20a%20term,UCC)%20Commission%20for%20Racial%20Justice.

5 The phrase "disposable people" has been borrowed from Rob Nixon, *Slow Violence and Environmentalism of the Poor.* Harvard University Press, 2011, p. 71.

Works Cited

Auty, Richard. *Sustaining Development in Mineral Economies: The Resource Curse Thesis*. Routledge, 1993.

Bullard, Robert D. "The Threat of Environmental Racism." *Natural Resources & Environment*, vol. 7, no. 3, 1993, pp. 23–56. JSTOR, www.jstor.org/stable/40923229. Accessed 13 April 2024.

Ekpali, Saint. "The Niger Delta Community Devastated by Yet Another Shell Oil Spill." *openDemocracy*, 12 December 2023, www.opendemocracy.net/en/shell-niger-delta-oil-spills-sabotage-equipment-failure/.

Guha, Ramachandra and Joan Martinez-Alier. *Varieties of Environmentalism: Essays North and South*. Earthscan, 1997.

Lester, James et al. *Environmental Injustice in the United States: Myths and Realities.* Westview Press, 2001.

Mbue, Imbolo. *How Beautiful We Were*. Canongate, 2021.

Nixon, Rob. *Slow Violence and Environmentalism of the Poor.* Harvard University Press, 2011.

Ross, Michael. *The Oil Curse: How Petroleum Wealth Shapes the Development of Nations*. Princeton University Press, 2012.

Szeman, Imre. "System Failure: Oil, Futurity, and the Anticipation of Disaster." *South Atlantic Quarterly*, vol. 106, no. 4, 2007, pp. 805–823.

Varma, R. Sreejith. "Resource Extractivism and Environmental Damage: An Analysis of Two Extractivist Fictions from Kerala." *ISLE: Interdisciplinary Studies in Literature and Environment*, vol. 30, no. 3, 2023, pp. 657–677, https://doi.org/10.1093/isle/isab052.

Watts, Michael. "Petro-Violence: Community, Extraction, and Political Ecology of a Mythic Commodity." *Violent Environments*, edited by N. Peluso and M. Watts, Cornell University Press, 2001, pp. 189–212.

Part IV

Disaster(s) and Dystopian Imaginaries

11 The Literary Dimensions of Pagan Spirituality in Fictionalising the Nuclear Tierratraumatic Experience

Inna Sukhenko

Introduction

Within the ecological aftermath of nuclear disasters, radiation has evolved from a scientific concept to a cultural phenomenon with spiritual and mythological components of a "radiation culture" (Romashko; Nakao), interpreting the mythological frameworks as a means to grapple with the daunting challenge of assessing nuclear risks and to seek solace and hope amidst devastation and despondency. The conceptualisation of an ecological disaster as a cognitive framework for religious encounters and the reevaluation of spiritual considerations through scientific paradigms illustrate the evolving interplay among religious convictions, ceremonial practices and human cognition within the realm of radiation. Nuclear disasters significantly impact religious beliefs and practices via fostering mythological thinking to explain these catastrophic events and prompting the consideration of radiation's influence on the relationship between science, spirituality and the long-term implications of nuclear technology on religious beliefs.

The Chernobyl nuclear power plant explosion (25 April 1986) and its aftermath catalysed the development of a "radiation" framework, transmitting nuclear radiation from a strictly scientific notion to a cultural construct while repleting with vivid beliefs and folklore, which contributes to altering global perceptions of nuclear hazards and the ramifications of radiation exposure, culminating in spiritual and mythological conceptualisations of radiation as a transcendent force necessitating esoteric explanations that transcend scientific comprehension – "the Chernobyl accident demands a spiritually transcendent explanation which neither science nor factual history can provide" (Romashko). Leaving a lasting impact on the physical landscape as well as engendering profound spiritual reflections and inquiries, resulting in "biological citizenship" concept (Petryna), the Chernobyl nuclear disaster not only resulted in widespread environmental contamination and human suffering but also contributed to prompting a reevaluation of spiritual beliefs and practices in the face of such devastation. Such a complex interplay of spirituality and the nuclear disaster can be regarded as an attempt of individuals, affected by this nuclear catastrophe, to seek solace and meaning

DOI: 10.4324/9781003602750-16

through their spiritual beliefs, while turning to faith, prayer and rituals in order to cope with the trauma and uncertainty, which in their amalgamation challenged traditional religious frameworks and prompted a deeper exploration of existential questions, related to human vulnerability, the fragility of life and the limits of human control over the contaminated nature. The aftermath of the nuclear disaster prompted a resurgence of religious observance and church attendance among many survivors – "I sing in the church choir. I read the Bible. I go to church – it's the only place they talk about eternal life. They comfort a person" (Alexievich 78), "Everyone prays. And those who don't know how to pray, also pray" (Alexievich 89) – while seeking solace and answers in their faith in the time of experiencing the desolation and suffering, followed the disaster. This led to reflections on themes of loss, resilience and the search for meaning in the face of overwhelming tragedy (Petryna 26), which gave rise to a unique spiritual narrative surrounding the concept of sacrifice, redemption and the interconnectedness of human/non-human beings, engaged in search for reconciling the spiritual dimensions of the nuclear disaster catastrophe with theological interpretations and ethical considerations as well as for reconsidering the spiritual narratives, offering a powerful testament to the human capacity for resilience, compassion in the time of unimaginable loss and uncertainty.

While questioning of faith and existential reflections and incorporating radiation into spiritual beliefs and practices, the nuclear trauma experienced communities go through reconsideration of nuclear-related human vulnerability, the fragility of life and the biological coexistence, which contributes to realising the Chernobyl disaster's impact on the spiritual beliefs and practices of survivors and their communities, leading to increased religious observance, reinterpretations of the disaster through spiritual frameworks, and the envision of radiation within cultural and spiritual beliefs.

Tierratrauma and the Chernobyl Disaster

Regarding spirituality as a societal value, contributing to framing the communication of "nuclear energy" in meaningful ways, includes fostering a deeper understanding of the environmental impact of nuclear energy-related activities from the intersectional perspective on the Chernobyl nuclear disaster and its aftermath, accompanied by emotionalisation of experiencing environmental changes. Such "diversity of emotional connections to our environment" (Harrison) can be represented by the emotional distress associated with witnessing traumatic environmental changes (experiencing the distress of environmental changes, or "tierratrauma," defined by Glenn A. Albrecht and clarified in *Earth Emotions: New Words for a New World* (Albrecht), but firstly represented in his *Psychoterratica* (2017), as "this is acute Earth-based existential trauma . . . (that) describes that moment when you experience sudden and traumatic environmental change . . ." (Albrecht).

While referring to the physical, psychological and emotional consequences of such sudden environmental changes, caused to the Earth and its ecosystems due to human activity, it is tierratrauma that results from experiencing a nuclear disaster in the aspect of its impact on the environment and contributes to framing "earth trauma" via emotionalisation of environmental transformations, represented by Albrecht in concepts *eutierria*, *sumbiocracy*, *Symbiocene*, *tierraphilia*, etc. (Albrecht).

The Chernobyl nuclear disaster serves as a vivid example of the intersection of spirituality and tierratrauma – implementation of spiritual beliefs and practices on radioactivity and widespread radiation contamination resulted in immediate human/non-human casualties (Assessment of the Radiation Effects from the Chernobyl Nuclear Reactor Accident) and the emotionalisation of long-term environmental damage to the surrounding area. The lack of understanding and proper management of nuclear energy inadequate safety measures and human error contributed to the catastrophic meltdown of the reactor, leading to the release of radioactive materials into the environment and devastating consequences for both human health and the ecosystem (Environmental Consequences of the Chernobyl Accident and Their Remediation: Twenty Years of Experience 2006). The emotionalisation of the environmental aftermath of the Chernobyl disaster highlights the distressful effects of environmental changes, resulting from contaminating water sources, and rendering vast areas uninhabitable for humans/non-humans, as well as mutations in plant and animal species, radiation and ecological disruption via experiencing the long-term effects of the environmental catastrophe, and leading to psychosocial traumatic effects of the nuclear disaster (Barnett 47). The case of the Chernobyl disaster outlines, that the need for increasing multisided awareness of the environmental consequences of the nuclear disaster (Global Nuclear Awareness) via addressing the Chernobyl tierratraumatic experience contributes to envisioning the spiritualisation of the nuclear contamination ecological disaster, regarded as a tool of the emotionalisation of transmitting scientific knowledge to the public. Such effects are achieved via not only the amalgamation of fictional and factual narratives in storytelling about the nuclear disaster (Sukhenko 172) but also the address to the intersectionality of narrating the tierratrauma of the Chernobyl disaster in fictional writings. The appeal to the range of various media, weaving together, in fictionalising this nuclear disaster contributes to enhancing the spiritualisation of the disaster as the emotional impact of the Chernobyl narratives in unique and compelling ways via the interaction between different media formats (such as text, images, sounds and more, within a single storytelling). Such amalgamation for narrating the distress of environmental changes as the tierratraumatic experience not only evokes emotional responses from the reading audience (having the potential to deepen the audience's emotional engagement and create a more immersive and memorable experience of the Chernobyl disaster in the multileveled, including intergenerational, perspectives (Lindsay)) but also enhances the outline of tierratrauma as a tool of transmitting scientific knowledge on the radiation contamination on the environment in a wide context of spirituality of energy.

The Chernobyl Disaster and Pagan Spirituality

The ecological-ethical framework of communicating the Chernobyl disaster (Plokhy) can be implemented via a unique combination of societal values, framed by the industrialisation of the Soviet Ukraine, Christianity and paganism, reflecting a complex interplay of historical, cultural and environmental influences. This multifaceted framework encompasses the impact of industrialisation on the natural landscape, the spiritual beliefs and practices of the nuclear trauma-experienced communities, and the traditional connections to the land and the divine.

The Chernobyl spirituality is framed with industrialisation and its significant force in implementing profound effects on the country's ecology, shaping the Soviet Chernobyl landscape. The rapid industrial development and technological advancement with its "industrial hopes of the common people" (Romashko) led to the construction and operation of nuclear power plants like the Chernobyl NPP, while bringing about extensive changes to the environment, including pollution, deforestation and habitat destruction of the region. The Chernobyl disaster happens to highlight the dark side of industrialisation, exposing the risks and consequences of prioritising progress over safety and environmental concerns. The juxtaposition of industrialisation impacts the Chernobyl NPP's nearby ecosystems but highlights the need for sustainable practices and environmental conservation efforts to mitigate the ecological consequences of industrial expansion and prompt a reevaluation of societal values.

Communicating the spirituality of Chernobyl from the perspective of "faith radiating from Chernobyl" (Welton) and "apocalyptic vision of the nuclear disaster" (Mattingly) appeals to Christianity as the predominant religion in Ukraine, where Orthodox Christian beliefs play a central role in shaping the spiritual and ethical values of the Ukrainian population of the region while guiding their interactions with the environment and influencing their attitudes towards nature. Interpreting the spiritual frames of the Chernobyl disaster addresses Christian teachings, emphasising stewardship of the Earth and the importance of caring for the surrounding world, fostering a sense of responsibility towards ecological preservation and well-being. Also the spirituality of the Chernobyl disaster can be envisioned via pagan beliefs and spiritual practices of the ancient polytheistic religion of East Slavic tribes (Ukraine's Pagan History 2024), where the

> development in the doctrines of paganism from animism to [monotheism] does not mean that all other pagan gods were forgotten and not

> worshiped anymore. Development in doctrines did not eliminate the previous beliefs, but added to them the supremacy of a new god without neglecting to worship to the old ones.
>
> (Proshak 141)

Such appeal to outlining the spirituality of the Chernobyl disaster from the perspective of paganism, with its deep roots in Ukrainian folklore and spiritual practices (Lesiv), offers a complementary outline within the ecological-ethical framework of the nuclear disaster, interpreted also from the position of pagan beliefs, which emphasise a harmonious relationship with nature and the divine, provide a holistic worldview that recognises the interconnectedness of human/non-human beings within the radiation-contaminated environment. The pagan spirituality of the Chernobyl disaster in this case cooperates with the reverence for natural elements, spirits of the land and seasonal rituals in pagan traditions, underscoring a deep respect for the Earth and a recognition of humanity's place within the larger ecological system. The spirituality of the Chernobyl disaster can be regarded from the perspective of amalgamation of the dominating Orthodox Christianity and paganism – "paganism survived and adjusted to this new religion after its coming" (Proshak 146) – by being incorporated into Ukrainian folk traditions, storytelling and customs and offering a spiritual framework, rooted in the land and the natural world, with beliefs of agrarian communities of East Slavic tribes (Ukraine's Pagan History 2024), but promoting an alternative vision for the nuclear disaster from the perspective of beliefs and spiritual practices, offered by neo-paganism (Lesiv 45) and contributing to the motifs of continuity, adaptation and revival as essential components of envisioning the spirituality of the Chernobyl disaster as a form of spirituality that emerged in the post-Chernobyl world. Under such circumstances narrating the Chernobyl disaster with references to Ukrainian folklore and pagan spirits emphasises a connection with nature and the divine as well as frames a broader context of spiritual interconnectedness with the environment, while reflecting a holistic worldview that recognised the interdependence of humans and the environment as well as contextualising the spiritual dimensions of environmental disasters.

Fictionalising the Chernobyl Disaster for Young Adults

Andrea White's *Radiant Girl* (2008) depicts the Chernobyl disaster and its aftermath by emphasising the pagan beliefs in narrating the nuclear disaster, while appealing to the ancient folklore and traditions, contributing to shaping the understanding of the spiritual landscape of the Exclusion Zone and serving as a haunting reminder of the interconnectedness of the power of the land and the spirits that dwell within it and go through the tierratraumatic experience as well.

Andrea White (b.1953), the writer, is known for her works of fiction that often explore themes of identity, culture and the human condition. As she comments to "Inspiring Houston Women,"

> I grew up in Houston, TX, though I only moved here from Louisiana when I was eight. After college, I went to law school and was a partner at a law firm here in Houston at a time when not many women were partners. But I always wanted to be a writer and I'm so grateful to have been able to walk down that path.
>
> (Leech)

Being the author of books of middle-grade fiction and having her first book "Surviving Antarctica" (2006) awarded the Golden Spur Award (2006) by the Texas State Reading Association (White), Andrea White currently serves on the editorial board of the Houston Chronicle. Nowadays she lives in Houston, TX with her husband, Bill White, the former Mayor and formerly an energy official in the Clinton administration, while being an education activist and a community volunteer (Leech).

Andrea White's *Radiant Girl* is a fictional story for young adults, which appeared after the writer's visit to the Chernobyl Exclusion Zone in the capacity of the *Huston Chronicle*'s journalist. The novel covers the story of Katya Dubko, an 11-year-old Ukrainian girl from the city Prypiat, whose life is profoundly impacted by the Chernobyl disaster. The writer manages to show Katya's life story before the Chernobyl NPP explosion and the influence of its aftermath on the life of Prypiat's residents' changes. The writer does not emphasise the technological nature of the Chernobyl disaster within the authoritarian regime but highlights the shifts of the teenager's value system in the context of the political, ecological, social and cultural transformation of the transitive society via emphasising the cultural component of the historical context – from the Cold War rhetoric to Perestroika's time and to the final moment of the station's functioning when it was closed under Ukraine's independence in 2000. The variety of everyday life details and objects is presented as the supplement of comments at the end of the book, which helps young readers realise a single person's life situation against the premises and the stages of Ukrainian society's transformations on its way from being one of the Soviet republics to the independent state. In her novel White emphasises the necessity to reread the event from the intergenerational perspective in order to remember the details which can help us prevent erasing the boundary between the true event and the myth, built around the Chernobyl disaster.

The intermediality of narrating a Chernobyl tierratraumatic experience in *Radiant Girl* is emphasised via appealing to the different media forms (textual, factual, visual components) in their amalgamation to explore how the themes, narratives or characters interact with environmental issues within narrating the environmental changes and experiencing the nuclear tierratrauma. The intermediality of the novel addresses blending the factual and fictional narratives to create a compelling narrative that intertwines real-world event – the Chernobyl disaster – with elements of folklore. The protagonists' tierratraumatic experience is shown here as addressing the Polissian (the regional) folklore stories, in interpreting the environmental changes, impacted by the nuclear disaster, via emotionalisation of these transformations against the teenager's considerations (uncertainty, distrust, doubt, love, anxiety), as well as personification of interpreting the distress of environmental changes within historical, ecological and cultural significance. White's focus on cultural identity and the intersection of traditional practices with modernity is a recurring theme in *Radiant Girl*, where the protagonist Katya navigates the complexities of her Ukrainian heritage in the aftermath of the nuclear disaster. The novel depicts the nuanced portrayal of characters via capturing the subtleties of human emotion, and exploring themes of family, community and the struggle to maintain cultural traditions in the face of societal changes, stemming from experiencing the Chernobyl disaster.

The Spirituality of Paganism in White's *Radiant Girl*

The spatiotemporal modality (Elleström 48) of *Radiant Girl* outlines the novel's story, primarily in Ukraine, with specific locations such as Yanov in 1986, Kyiv in 1986 and Slavutych in 1990, which helps deeply intertwine the plot of the novel with the aftermath of the Chernobyl nuclear power plant disaster, which serves as a backdrop for the story and reveals as well as influences the characters' lives and beliefs. The rural and natural settings depicted in the novel, such as the forest trails, wooden cottages and gardens, evoke a sense of closeness to nature and frame the spiritual realm via integrating pagan spirituality. The description of the Earth glowing at sunset juxtaposed with the artificial light of the Chernobyl nuclear power station, highlights the tension between natural and man-made forces via an appeal to the Earth as a sacred embodiment of life and energy: "In my village, the Earth glowed at sunset. But tonight as I headed home, I walked in the direction of a man-made brightness – the Chernobyl Nuclear Power Station" (White 22). The contrast with the nuclear facility serves as a reminder of humanity's disruption of the natural world, prompting a reevaluation of spiritual practices in light of environmental devastation

> Mama pointed at the wooden hut . . . behind the cottage next to a fenced garden and the outhouse. Nearby, a wooden table sat under an expansive old oak tree. The red barn and a chicken coop were on the other side of our yard.
>
> (White 13)

It encapsulates the pagan ethos of reverence for nature while acknowledging the challenges posed by modernity and technological advancements.

The interactions with the environment provide a rich background for exploring the intersection of pagan beliefs with the characters' experiences and worldview, where the rural landscapes, traditional wooden cottages and the wild nature contribute to the novel's exploration of cultural heritage, spirituality and the enduring presence of pagan beliefs in the characters' lives. The novel appeals to the introduction of pagan rituals and traditions, represented by Granny Vera's traditional Ukrainian costume and her way of telling stories by using the traditional opening phrase, reflecting the importance of pagan cultural practices in the context of connecting the protagonists to ancestral heritage:

> Unlike most storytellers who started with, Once upon a time, she began her stories with, "Now at the beginning . . ." It was as if I could hear her saying this phrase, and I sensed that I was at the beginning of something new.
>
> (White 20)

Granny Vera's attempts to pass on her pagan stories and beliefs to Katya demonstrate the importance of intergenerational transmission of these traditions within the family, connection to nature and engagement with these traditional spiritual beings, even in the face of modernisation and societal changes. With this strong appeal to pagan spirituality in narrating the Chernobyl disaster, the writer depicts the conflicts and challenges, arising from the characters' practice of pagan beliefs in the novel. Thus, the novel depicts a contrast between Katya's openness to pagan beliefs and her father Ivan's dismissal of them as "superstitious myths" – "She [Granny Vera] would never forget when her son, Ivan, a boy of about seven and full of the patriotic talk fed to him by his schoolteacher, informed her that her creatures were superstitious myths. He had said, 'Mama, for my country to make progress, we must forget the old ways." He gazed squarely into her eyes before adding, "We must denounce them'" (White 10) – which reflects the broader societal tensions surrounding the role of pagan traditional spirituality in the face of Soviet-era secularisation and progress. Although the protagonist Katya understands the importance of intergenerational transmission of these pagan traditions and beliefs, inherited from her Granny Vera's storytelling, this is challenged by the changing societal landscape (moving to Kyiv as a result of forced replacement from the radiation-contaminated zone) and Katya's own doubts about the reality of the pagan spirits, leading to growing scepticism towards pagan beliefs (after Granny Vera's death, Katya begins to suspect that her [Granny's] wood sprite may not be real, as the characters grapple with the realities of the modern world) or even to potential dangers associated with pagan beliefs (Katya starts to believe that the encounter with pagan spirits in the woods implies that there may be risks or consequences associated with engaging with pagan spirits). These all together create the protagonist's sense of unease and uncertainty around the practice of these beliefs as well as reflect the broader societal and personal struggles faced by the characters of the novel as they navigate their cultural heritage and spiritual beliefs.

Domovyk in White's *Radiant Girl*

The appeal to pagan beliefs and spiritual practices in the novel occurs through the portrayal of traditional Slavic folklore and the introduction of pagan spirits. The presence of the domovyk, a household spirit in Slavic mythology, introduced by Granny Vera's stories about encountering the domovyk Vasyl, the non-human being – "Have I ever told you about the day I saw our domovyk? Granny Vera asked. 'Domovyks were house elves'" (White 8) – symbolises the characters' ties to their cultural heritage and pagan traditions, as well as reflects a belief in the supernatural and the interconnectedness of the natural world.

The writer provides readers with an explanation of this pagan spirit – "Domovyk: a house elf is part of folklore. It protects the house from evil and trouble and alerts the family or a specific individual to impending danger. The elves are known to play tricks on people" (White 183) – as well as introduces his appearance:

> "What did he look like?" "He was dressed like a young boy, in brown pants and an embroidered shirt. . . . I thought he was a villager, but then he turned, and I saw his profile. His blue eyes were a little too round, almost like an owl's. His blond hair was low on his forehead, and there was much hair on the back of his hands. I knew he wasn't human."
>
> (White 10)

Such detailed description of the domovyk, depicted in the form of a wooden figurine in household pieces, together with the protagonists' fascination with the domovyk's significance makes this spiritual non-human being among central protagonists in the novel via unfolding the recognition of Vasyl as their domovyk, indicating a personal bond and a sense of familiarity with the spirit – "I was now convinced that Vasyl was my domovyk. For some reason that I didn't understand, my house elf had chosen to appear to me in the woods instead of my cottage" (White 157). The novel introduces Vasyl as a mystical creature from forklore stories, the conversations with whom helped Katya clarify the internal and external transformations of experiencing the growing connection with the environment via appealing to the pagan stories:

> His face appeared translucent, as if I could strain my eyes and see through him. That blond, shaggy hair curling over his ears also covered the back of his hands. It shone in the yellow light of the afternoon. And his eyes were so blue that they kept me from noticing anything else about them. "Are you a wood sprite?"
>
> (White 38)

The encounter with this being reminiscent of pagan folklore evokes a sense of enchantment intertwined with mystery and scepticism, reflecting the tension between belief and doubt, particularly following the death of Granny Vera, who served as a link to the world of forest creatures. Vasyl's denial of being a wood sprite, coupled with his knowledge of the narrator's name – '"I know your name." I was puzzled, though not yet afraid. "How could you know my name?" I asked. Vasyl smiled brightly. "You are Katya" (While 21) – enhances his enigmatic presence, suggesting a deeper connection to the spiritual realm. His hollow laughter, akin to the wind, symbolises the fragility of belief in a world increasingly dominated by rationality, yet still yearning for the magic of nature and its mythological inhabitants. The domovyk is portrayed as a creature that can be carried in an empty box when families move or travel, highlighting its connection to domestic spaces and the rituals associated with its presence:

> After I had packed a change of clothes, my brush, a few pencils and a notepad, I took another quick glance around the room and noticed the mirror carved with the little domovyk. I would have liked to take him with me, but the frame was too big. I remembered Granny Vera had told me when families wanted to take a domovyk with them on a move or a trip, they carried an empty box and invited the house elf to jump in.
>
> (White 67)

The interaction with the domovyk is imbued with ritualistic significance, as seen in the protagonist's attempt to invite the spirit to accompany them by placing an empty box in their suitcase – "I stuck the empty blue box in my suitcase and said out loud, 'Come with us, domovyk.' Nothing happened, but then, I was too nervous to wait more than a few seconds" (White 64) – by emphasising this ritual gesture, underscoring the bond between the human protagonist and the domovyk, for highlighting a sense of mutual respect and cooperation between human and spirit. On the other hand, the domovyk's ability to manifest outside the confines of the cottage suggests a fluidity in its presence and a capacity to appear where needed, transcending physical boundaries. This emotional resonance adds a poignant layer to the domovyk's character, portraying it as a sentient being attuned to the world's upheavals – "If he doesn't feel welcome in his house, a domovyk can travel anywhere. He warns his people of danger or misfortune. His visits can be like nightmares," Margarita shrugged, "or just interesting" (White 140).

Furthermore, the domovyk is depicted here as a harbinger of events, warning of impending danger or misfortune to those it watches over – "There's going to be a thunderstorm tonight, the domovyk warned us" (White 10). The novel also delves into the emotional depth of the domovyk, as seen in

Margarita's recollection of the spirit shedding red tears before a catastrophic event, symbolising its awareness of impending tragedy –

> Before the accident. I looked out the window of my apartment. My domovyk sat on the curb. . . . He was crying red tears. Bright red. . . . He knew that something terrible was about to happen. . . . But I never dreamed that it would be a nuclear accident.
>
> (White 140)

Its role as a protector is underscored by its ability to communicate through subtle gestures and expressions, such as the half-smile on its wooden lips, hinting at deeper insights and premonitions:

> As I looked into the mirror, I glanced at the figure of the domovyk. He had the same half-smile as always on his wooden lips, and remembered again the horrible impression that I had in the instant before Vasyl said, Our world is going to be destroyed.
>
> (White 41)

The novel demonstrates how experiencing the distress of environmental changes contributes to emotionalization of teenagers' value paradigm transformation and enhancing uncertainty and fear, resulted from these transformations:

> As I moved deeper into the woods, my eyes adjusted. I found that the darkness had transformed the familiar landscape. Now, the tops of the trees disappeared into inky darkness. Spiderwebs sparkled like bright silver ladders. Patches of white moonlight checkered the otherwise black leaves.
>
> (White 36)

The environmental changes following the nuclear disaster symbolise the loss of safety and the intrusion of unnatural elements into nature as well as reflect the disruption of ecological balance and the haunting presence of radiation. The narrator's fear, heightened by the sound of heavy footsteps – "Footsteps. And not a boy's, a man's. They were moving heavily through the trees. I didn't know who would be out this late at night, but fear, for myself and for Vasyl, came over me" (White 36) – signifies a pervasive anxiety about the unknown threats posed by the disaster. This atmosphere of tension and uncertainty encapsulates the broader emotional and environmental turmoil experienced in the aftermath of the Chernobyl disaster.

And as a result of such emotionalised tierratraumatic experience, closely connected to the motif of survival, the protagonist appeals to the folklore images – "The boy in the woods! I need to see you again. . . . Come back to the boulder" (White 35) – in the attempt to find the interpretation of environmental and personal transformations. The protagonist's encounter with Vasyl, a mysterious figure in the woods, mirrors the unsettling changes in the surrounding landscape. The rustling leaves and shifting shadows symbolise the instability and uncertainty that permeate both the natural world and the protagonist's reckonings – "I mustered the faith to whisper, 'Vasyl?' I heard a rustling sound as though an invisible curtain were being pulled back – more likely an animal crunching the leaves" (White 38). Katja's shift from fear to empathy towards Vasyl reflects a personal transformation, a growing awareness of the vulnerability and isolation experienced by those affected by the disaster:

> Then, out of the darkness, a small figure emerged. As I looked at the boy gliding towards me, I wasn't afraid. I felt sorry for him. Just a boy, he was alone in the woods. But also, once again, I was overwhelmed by the sensation that he was a magical being.
>
> (White 38)

This interplay between environmental and emotional upheaval underscores the profound impact of nuclear catastrophe on both the physical and psychological realms.

The novel introduces readers with the character of the domovyk, this pagan spirit, by depicting it as a multifaceted entity with protective, prophetic and emotional dimensions, embodying the rich folklore and spiritual beliefs surrounding house spirits in Slavic pagan culture. Through its interactions and manifestations, the domovyk emerges as a symbol of connection to the home, a guardian of the household, and a bearer of warnings and insights that transcend the physical realm, exhibiting both physical and metaphysical attributes that set it apart as a unique entity in the pagan cultural context within the environmental disaster. The domovyk represents here the significance of pagan beliefs and spiritual practices for the protagonist's connection to her heritage and the enduring power of traditional beliefs within the societal and environmental transformations, stemming from the nuclear disaster.

Conclusion

The intention to unveil the spiritual interpretations of radiation brought the realms of science and spirituality, which resulted in conceptualising spiritual radiation as well as "religion as radiation," delving into the impact of radiation on religious beliefs, reshaping sacred locations and spiritual encounters, leading to a "radiation culture," where nuclear radiation has transcended the scientific realms to become a cultural phenomenon with its spiritual and mythological interpretations. This highlights the evolving relationship between science, spirituality and the enduring implications of nuclear technology on religious beliefs, reshaping religious practices through new spiritual interpretations of radiation and mythological frameworks to comprehend the ecological, including nuclear, disasters. The Chernobyl nuclear power plant explosion and its aftermath catalysed the development of a "radiation culture," transforming nuclear radiation from a strictly scientific concept into a cultural construct replete with vivid beliefs and folklore. This transition has contributed to the alteration of global perceptions regarding nuclear hazards and the ramifications of radiation exposure, culminating in spiritual and mythological conceptualisations of radiation as a transcendent force. The complex interplay between spirituality and the nuclear disaster can be regarded as an attempt by individuals affected by this catastrophe to seek solace and meaning through their spiritual beliefs, turning to faith, prayer, and rituals to cope with the trauma and uncertainty. This process can challenge traditional religious frameworks and prompt a deeper exploration of existential questions related to human vulnerability, the fragility of life and the lack of ecological balance within the contaminated environment. The Chernobyl nuclear trauma, experienced by these communities, has contributed to a reconsideration of nuclear-related human vulnerability, the fragility of life and biological coexistence, leading to increased religious observance, reinterpretations of the disaster through spiritual frameworks, and the envisioning of radiation within cultural and spiritual beliefs. The amalgamation of industrialisation, Christianity and paganism beliefs establishes the ecological-ethical framework of communicating the Chernobyl disaster, influencing the strategies towards environmental preservation, religious beliefs and cultural legacy as well as reflecting the complex interplay of historical, cultural and environmental influences on the region. Narrating the Chernobyl disaster within the coexistence of Christianity and paganism in the ecological and ethical dimensions contributes to diverse and inclusive debates about the amalgamation of environmental conservation and spiritual values. This interplay of belief systems highlights the multifaceted relationship between human society, spirituality and the natural world, emphasising the importance of integrating diverse perspectives and traditions in addressing ecological challenges.

The envisioning of the tierratraumatic experience within the lens of spirituality, exemplified by the Chernobyl nuclear disaster here, showcases the emotionalisation of situating environmental changes and the profound impact of nuclear energy-related activities on ecosystem of the region. This emotional connection to the environmental changes, termed "tierratrauma" (by Glenn A. Albrecht), signifies the acute existential trauma, resulting from sudden and traumatic environmental shifts, such as those experienced in the aftermath of nuclear disasters. The Chernobyl catastrophe underscores the emotional and psychological consequences of environmental devastation, highlighting the distressful effects of radiation contamination on water sources, habitat destruction and long-term ecological disruptions. The need for increased awareness of the environmental repercussions of the nuclear disaster, including proper waste management and policy development, sounds crucial for addressing global sustainability challenges and climate change. By emotionalising scientific knowledge through storytelling and intermedial narratives, the emotional impact of the Chernobyl disaster can be effectively conveyed to the public, fostering deeper engagement and understanding of the spiritual and environmental dimensions of energy and radiation contamination.

White's *Radiant Girl* is a bright example of how narrating the Chernobyl disaster with its nuclear tierratraumatic experience addresses Slavic pagan beliefs, regarding the environment as a sacred universe, with spirits inhabiting the land, water and sky. Envisioning the Chernobyl Exclusion Zone, with its overgrown forests, abandoned villages and lingering sense of unease, via appealing to pagan spiritual creatures, believed to wield great power and influence over human's life, seemed to be a manifestation of the power of these ancient spirits, said to be the guardians of the forest and protectors of wildlife and get awakened or angered by ruining the ecological balance, which, in some concern, led to the tragic event in the region. *Radiant Girl* profiles a pagan spirit as a medium for decoding humans within the environment in times of crisis and uncertainty, while turning people to the supernatural or spiritual for seeking the sense of events with its widespread destruction and long-lasting effects, as a reminder of the awesome power of nature and the fragility of human civilisation. Such focus on pagan spirituality of narrating the Chernobyl disaster, in the amalgamation with the spiritual frames of industrialisation and Christianity, allows envisioning the ecological-ethical framework of the Chernobyl disaster's area and the Chernobyl Exclusion Zone in particular, which reflects a dynamic interplay of cultural, historical, spiritual and environmental factors of the protagonists' transformations. Such a framework of framing the pagan dimensions of Chernobyl's spirituality in the novel highlights the intricate relationship between the tierratrauma-experienced human society, technology and the environment, emphasising the need for a balanced approach to ecological stewardship that integrates ethical, spiritual and practical considerations. By drawing on the diverse traditions and beliefs that shape the spirituality of the Chernobyl in a wide concern, *Radiant Girl* offers a multifaceted perspective on the interconnectedness of humanity and the environment of the tierratrauma-experienced community. Among narrating other issues on the Chernobyl disaster and its aftermath, White's *Radiant Girl* emphasises the significance of pagan beliefs in shaping the characters' identities and worldviews by appealing to pagan spirits (domovyk), serving as symbols of the protagonists' connection to their cultural heritage and the enduring power of traditional beliefs within societal and environmental transformations. Such perspective on the spirituality of the Chernobyl disaster via the pagan spirituality not only offers the complementary outline of narrating the nuclear event with a focus on framing the nuclear tierratraumatic experience of communities but also contributes to gaining a deeper appreciation for the interconnectedness of human/non-human beings for envisioning the amalgamation of the spirituality of science and the spirituality of the environment.

Works Cited

Albrecht, Glenn A. "Tierratrauma." *Psychoterratica*, 2017, glennaalbrecht.wordpress.com/category/tierratrauma/.
———. *Earth Emotions: New Words for a New World.* Cornell University Press, 2019, p. 256.
Alexievich, Svetlana. *Voices from Chernobyl: The Oral History of a Nuclear Disaster.* Translated by Keith Gessen, Picador, 2005, p. 256.
Assessment of the Radiation Effects from the Chornobyl Nuclear Reactor Accident. United Nations Scientific Committee on the Effects of Atomic Radiation, unscear.org/unscear/en/areas-of-work/chernobyl.html.
Barnett, Lynn. "Psychosocial Effects of the Chernobyl Nuclear Disaster." *Medicine, Conflict and Survival*, vol. 23, no. 1, January–March 2007, pp. 46–57, jstor.org/stable/10.2307/27017749.
Elleström, Lars. "The Modalities of Media II: An Expanded Model for Understanding Intermedial Relations." *Beyond Media Borders: Intermedial Relations among Multimodal Media*, vol. 1, edited by Lars Elleström. Basingstoke: Palgrave Macmillan, 2020, pp. 3–91.
"Environmental Consequences of the Chernobyl Accident and Their Remediation: Twenty Years of Experience." Report of the Chernobyl Forum Expert Group 'Environment'. International Atomic Energy Agency. Vienna, 2006, iaea.org/MTCD/Publications/PDF/Pub1239_web.pdf.
"Global Nuclear Awareness." globalnuclearawareness.org/.
Harrison, Hannah. "Generation Climate Europe." 2024, gceurope.org/putting-on-a-brave-face-what-is-eco-anxiety-and-climate-care/.
Leech, Caroline. "Andrea White." *Inspiring Houston Woman*, Monday, 14 October 2013, inspiringhoustonwomen.com/2013/10/andrea-white.html.
Lesiv, Mariya. *The Return of Ancestral Gods: Modern Ukrainian Paganism as an Alternative Vision for a Nation,* McGill-Queen's University Press, 2013, p. 254, jstor.org/stable/j.ctt32b80f.
Lindsay, Stuart. *Reading Chernobyl: Psychoanalysis, Deconstruction, Literature.* PhD thesis, University of Stirling, 2014, p. 267.
Mattingly, Terry. "Apocalyptic Visions about Chernobyl." *On Religion*, 8 October 2012, tmatt.net/columns/2012/10/apoc-alyptic-visions-about-chernobyl.
Nakao, Maika. "Radium Traffic: Radiation, Science and Spiritualism in Early Twentieth-Century Japan." *Medical History*, vol. 65, no. 1, 2021, PMCID: PMC7739056, pp. 32–45, https://doi.org/10.1017/mdh.2020.47, ncbi.nlm.nih.gov/pmc/articles/PMC7739056/.
Petryna, Adriana. *Life Exposed: Biological Citizens after Chernobyl.* Princeton University Press, 2013, p. 304.
Plokhii, Serhii. *Chernobyl: The History of a Nuclear Catastrophe.* Basic Books, 2018, p. 432.
Proshak, Vitaliy. "Paganism in Ukraine: Its Beliefs, Encounter with Christianity, and Survival." *Theological Reflections*, no. 7, 2006, pp. 140–148.
Romashko, Elena. "Religion and 'Radiation Culture': Spirituality in a Post-Chernobyl World." *Material Religions*, Web blog post, 1 June 2016, materialreligions.blogspot.com/2016/05/religion-and-radiation-culture.html.
Sukhenko, Inna. "Implicating "Factual" in U.S. Nuclear Fiction: Nuclear Criticism's Perspective." *Energy Humanities and Energy Transition. Current State and Future Directions,* edited by M. Mišík. Switzerland: Springer Nature, 2021, pp. 169–185.
"Ukraine's Pagan History." 2024, ukraine.com/blog/ukraines-pagan-history/.
Welton, Gary L. "Lessons of Faith Radiating from Chernobyl." *Faith and Freedom*, 20 January 2016, faithandfreedom.com/lessons-of-faith-radiating-from-chernobyl/.
White, Andrea. *Radiant Girl.* Bright Sky Press, 2008.
———. Harper Collins Publishers, 2024, harpercollins.com/blogs/authors/andrea-white.

12 Some Things Are More Equal Than Others

Or, How to Read *On the Beach*

Anna Frieda Kuhn

The Anthropocene?

As is evidenced by the plethora of literature published on climate change – literary scholarship is greening. The burgeoning interest in ecocritical themes can be traced to a geocultural shift that occurred after World War II, widely accepted as the Anthropocene. Critiqued by posthumanist scholars such as Donna Haraway because of the Enlightenment-humanist-based conceptions of the anthropos it perpetuates, we are, however, made acutely aware of the perceived limitations of this metaphor. Under the aegis of this sentiment, my article will aim to look at the metaphors and classification systems used when dealing with issues of environmental concern. These considerations will centre on Nevil Shute's post-apocalyptic Cold War novel *On the Beach* (1957), which offers several points of reference to modern debates on climate change. Set in the aftermath of a nuclear World War III, the novel highlights the final days of an apparently heterogeneous group of people living in the Melbourne area.

In considering the often-neglected relationship between Cold War politics and anthropogenic climate change in literary Anthropocene scholarship, a brief genealogical consideration of these debates is necessary. Jacob Hamblin in his *Arming Mother Nature: The Birth of Catastrophic Environmentalism* (2013) points out that much of the impetus behind current climate-related debates can be traced to the Cold War. Military endeavours and environmental thinking, he claims, are more closely related than thus far recognised. The main factors behind environmental change can, for the most part, be traced to defence economics, with economic production comprising one of the main areas of competition during the Cold War. Indeed, both systems, communist and capitalist, tried proving their superiority by endeavouring to outproduce one another. These contests led to hitherto unwitnessed impacts and disruptions on ecology and geology (McNeill 423). This, coupled with factors such as agricultural change, population growth and technological advances, then laid the groundwork for unforeseen, anthropogenic climate change. Consequently, environmental discourses became matters of global concern during the Cold War, with the ecological effects of human interference on ecology

DOI: 10.4324/9781003602750-17

being debated across the globe. Particularly as the Cold War experience has majorly impacted how human subjectivity and the earth are thought of, it may prove helpful to look to *On the Beach* as a tool for thinking about environmental change and changes about how "the human" is thought – especially considering the remarkable similarities between Cold War discourses and current debates on climate change.

Postcolonial Eco-Imaginaries

With *On the Beach* set primarily in Australia, a brief consideration of postcolonial ecology is at this point necessary. Combining postcolonial studies and ecocriticism, however, proves a somewhat thorny endeavour, with both fields often having strongly contrasted interests. One of the major differences between the two is the apparent anthropocentricity of postcolonial studies; an approach that is decidedly rejected, particularly by second- and third-wave ecocritics (Huggan and Tiffin 2–3). A further issue that arises within the field of postcolonial ecocriticism, is, as Graham Huggan and Helen Tiffin point out, that certain streams of ecocriticism privilege white, male, Western subjectivity. It is, therefore, possible to identify significant differences between "Northern environmentalisms" and "Southern environmentalisms" – especially as it relates to the socio-economic positioning of these spaces (2). This dichotomisation has its shortcomings, of course, but it does nonetheless seem necessary to make this distinction as the effects of climate change, though universal, affect subjects differently. Tying in with these observations, Huggan and Tiffin propose the idea of "Hegemonic centricism." Hegemonic centricism is closely linked with species thinking and the inability to ethically situate human and non-human subjectivities. Higgins and Tiffin explain:

> "Hegemonic centricism" . . . accounts not only for environmental racism, but also for those forms of institutionalised speciesism that continue to be used to rationalise the exploitation of animal (and animalised human) "others" in the name of a . . . human- and reasoned-centred culture that is at least a couple of millennia old.
>
> (5)

Strikingly, though set in Australia, Southern environmentalisms are conspicuously absent in Shute's narrative. In this way, the novel points to what Jacques Derrida says elsewhere about the way knowledge production is entirely in the hands of an Anglo-American hegemony (122), at the same time as inadvertently highlighting the prominence of a peculiarly Western ecological discourse during the Cold War. The characters narrating the causes and effects of nuclear apocalypse in *On the Beach* are primarily white, male, Western subjects: Dwight Towers, Peter Holmes and John Osborne are the most prominent voices in the novel, with John Osborne's scientific insights and Dwight Towers' Americanist narrative forming the backbone of the story. Dwight Towers, the captain of an American nuclear submarine (6), for example, had been in his submarine when the war began. And we are told that the Lieutenant Commander, Peter Holmes, who also serves on the nuclear submarine "had read the American's report" (6). This seemingly minor detail draws attention to the way the narrative of the war is constructed, as the characters' knowledge thereof is based, albeit only partly, on Dwight Towers' reports. What is understood about the war, in other words, is based on the narrative as put forth by the American captain and, as a consequence, constructed by way of an Anglo-American discourse. Such a reading is further supported when considering that in crisis the Australian John Osborne (who works for the Commonwealth Scientific and Industrial Research Organisation) is called upon for help because "he's a scientist and he understands radiation" (25). His narrative is, then, to a certain extent believed solely because "he is a scientist." The discourse about the world and the understanding of it, is, in other words, based within scientific, particularly Anglo-American scientific, discourses. An example of what Derrida would consider to be the way an Anglo-American hegemony imposes itself on all techno-scientific exchange. Although there would be much more to unravel here, due to space constraints, I will focus primarily on questions regarding classification thinking that the latter quotation inevitably evokes.

The Order of Things

From Plato through to Claude Lévi-Strauss and Michel Foucault scholars have highlighted the way classification thinking forms the heart of human subjectivity. This has been aptly summarised by the literary critic John Frow who contends that "[n]o speaking or writing, or any other symbolically organised action takes place other than through the shapings of generic codes, where 'shaping' means both 'shaping by' and 'shaping of': acts and structure work upon and modify each other" (10). The importance of classification thinking, specifically as it relates to questions of genre in ecocriticism, is also considered by Amitav Ghosh. In *The Great Derangement* (2016), Ghosh asks the pertinent question of how to narrativise climate change. This need to find modes of classification in eco-narratives is evitable, not only inside but also outside of academic circles, pointing to the acute need to come to terms with rapid environmental change. That *On the Beach* – as a dystopian science-fiction novel that traces the slow extinction of the human subject by radiation – also resonates with current debates around climate change is, then, no surprise. In comparing the threat of nuclear war in the 1950s to contemporary debates, Australian scholar Peter Christoff states that "[l]ike characters in On the Beach, we are . . . suffering from a radical failure of imagination" (qtd. in Milner 257). The question, then, seems to be why, like the characters in *On the Beach*, we have difficulties narrativising climate change. In keeping with these considerations, I will aim to argue that *On the Beach* offers a paradigm, a metaphor, of how to imagine the unimaginable. And, in particular, how to think about climate change.

The Order of "Things"

With the (re)emergence of materialist ontologies, these observations also necessitate a brief reflection on the ways agency is thought beyond the anthropos. Such forms of philosophical inquiry have, in recent years, been widely discussed in the fields of new materialism (or neo-materialism) and object-oriented ontology. Largely informing this "material turn" is Jane Bennett's *Vibrant Matter* (2010), which introduces the concept of "vital materiality" as a means of cancelling out the modern enterprise of dichotomising the human and non-human. In doing so, she shifts the idea of attributing agency only to the human subject to all things: human and non-human alike. This material turn, however, also has its shortcomings. As Benjamin Boysen points out in his "The Embarrassment of Being Human: A Critique of New Materialism and Object-Oriented Ontology" (2018), these movements are marked by an acute semiophobia: "an unease and malaise with the idea of human reality as being semiotic" (225). He further argues that the paradox of this radical critique is the large-scale anthropomorphisation of material objects it advances. Hence, the basis of the approach privileges human subjectivity, even though its basis should be to forego such a privileging. What needs to be considered here are particularly also the ethico-political consequences of moving beyond the cultural/linguistic turn to the material. As already mentioned earlier, both object-oriented ontology and new materialism have a heightened awareness of human subjectivity's embeddedness in material contexts.

With this discussion centring on what classifies as human and non-human, these debates will be considered in the larger framework of biological classification and species thinking next. Richard A. Richards takes this up in his *Biological Classification: A Philosophical Introduction* (2016). Here he delineates the history of classification and considers how they link to current debates in the life sciences, effectively asking whether classification would exist beyond the framework of human discourse. As Richards points out in regard to the constructedness of species categorisations: "[w]hat counts as a species depends on the theoretical interests and needs of the investigator" (191). The problem of species categorisation and taxonomical order is also considered in Haraway's *Staying with the Trouble: Making Kin in the Chthulucene* (2016). Here she aims at countering the dominant discourse, particularly in the life sciences, associated with hierarchisation and competition. Albeit from a very different angle, this hierarchisation within the "natural" world is also delineated in Sean B. Carroll's consideration of the rules and laws in "nature" in his aptly titled "Some Animals Are More Equal than Others" (111). In this chapter Carroll addresses the way animal populations are regulated, with the premise being that population regulation is effectively "unequal." This observation clearly weakens the metaphors Haraway introduces in her *Staying with the Trouble*. Her volume nonetheless rightly points out that species are constituted by way of symbiotic associations. She is also correct in pointing out that in scientific discourses too much emphasis has been placed on competition rather than symbiotic relationships. It is, however, arguable that she heavily downplays the role of competition and species hierarchisation in a somewhat myopic perpetuation of a symbiosis/sympoiesis metaphor (Trouble 58). Although both Carroll and Haraway assert that all forms of life are inextricably connected (entangled), the question that needs to be answered is to what extent, as Haraway contends, metaphors such as symbiosis/sympoiesis are indeed fruitful. Indeed, the different interests in debates surrounding environmental crises then mean that ecocriticism is also particularly concerned with analysing and reflecting on these debates. It is, therefore, necessary to reflect critically on the pragmatics of the metaphors that scholars such as Haraway introduce to the field. And it seems, especially if one considers Carroll's work in relation to Haraway (though there are many points of overlap) that Haraway's utopia of symbiosis simply does not hold as a productive metaphor for an ethics in an age of universal climate change.

Thus, the question that remains is how to introduce metaphors that do not gloss over species hierarchisation. According to a new materialist view only if one considers objects apart from human subjectivity, then only is a truly dis-anthropocentric reading of climate change possible. These objects, however, as has been considered earlier, also exist hierarchically. In light of this, we have to discuss whether a dis-anthropocentric, new materialist view really is the solution to the continued practice of sustaining the human as a problem. We also have to consider which metaphors are best used to deal with climate change. As has been shown earlier, the metaphor of the Anthropocene consolidates the human as a problem. Going beyond the human, however, is an immense source of anxiety, as can be seen when reflecting on the tremendous popularity of pieces, such as Alan Weisman's *The World Without Us* (2007), which depicts a world in a posthuman epoch. Such fears are exacerbated by the advances in science and technology. From an evolutionary standpoint, one should, however, be wary of this narrative and be reminded, as Helena Feder puts it, "that 'we' are not the world" (2). But although we are not the world, we are nonetheless still part of this "thing" called the world.

This ties in particularly well with a new materialist reading of *On the Beach*, as new materialism considers fundamental questions about the position of the contemporary anthropos in a material world. Such a reading means looking at the ways in which the material environment is currently reproduced and consumed. What is of particular importance here, it deals with changes in the bio- and ecosphere. New materialism, in other words, is a means of going beyond normative assumptions about the human and the way in which our interaction with nature takes place. An example of how both anthropocentric and new materialist ideas figure in *On the Beach* is reflected in a debate between Dwight Towers, John Osbourne and Peter Holmes on the impossibility of imagining a post-apocalyptic world:

> 'You can't imagine anything,' said the scientist. 'All army and navy people are the same. You think it can't happen to me.' He paused. 'But it can and it certainly will.'
>
> [Peter Holmes:] 'I can't imagine the end of the world. It's too difficult.'
>
> 'It's not the end of the world,' John Osborne said. 'It's only the end of us. The world will go on just the same. But we shall not be in it. Perhaps it'll be right without us.'
>
> (38–39; emphasis added)

Interpreting the latter through an anthropocentric ecocritical lens, it would seem that the impossibility of imagining the end of the world is rather the impossibility of imagining a post-anthropocentric world. The "world," as described by Peter Holmes, is seen anthropocentrically; a view that conflates the world with the human subject. The voice of John Osborne, on the other hand, can be read as a new materialist figure that questions the narrative that treats the human subject as a superior agent. Osborne's understanding of the "world" can, then, be pitted against that of Holmes, with him possibly questioning anthropocentric assumptions about the interaction between the human subject and the material world. It is indeed possible that he sees himself not as the world but as an agent within a larger material reality. To overcome anthropocentricism, in keeping with the latter analysis, then seems to mean the ability to "imagine" – to imagine both "anything" and "any thing." Such a reading, therefore, proposes that it is not only necessary to "imagine" if one is to come to terms with ecological crises, but also to imagine "things." Thus, taking what could be considered a new materialist view of the world. In a sense, an ontological shift occurs in the way in which things are classified through generic codes. The "perhaps" in Osborne's comment, however, also points to the possibility that no such shift has occurred after all. The "[p]erhaps it'll be right without us" may then also be read quite differently, as it may reflect on the way agency is placed primarily in the agential sphere of the anthropos. One could, in other words, argue that Osborne here dichotomises the "us" and the "it," hereby placing the agency with human subjectivity rather than the larger material world. This understanding of the world as thing is also considered by Jane Bennett who, in *Vibrant Matter*, argues that the philosophical project is marked by the treatment of matter as passive and of "parsing the world into dull matter (it, things) and vibrant life (us, beings)" (vii). Osborne may indeed be consolidating the philosophical project by viewing the world thus. The ambiguous "perhaps," however, sets these statements in a Derridean suspension of sorts and opens up possibilities of reading his statement in different ways.

"Let's Invent New Metaphors!"

With the plethora of dystopian fiction published of late, the question necessarily arises as to the usefulness of reading *On the Beach* for ecocritical purposes. The Australian critic McKenzie Wark may hold some answers here: "[N]ew experiences," she says, "often have to be thought within the basic metaphors that already exist" (223). Arguably, *On the Beach* can be treated as a metaphor for thinking and reading contemporary climate change. It is a metaphor and frame of reference that seems to be particularly useful when keeping in mind that the impetus behind contemporary thinking about environmentalism can be traced to the Cold War. With a further aim of ecocriticism being the analysis and reflection on current debates surrounding environmental concerns, the efficacy of common present-day concept metaphors needs to be looked at further. It is particularly the usefulness of alternative metaphors as the ones introduced by Haraway, or even such metaphors as mis- or disanthropocene when treating topics of environmental concern, as does, for instance, Wark, who also says that we should "invent new metaphors!" (223), that need be considered.

This short analysis of *On the Beach* has, then, attempted to show that the narrativisation of ecological themes pivots on classification. Ecocriticism will thus have to focus particularly on classifications if it is to adequately deal with environmental concerns. Despite the semiophobia of many new materialist streams, I would still contend that the world is largely structured by discourses. It is, therefore, necessary to be critical of the language used when discussing environmental concerns, with classification continuing to be at the heart of human subject and discourse formation. It is indeed not a secondary matter whether we call the current geological epoch "Anthropocene," "Chthulucene" or perhaps "Misanthropocene"; because these classifications constitute worlding, they will construct the discourse on climate change. Arguably, rather than pitting "the linguistic" against "the material," we may need to look for the material in the linguistic and the linguistic in the material – to see that both have been "the world" all along.

Works Cited

Bennett, Jane. *Vibrant Matter: A Political Ecology of Things*. Duke University Press, 2010.

Boysen, Benjamin. "The Embarrassment of Being Human: A Critique of New Materialism and Object-Oriented Ontology." *ORBIS Litterarum*, vol. 73, no. 3, 2018, pp. 225–42, www.onlinelibrary.wiley.com/doi/epdf/10.1111/oli.12174. Accessed 21 January 2022.

Carroll, Sean B. *The Serengeti Rules: The Quest to Discover How Life Works and Why It Matters*. Princeton University Press, 2016.

Feder, Helena. "Ecocriticism Posthumanism and the Biological Idea of Culture." *The Oxford Handbook of Ecocriticism*, edited by Greg Garrard, Oxford University Press, 2013, www.oxfordhandbooks.com/view/10.1093/oxfordhb/9780199742929.001.0001/oxfordhb-9780199742929-e-006. Accessed 22 January 2022.

Foucault, Michel. *The Order of Things: An Archaeology of the Human Sciences*. Routledge, 2002.
Frow, John. *Genre*. 2nd ed. Routledge, 2015.
Hamblin, Jacob Darwin. *Arming Mother Nature: The Birth of Catastrophic Environmentalism*. Oxford University Press, 2013.
Haraway, Donna. "Anthropocene Capitalocene Plantationocene Chthulucene: Making Kin." *Environmental Humanities*, vol. 6, 2015, pp. 159–165.
———. *Staying with the Trouble: Making Kin in the Chthulucene*. Duke University Press, 2016.
Huggan, Graham and Helen Tiffin. *Postcolonial Ecocriticism: Literature Animals Environment*. 2nd ed. Routledge, 2015.
McNeill, J. R. "The Biosphere and the Cold War." *The Cambridge History of the Cold War*, edited by Melvyn P. Leffler and Odd Arne Westad, vol. 3. Cambridge University Press, 2010.
Milner, Andrew. "On the Beach and The Sea and Summer: Two Paradigmatic Australian Dystopias?" *Spectres of Utopia: Theory Practice Conventions*, edited by Artur Blaim and Ludmila Gruszewska-Blaim. Peter Lang, 2012, pp. 256–268.
Nelson, Alondra. *The Social Life of DNA: Race Reparations and Reconciliation after the Genome*. Beacon Press, 2016.
Richards, Richard A. *Biological Classification: An Introduction*. Cambridge University Press, 2016.
Shute, Nevil. *On the Beach*. 1957. Longman, 1972.
Wark, McKenzie. *Molecular Red: Theory for the Anthropocene*. Verso, 2015.

13 Gender, Famine and Masculinities

An Ecofeminist Insight into the Irish Great Hunger

Asmae Ourkiya

The Great Irish Hunger: An Overview

The Great Hunger, commonly referred to as the Irish Potato Famine, represents a monumental tragedy in Irish history that transpired between the years 1845 and 1852. This period was characterised by an immense loss of life, extensive forced displacement, rampant disease and mass emigration, resulting in the death of over a million people and the displacement of millions more. The often-debated reasons behind this calamity include factors such as the failure of the potato crop due to the devastating effects of late blight, an excessive reliance on the potato as a primary food source, the colonial policies of the British government and an inadequate response from authorities to provide relief to the affected population. In this chapter, I argue that The Great Hunger, in Irish "an Gorta Mór," was not solely a natural disaster but rather a complex historical event that was shaped by political, social and economic factors and remains a significant and contentious topic in Irish history.

Before the Great Hunger reached its peak in Ireland, the country had already endured decades of British colonial control. In early 1801, the *Act of Union* came into effect and resulted in the official creation of the United Kingdom of Great Britain and Ireland. This led to the aggravation of the Catholics' conditions and oppression since they did not have any presence among the hundred Irish representatives in the Parliament at Westminster, England. Given the fact that most of the Irish people were Catholics, the results of such an occurrence were inevitable. The Catholics did not give up fighting British rule, and men like Daniel O'Connell made changes in Ireland's religious history. These changes include one of the most significant achievements of Daniel O'Connell, known as the Liberator, Catholic Emancipation (1829). O'Connell fought for the rights of the Catholics, and despite that he was a unionist and did not work on freeing Ireland from British rule, he successfully managed to grant the Catholics the right to sit in the parliament.

In exploring the historical impact of the famine, I also consider its continued cultural resonance, as seen in various literary and filmic interpretations. One such work is the movie *Black 47* (2018), a film that dramatically portrays the suffering of the Irish during the famine, intertwining personal

DOI: 10.4324/9781003602750-18

loss with the broader backdrop of imperial neglect and exploitation. As we move forward, this chapter will extend the discussion of the famine's causes and consequences by examining how the film positions the historical tragedy within the framework of global ecocatastrophe. Black 47 serves as a contemporary cultural reflection on environmental disaster, food security and colonialism, which makes it a valuable text for understanding the intersection of natural and human-made crises.

The mid-1840s saw the plagues of Ireland – hunger, disease and governmental neglect – compound each other like a battleground of contending dooms. Fragile lifelines of aid reached only a minority of the population (Scally). In 1835, Gustave de Beaumont, a French sociologist, paid a visit to Ireland and described it in his book *L'Irlande: Sociale, Politique et Religieuse* as follows:

> I have seen the Indian in his forests, and the negro in his chains, and thought, as I contemplated their pitiable condition, that I saw the very extreme of human wretchedness; but I did not then know the condition of unfortunate Ireland. . . . In all countries, more or less, paupers may be discovered; but an entire nation of paupers is what was never seen until it was shown in Ireland.
>
> (De Beaumont)

A visitor's thoughts on the country reflect the horrid situation that the Irish were living in prior to the Famine. Most of the focus has been on freeing the country and liberating it from Protestant rule, which has resulted in the neglect of a number of sectors, including agriculture and the empowerment of the politics of food marketing. It is undisputed that the blight has caused the failure of the only crop that the majority of the Irish were relying on for survival: potatoes. Yet Ireland could have done more work in preventing the disastrous consequences of the blight. The following subsections will discuss the agricultural practices in Ireland before the Great Hunger, as well as the food export system based on Christine Kinealy's findings in her book *A Death-Dealing Famine* (1997).

The Ecology of Agricultural Practices: The Rundale System and the Overreliance on Potatoes

The rundale system was a communal agricultural practice that was widely used in Ireland in the 1800s. This system involved the sharing of land among multiple families for farming, with each family being responsible for a small portion of the land. While the rundale system allowed for the cultivation of small plots of land by families who would not have otherwise been able to afford their own, it also contributed to ecological problems in the country. One of the major issues with the rundale system was its inefficient use of land. Since each family was only responsible for a small portion of the land, the system led to a lack of individual ownership and a lack of incentive to maintain the land. This ultimately led to overgrazing, soil depletion and other ecological problems (Kussmaul).

Furthermore, the rundale system made it difficult for farmers to adopt sustainable agricultural practices, such as crop rotation. This is because each family only had a small plot of land and did not have enough space to rotate their crops effectively. As a result, the soil became depleted of nutrients, which made it more vulnerable to disease and erosion (Kussmaul). The ecological problems caused by the rundale system had devastating consequences for the Irish people. The Great Famine of the 1840s, which resulted in the deaths of over a million people and the emigration of millions more, was in part caused by the ecological damage caused by the rundale system and the overreliance on potatoes as a staple.

The unsustainable agricultural practices contributed significantly to the mass starvations that occurred in Ireland due to the British plantations that go back all the way to the sixteenth century, and the power and ownership of land in Ireland. The Catholics have been deprived of land ownership. The economy of the country was predominantly agricultural considering that it was not producing food for its people only but was exporting it as well. Yet the agricultural practices were regarded as underdeveloped and unsustainable. The number of farmers was too few compared to the population, and they lacked the tools and the ideas to modernise and cope with the increase in the population. Between 1700 and the Great Hunger, the population of Ireland grew rapidly.

In analysing the factors that contributed to the vulnerability of Ireland during the famine, it is crucial to understand the socio-economic conditions of the pre-famine years. One key element was the demographic shift, which saw a significant population increase alongside the country's increasing dependence on the potato crop as a staple food. This reliance on a single crop, coupled with economic changes in agricultural practices, made the country particularly susceptible to disaster when blight struck. As historian Connell explains, the population growth during these years can be attributed to "the factors of potato dependence, and the swing from pasture to arable in the wake of rising wartime grain prices toward the end of the eighteenth century" (Flaherty). These shifts not only created a precarious economic foundation but also intensified the effects of the potato blight, as much of the population had come to rely on the potato for survival. According to Connell, the explanation for Ireland's population increase throughout the pre-famine years resided in the factors of potato dependence, and the swing from pasture to arable in the wake of rising wartime grain prices towards the end of the eighteenth century (Flaherty). This growth led farmers to seek available lands around the Atlantic coast (Connaught especially where the lazy bed was invented). These lands were poor due to the lack of good soil; thus, they were only capable of growing either oats or potatoes (Whelan).

Despite the lack of modernisation, the *Act of Union* opened Ireland to an international market which was for obvious reasons a means to develop the English economy. A few Irish landlords, influenced by the changes in English agriculture, became filled with the desire to "improve" their properties and thus take advantage of new technologies and processes. However, this also required the consolidation of property which, in the context of Ireland, meant the ending of subdivision. The transition to a more capitalised system of agriculture represented a threat not only to the livelihood of the small tenants and cottiers but also to the society and culture that was an integral part of it (Kinealy).

Adopting the rundale agricultural system, characterised by periodic land redistribution among small groups or families, was essential for farmers to increase production and meet market demands. This system was ecologically sustainable until it could no longer accommodate the heightened productivity demands driven by the pursuit of profit and a consistently growing population. The rundale system, associated with marginal lands, is considered a resource management strategy well-suited to such conditions. Consequently, many rundale settlements employed "lazy beds," a labour-intensive method of spade husbandry involving ridge and furrow planting, which provided drainage, aeration and wind protection while enabling selective fertiliser application (Flaherty). This shift in Irish agriculture benefited both farmers and colonisers, who ignored the dietary needs of marginalised populations. Despite relying on potatoes as their primary affordable crop, the English government made no effort to diversify their diet. Instead, they continued to exploit the country's food resources for market purposes without making other crops more accessible to Ireland's poor. The system lacked a food security plan, and with no measures in place for unexpected crop failures, neither the Irish nor the English were prepared for the potato blight.

The Earl Grey Scheme: The Human Trafficking of Teenage Girls

Although the debate will perhaps always remain as to whether the Great Hunger was in fact a natural disaster or an active genocide from the British government for failing to provide adequate relief, archives and history prove abundant proof that it was in fact an ethnic cleansing and an abuse of human rights on different levels. One of the most savage violations of human rights in Irish history is the trafficking of Irish teenage orphan girls from Ireland to Australia.

In 1848, the British government initiated the Earl Grey Scheme, ostensibly to alleviate Irish suffering by offering opportunities to emigrate to Australian colonies. Named after Charles Grey, the second Earl Grey and Secretary of State for the Colonies, the scheme promised free passage, land and other incentives to willing Irish emigrants. However, the scheme had a darker aspect, functioning as a means of human trafficking. Irish teenage girls who came mostly from impoverished families, were recruited and sent to Australia against their will, under false promises of better opportunities. Many ended up working as domestic servants or prostitutes. Daly and Grayson describe the scheme as a way to "dispose of the pauper girls of Ireland" by sending them to Australia (111). Though intended to relieve the Irish population during the Great Hunger, it also facilitated human trafficking.

Recruiters, known as "soul drivers," played a crucial role in persuading young Irish girls to emigrate by using deceitful promises of jobs as governesses or ladies' maids. These recruiters travelled across Ireland, convincing families to let their daughters go. Upon arrival in Australia, the girls were sold to the highest bidder and forced to work for little or no pay. Daly and Grayson reveal that the British government was aware of the trafficking under the Earl Grey Scheme, leading to a parliamentary inquiry in 1852. The inquiry found that many girls had been subjected to "fraudulent and deceptive practices" and "gross immorality" during their voyage (111). Local officials and clergymen, eager to reduce Ireland's surplus population, often ignored these abuses. Cullen, in "Irish Women and the Creation of Modern Catholicism," also discusses the parliamentary inquiry, noting that many girls were "deceived and betrayed" by recruiters who promised a better life (41). The inquiry highlighted the "gross immorality" and abuses faced by the girls during their journey, with local authorities complicit in these practices.

Between 1849 and 1851, over 4000 Irish orphan girls were transported from workhouses in Ireland to Australia, primarily to Sydney, Port Phillip and Adelaide. Véronique Molinari notes that this "emigration" was perceived as a way to alleviate overcrowded Irish workhouses, address Australia's gender imbalance and tackle labour shortages (Molinari). However, the term "emigration" misleadingly implies voluntary relocation, which was not the case for these underage girls, who were too young to make such decisions. Joseph A. Robins, in his 1968 article "Irish Orphan Emigration to Australia 1848–1850," argues that the term "emigration" inaccurately describes the coerced journey of these girls. The scheme, initiated by Henry, the third Earl Grey and Secretary of State for the Colonies (1846–1852), was politically supported by Britain to reduce poor law union expenses, effectively contributing to a covert ethnic cleansing by the British government.

The Australian government had strict eligibility criteria for who can join the scheme. The girls had to be "imbued with religion and morally pure," meaning they had to be virgins and obedient, ready to serve as wives to the men they were about to marry. In the twenty-first century, this would be the definition of human trafficking for sexual exploitation. What seems to be a relief programme from Australia and what some might consider as a support from the Australian government, was in fact an exploitation of the young girls' bodies to serve a heteronormative society's reproduction goals. The girls were referred to as "breeders," a term that commodifies their genitals and their reproductive abilities. "Young marriageable women were also (although only secondarily) needed to bring a stabilizing moral influence to rough masculine colonial societies" (Molinari).

In 1848, 110 orphan girls were sent from the Skibbereen Union Workhouse (Skibbereen Community Hospital campus), where it was estimated that around 60 people died per day due to hunger and disease. Toma McCullim, Irish artist in residence in Skibbereen's Ludgate Hub and awardee of grants from Create/Arts Council and the Cork County Arts Department, worked on a series of 110 bronze spoons, each representing one of the 110 Skibbereen girls. Her artwork 110 Skibbereen Girls was a result of McCullim's work with the Skibbereen Community Hospital campus staff, service users, residents and visitors to develop a permanent site-specific artwork on the campus grounds.

Earl Grey's sexist and paedophilic plan aimed at hitting two birds with one stone: Relieving the crowded Irish workhouses and supplying female labour and bodies to Australia. Sending underage girls on months-long journeys across the world accompanied by men leading the ships has resulted in what was inevitably traumatic on-board and post-arrival experiences.

As it was, various cases of abuses were reported: the first and second mates as well as some other officers on the Earl Grey were accused of having paid "improper attentions to the young girls on board" and the crew of having had "unrestricted intercourse with them"; the officers and seamen on board the James Gibb, in 1849, were reportedly guilty of similar conduct and six of the females who sailed to Port Phillip on the Manchester were said to have been hired by brothel keepers the moment they arrived on board; again, the surgeons of the Waverley, of the Lysander and of the Thomas Arbuthnot were accused of having treated unmarried females "improperly" (Molinari).

This scheme had everlasting impacts on the girls and their descendants. Some describe the female orphan survivors as women who flourished. By flourished they mean married and had children from the men they most likely had no choice but to marry. Forced marriages, lack of or absence of contraception and unreported domestic abuse are thus what the Irish girls were given as an alternative to the risk of starvation and disease. Yet not all girls were lucky enough to "flourish." From suicide stories to being admitted to mental asylums, not all Irish girls had a safe journey. The conditions in mental asylums during this time were poor, and patients were subjected to harsh treatments, such as straitjackets, restraints and even lobotomies (Kenny). Many of the patients in these asylums were not actually mentally ill but rather were suffering from other conditions, such as epilepsy or tuberculosis (Kenny). While the exact number of Irish teenage girls who were admitted to mental asylums during the Great Hunger cannot be determined due to the lack of available records, it is evident that this was a challenging and traumatic period in Irish history, with a significant impact on the mental health of many individuals, particularly those who were already vulnerable due to poverty and social stigma (Connolly). Famine, whether perceived in this case as an intended genocide or a natural disaster, does indeed impact people disproportionately merely because of their genitals and the gender they were assigned to at birth.

Irish (Un)Industrial Breadwinner Masculinities: Food Supplies and Food Exports

Weary men, what reap ye? – Golden corn for the stranger. What sow ye? – Human corpses that wait for the avenger. Fainting forms, hunger-stricken, what see you in the offing? Stately ships to bear our food away, amid the stranger's scoffing. There's a proud array of soldiers – what do they round your door? They guard our masters' granaries from the thin hands of the poor (Wilde 1847).

Insufficient stocks of food never cause famines. The unjust politics of food distribution are the major cause behind deprivations and starvations. During the nineteenth century, the United Kingdom was one of the most advanced economies in the world, yet it failed to respond to the Irish food crisis. This latter did not receive appropriate responses and thus resulted in the death and immigration of millions of Irish people. Knowing that around half of the country's population back then relied solely on the potato crop for survival, the British government did not meet the needs of the people during the Famine and was not able to protect the rural underclass which was the main victim of the mismanagement of food during the Hunger. It is certainly the role of economists and historians to investigate the free trade and food exports' impacts on the British government's economy and whether their assumptions behind their trade policies worked in fact or not.

In the mid-nineteenth century, the crops failed throughout several European countries; nevertheless, Ireland was the only one with long-term suffering and high mortalities. Alas, when the blight hit the country, the British government was still in control, and the free-trade policies of the Whig and Peel governments undoubtedly affected the food supplies to Ireland leading to a tragic cataclysm. Britain was not to suffer financially because of Ireland, even if that meant the death and sickness of millions of people: "the problem in Ireland was not lack of food which was plentiful, but the price of it, which was beyond the reach of the poor" (Woodham-Smith 1991, 165). In order to understand the politics of food exports during the Great Hunger, it is important to understand the political systems of Great Britain in the mid-nineteenth century and the decisions that were made by the government regarding the situation in Ireland.

From a postcolonial view, though some may blame the British and accuse them of neglect, others would blame the Irish for their misfortune and accuse them of laziness. Nonetheless, the blame should be put aside and the ethical judgments, prejudices and presumptions that are based on biased ideologies should be disregarded. Instead, the major reasons why the deaths occurred despite that the food was plentiful should raise questions among researchers who seek answers beyond what has already been written on the subject. Therefore, it is worth incorporating theories on ecological masculinities and how Breadwinner (Hanlon) and industrial masculinities can affect a whole nation, especially in times of crisis. Drawing from Martin Hultman's theories on breadwinner masculinities, the following passages will examine the masculinities of Ireland that are not so industrial but more agricultural, capitalistic and catholic. Hultman and Pulé describe "Industrial/Breadwinner Masculinities" in their 2018 book as similar to hegemonic, patriarchal and normative masculinities that are more often than not adopted by men. It is a fact that industrialisation and capitalism have caused severe environmental and social damages. In mid-nineteenth century Ireland, market growth, the distribution of resources and the way labour was paid and provided were imposed by the British colonists and had drastic impacts on Ireland.

In Alvin Jackson's *Ireland, 1798–1998: War, Peace and Beyond* (2010), the fourth section of the book, The Ascendancy of the Land Question 1845–91, starts with the chapter "Guilty Men and the Great Famine." In the chapter, the author described the Great Hunger as the most costly ecological disaster of modern times and took an interesting angle in analysing the apocalyptic event. Initiating with a discussion of food arrival to Ireland during the period when Robert Peel was minister, he describes the government's measures in 1845 and 1846 as the *Peelite* measures that were designed to control the private traders and not replace them. Jackson briefly describes the three strategies that were adopted by the Peel government, which were grain purchases from abroad, local relief committees and work schemes. According to Jackson, the pragmatic economic liberal Robert Peel used the Great Hunger as an excuse to repeal the Corn Laws as he was a firm believer in the laissez-faire system. The Peel government attempted to establish effective famine relief measures. Loads of goods, especially Indian maize and oatmeal, were shipped to Ireland in 1846, which contributed effectively in helping the Irish during the beginning of the Famine. Peel's administration came to an end in 1846 and was succeeded by a more damaging one. The Famine started to worsen in 1846 due to the second potato crop failure. When the Peel government was brought down mainly because of his plans to repeal the Corn Laws, Robert Peel resigned. This was followed by the formation of the Whig government under Lord John Russell along with his chief economic advisor Sir Charles Trevelyan. This government was a firm believer in laissez-faire policies since Trevelyan, who was in charge of administering Irish Famine relief policies, believed that protecting the British economy comes before helping the Irish. Whether he was a true believer in the "providence" of the Famine or whether he used it as an excuse to explain his absence of support and relief measures, he claimed that: The judgment of God sent the calamity to teach the Irish a lesson, that calamity must not be too much mitigated. The real evil with which we have to contend is not the physical evil of the Famine, but the moral evil of the selfish, perverse and turbulent character of the people (McCourt 2015, 117). Trevelyan played an important role in increasing the number of deaths and starvations by providing little to almost no help to the Irish. In one of his letters that were addressed to Sir R. Routh in 1846, he explained why the British government should no longer follow the Peel government's relief plan and why the Irish should find a way to solve their problem. He says in the letter:

> The greatest improvement of all which could take place in Ireland would be, to teach the people to depend upon themselves for developing the resources of their country, instead of having resource to the assistance of the Government on every occasion.
>
> (H.M. Stationery Office 1846)

The worst year of the Famine was 1847, which is why many researchers and historians refer to it as Black 47. Nevertheless, Ireland kept being the source of several goods to feed the British economy. Seventy-five ships sailed from Tralee to Liverpool, most of which were carrying grain. In the same period, six vessels sailed from Kilrush, County Clare, to Glasgow carrying a total of 6624 barrels of oats. Throughout 1847 both Indian corn and potatoes were exported from Ireland. A wide variety of other foodstuffs left Ireland apart from livestock – vegetables and pulses (particularly peas, beans and onions), dairy products, fish (especially salmon, oysters and herrings) and even rabbits. In February 1847, 377 boxes of "fish and eggs" and 383 boxes of fish were imported into Bristol alone. The butter export trade was particularly buoyant. In the first week of 1847, for example, 4455 firkins of butter (a firkin equals 9 gallons) were exported from Ireland to Liverpool. In the following week, this has risen to 4691 firkins (McGowan 2017, 16).

On Irish (Hyper)Masculinities in the Movie *Black 47*

"When I kill someone they call it murder. When they do it they call it war." (*Black 47* movie, 2018)

Understanding Irish masculinities is perhaps as complex as understanding the complex gender dynamics of the country's population. Masculinity is constructed, claimed and distributed by all people regardless of their gender identities or biological sexes. In Ireland, this masculinity has long been naturally linked to maleness and therefore political and religious hegemonies. The Great Hunger in Ireland was a period marked by a profound association between masculinity and violence. This association manifested in a series of rebellions, agrarian and sectarian violence, which were driven by a desire to assert Irish identity and autonomy in the face of British colonialism. Concurrently, the period was marked by a deeply ingrained culture of masculinity that prioritised physical prowess, courage and aggression. Men who were deemed weak or cowardly were ostracised or subject to ridicule, while those who engaged in violent acts were celebrated as heroes and martyrs. The intertwining of masculinity and violence during the Great Hunger can be understood through a variety of theoretical frameworks. For instance, Joseph A.

Kestner's analysis of British adventure fiction during the late nineteenth and early twentieth century (Kestner) highlights how anxieties about masculinity, empire and modernity were played out in representations of violent heroism. These anxieties were not unique to Britain, however, as they were also present in Irish society during the Great Hunger. Helena Goodwyn's exploration of the portrayal of violence and masculinity in Frank McCourt's *Angela's Ashes* underscores how these themes are integral to the construction of Irish identity in the context of British colonialism.

The issue with masculinity in Ireland is that it has only begun to be questioned as a gendered issue less than a few decades ago. Prior to the multiple publications that put this masculinity under the spotlight such as Conn Holohan and Tony Tracy's *Masculinity and Irish Popular Culture: Tiger's Tales* (2014) or Diane Negra's *Gendering the Recession: Media and Culture in an Age of Austerity* (2014), masculinity has long been coupled with maleness, and males with normativity. Ideally, this book's first chapter focuses on analysing Irish masculinities in the eighteenth century, providing an overview of the gendered power relationships in pre-Famine Ireland. An analysis of the coming of age is based on Kevin McKenna's writings on the coming-of-age ceremonies. McKenna believes that such rituals, which took place in Co. Galway in 1855 and 1890 to celebrate the coming of age of the heir to the Clonbrock estate, were used both to reaffirm the family's position "at the apex of the social order" and to "maintain a bond of identification" between the superordinate and subordinate" (McKenna 2013). A well-detailed discussion on primogeniture in Ireland is in fact written by McKenna himself in the fifth chapter of the 2019 book, where the problematics of such rituals and their masculinities are discussed thoroughly. The main issue with these celebrations, which ended up fading away in the late nineteenth century, is the reinforcement of a paternalistic social contract between landlords and tenants (McKenna 2019). The landlord-tenant relationships are of massive importance in analysing the Great Hunger's narrative through an ecofeminist lens, considering that in times of crisis massive evictions occurred, which I argue were conceived due to such masculinities that have been nurtured for decades, if not centuries, by the historically understudied Irish masculinities. The weakening of these celebrations has only begun to occur a couple of decades after the Famine ended.

Understanding hypermasculinity is crucial for comprehending the predominant form of masculinity that dominated Irish society during the height of the Great Hunger. Hypermasculinity denotes an exaggerated adherence to traditional male gender roles rooted in outdated societal norms. It is associated with aggressive behaviour towards women and individuals who challenge these gender norms, increased engagement in risky behaviours such as substance abuse and multiple sexual partners, and lower academic achievement (Zernechel and Perry). This concept portrays a distorted form of masculinity characterised by overt displays of toughness, hostility towards women and LGBTQIA2S+ individuals, and a rejection of non-traditional male behaviours.

The link between hypermasculinity, hegemonic masculinity and performativity has been explored by scholars such as Judith Butler and Erving Goffman. Their studies highlight that gender is a social performance rather than a biological given, compelling some men to exaggerate these performances to affirm their manhood.

Rebecca Anne Barr, Sean Brady and Jane McGaughey edited the volume *Ireland and Masculinities in History* (2019), which delves into various facets of masculinity in Irish history. The collection includes essays on contractarian masculinities (Ward 2019) and Protestant masculinity (Ó Gallchoir 2019), examining normative gender ideas and Irish masculinist ideologies. The authors address gender and power dynamics in Irish historiography, encompassing sexual orientation, class, religion and politics. The introduction to the book focuses on the challenge of gender in Irish historical studies, historically framing it as primarily a women's issue and noting the marginalisation of women's experiences. While efforts have begun to integrate women into Irish history, they are usually segregated rather than integrated into the broader historical narrative (Barr et al. 2019). Christine Kinealy's work, for example, has dedicated specific analysis to women's experiences, exemplified by her 2017 publication *Women and The Great Hunger*, which explores their long-overlooked roles.

The dominance of men's experiences as the norm in Irish writing has led the authors to advocate for a deeper exploration of Irish masculinities in historical research, arguing that persistent conservatism has relegated women to exceptional status while men remain the default. The collection characterises Irish masculinities as "religiously orientated", aiming to expand and redefine existing knowledge at the intersection of gender and society. In examining the portrayal of masculinity and its implications in nineteenth-century Ireland, the absence of femininity in men due to societal and religious constraints has deprived them of nurturing qualities. Imagining a scenario where queerness was socially accepted during colonisation suggests a departure from toxic masculinities and heteronormative pressures, potentially fostering a culture of care and nurturing.

The film "Black 47," directed by Lance Daly and featuring Hugo Weaving, James Frencheville and Stephen Rea, has achieved significant commercial success, grossing over 1 million euros at the Irish Box Office. It depicts hypermasculine representations of Irish men, characterised as aggressive, violent and stoic in their resistance against British oppression. This portrayal reflects societal expectations of Irish men as primary providers whose masculinity and societal worth are intertwined (Connell).

The narrative centres on Feeney, an Irish mercenary who deserts the British army to seek vengeance for his family's demise. Feeney embodies stoicism and physical prowess, employing violence to confront the British occupation (Sugars). His relentless pursuit of justice in a famine-stricken Ireland, where landlords prioritise rent collection over humanitarian concerns and the British prioritise economic gain at the expense of Irish lives, underscores themes of power dynamics and revenge.

Throughout the film, oppression manifests in various forms: suppression of the Irish language, discrimination based on religion, and exploitation of land and resources by British authorities. The narrative highlights a struggle against British soldiers, clergy and authoritative figures who dehumanise and subjugate the Irish under the guise of empire and religion.

Despite its title suggesting a focus on the peak year of the Great Hunger, "Black 47" primarily explores themes of individual vengeance and resistance against colonial powers. Female characters are largely portrayed as victims of the famine's horrors, while male characters are depicted as fighters who migrate, betray and resort to violence in their quest for justice.

The film underscores the pressure on men to conform to hypermasculine ideals, exemplified when young Hobson, influenced by older, more aggressive men, is coerced into committing violent acts to prove his manhood. This underscores the performative nature of hypermasculinity, where adherence to traditional masculine norms is enforced through peer pressure and mentorship.

Even recent media productions, such as the 2020 documentary series "The Hunger: The Story of The Irish Famine," narrated by Liam Neeson and co-produced by RTÉ and University College Cork to commemorate the Great Hunger's 175th anniversary, briefly touch upon historical topics like the Earl Grey Scheme's shipment of girls. This reflects ongoing narratives where women's experiences remain marginalised, while men's roles as protagonists and agents of historical change persist as the norm.

Conclusion

During the Great Hunger, societal gender expectations dictated that men assume the role of primary providers for their families, thereby linking their ability to do so with their masculinity and worth. This expectation became particularly arduous during the famine period when an increasing number of men were unable to secure work or provide for their families, leading to a sense of shame, despair and hopelessness (Ó Gráda 1999). These feelings may have contributed to the elevated rates of alcoholism and suicide among men, as well as a sense of powerlessness and emasculation (Ó Gráda, 1999). Moreover, the pressure on men as primary providers resulted in a lopsided distribution of the famine's effects, with women and children bearing a disproportionate burden. The expectation that women were responsible for taking care of their families despite the limited resources led to many resorting to begging or prostitution to survive (Kinealy, King, and Reilly 2017). Children experienced increased rates of mortality as well, due to high levels of malnutrition and disease (Ó Gráda 1999). Therefore, the gender roles and expectations prevalent in Irish society during the Great Hunger had a prejudiced impact on women and children, who were left to suffer the consequences of toxic masculinity and a societal overemphasis on men as providers.

The earlier examination of the 2018 film *Black 47* highlights the destructive impact of these societal gender roles and expectations. The film's protagonist, an Irish soldier returning home during the famine, is confronted with the devastation of his family, mirroring the sense of failure experienced by many Irish men who were unable to protect or provide for their loved ones. The protagonist's turn to violence can be seen as a manifestation of the emasculation and hopelessness that Irish men endured during this period. In this way, *Black 47* captures the wider social implications of the famine, illustrating how the inability to fulfil traditional masculine roles led to despair and, in some cases, extreme responses. The film also demonstrates the role of British masculinities during the famine, depicting British officials as distant, indifferent and authoritarian. This portrayal reflects the patriarchal British perspective that viewed the Irish as inferior and in need of paternalistic control, a belief that shaped the government's inadequate response to the famine. The British reluctance to intervene in the Irish economy, coupled with their gendered approach to relief efforts, further reinforced this power dynamic. As the film suggests, British authorities focused on controlling the Irish population rather than addressing the root causes of the crisis, reflecting a gendered understanding of Irish society that dismissed the needs of Irish men while reinforcing traditional gender roles for women and children.

Parallelly, the role of British masculinities in shaping politics in Ireland during the Great Hunger was significant. The British government's approach to the famine was influenced by a patriarchal view of Irish society, which regarded the Irish as inferior and in need of paternalistic guidance. This approach was characterised by a reluctance to intervene in the Irish economy, with a focus on maintaining British authority and control over Ireland. The British view of their own masculinity as superior to that of the Irish was a key factor in shaping the government's response to the famine. It reflected a belief that the British were rational and capable, while the Irish were emotional and incapable of governing themselves. This gendered understanding of Irish society was evident in the British government's relief efforts, which focused on providing assistance to women and children, while neglecting the needs of men. The gendered approach to relief efforts was also reflected in the workhouse system, which aimed to provide relief to the destitute but was intended to promote a particular view of gender roles and family life. This system separated men and women, and imposed strict rules on family life, reinforcing the notion of women as passive victims in need of protection, and men as irresponsible and unable to provide for their families.

Works Cited

Butler, J. *Gender Trouble: Feminism and the Subversion of Identity*. Routledge, 1990.

Connolly, S. "The Irish Famine: The Forgotten Context of Mental Health Experiences." *History of Psychiatry*, vol. 20, no. 4, 2009, pp. 429–445.

Cullen, L. *Irish Women and the Creation of Modern Catholicism*. Indiana University Press, 2014.
Daly, M. and J. Grayson. *Crime, Justice and Society in Ireland*. Routledge, 2015.
Groganyy, E. "Women and the Irish Famine: A Case Study in Collective Memory." *Journal of Historical Sociology*, vol. 13, no. 3, 2000, pp. 257–285.
Hanlon, N. "Breadwinner Masculinities." *Masculinities, Care and Equality*. Palgrave Macmillan, 2012, https://doi.org/10.1057/9781137264879_6.
Kenny, M. *Irish Asylum Experiences: Inside the Nineteenth-Century Lunatic Asylum*. Irish Academic Press, 2012.
Kestner, J. A. *Masculinities in British Adventure Fiction, 1880–1915*. Palgrave Macmillan, 1999.
Kussmaul, A. "Community and Communal Ownership in Ireland." *Irish Economic and Social History*, vol. 31, 2004, pp. 40–60.
McCourt, J. "Famine." *The Oxford Handbook of Modern Irish History*, edited by Alvin Jackson. Oxford University Press, 2015, p. 117.
McGowan, M. G. "The Famine Plot Revisited: A Reassessment of the Great Irish Famine as Genocide." *Genocide Studies International*, vol. 11, no. 1, 2017, pp. 87–104.
McKenna, K. "Elites, Ritual and the Legitimation of Power on an Irish Landed Estate, 1855–1890." *Irish Elites in the Nineteenth Century*, edited by Ciaran O'Neill. Four Courts Press, 2013, pp. 68–82.
———. "Primogeniture, Strict Settlement and the Rituals of Masculinity on an Irish Landed Estate, 1855–90." *Ireland and Masculinities in History*, edited by Rebecca Anne Barr, Sean Brady, and Jane McGaughey. Palgrave Macmillan, 2019, pp. 68–82.
Ó Gallchoir, C. "'Whole Swarms of Bastards': A Modest Proposal, the Discourse of Economic Improvement and Protestant Masculinity in Ireland, 1720–1738." *Ireland and Masculinities in History*, edited by Rebecca Anne Barr, Sean Brady, and Jane McGaughey. Palgrave Macmillan, 2019, pp. 39–66.
Ó Gráda, C. *Black'47 and beyond: The Great Irish Famine in History, Economy, and Memory*. Princeton University Press, 1999.
Sugars, C. "'Fighting for Home': Masculinity and the Irish Revolution in *Black'47*." *New Hibernia Review*, vol. 23, no. 4, 2019, pp. 80–96.
Ward, J. "Caught in a Contract: Congreve, Farquhar and Contractarian Masculinities." *Ireland and Masculinities in History*, edited by Rebecca Anne Barr, Sean Brady, and Jane McGaughey. Palgrave Macmillan, 2019, pp. 19–38.
Woodham-Smith, C. *The Great Hunger: Ireland 1845–1849*. Penguin Books, 1991, p. 165.
Zernechel, A. and A. L. Perry. "The Final Battle: Constructs of Hegemonic Masculinity and Hypermasculinity in Fraternity Membership." *College Student Affairs Leadership*, vol. 4, no. 1, 2017, article 6. Available at: scholarworks.gvsu.edu/csal/vol4/iss1/6.

14 The Ecology of Reading Lithuanian Dystopia

The Cases of *Dorandobongas* by Jurgis Volandas and *Ėko* by Valdas Papievis

Indrė Žakevičienė

Introduction

The realm of ecocriticism is vast, and literary analysis from an ecocritical perspective can be more interpretive rather than strictly analytical. The problem of such analysis was discussed by Dana Phillips a quarter of a century ago. One of its aspects was defined distinctly:

> [W]ords like "ecosystem," "organism, " and "wilderness" are used metaphorically, with no acknowledgment of their metaphorical status, as if literary, ecological, and environmental ways of speaking were a lot more compatible than they are, and as if their differences could safely be overlooked.
>
> (Phillips 579)

An ecocritical perspective, in the broadest sense, emphasises relationships, and perhaps this could be enough to start thinking about the ecology of literature without attempting to imitate "ecological and environmental ways of speaking" (Phillips 579).

The relationship between the text and the reader is essential for any ecocritical problem to be highlighted via fiction, so it is worth considering the reception of the text and the ways the message of the fictional narrative could be delivered. According to Margaret Freeman,

> poetry (as is true of all the arts) is an attempt to break through or transcend the fact that the way we are constructed as human beings, and how that structure gets expressed in human language, constrains our view of reality; that the minute we become conceptually aware of the world around us, we have at the same time made that world a phenomenological one, rich in its overlayering leaves but poor in enabling us to experience its essence directly.
>
> (Freeman 425)

DOI: 10.4324/9781003602750-19

Supposedly the same could be said about prose genres, as they illustrate the same effort to enable the reader to experience the world's essence directly. Environmental narratives should be no exception. If such narratives are to be defined "as written texts focusing on the environment, often describing events and experiences that took place at a particular location or in a particular type of location" (Purves et al. 4), there would be a great number of them in Lithuanian fiction spanning over a hundred years, from the end of the nineteenth century, when Lithuanians started contemplating their modern national identity.

Environment refers to specific places and spaces; Gabriel Viehhauser states that

> space is a pre-condition of every narrative, but exactly because of this it does not have to be explicitly mentioned or described every time it is evoked. Characters move through and act in space, thus constituting space implicitly. Practically every object that appears in a narrative can be situated in space and therefore has potential spatial dimensions.
> (Purves et al. 38)

Having this in mind it could be said that every narrative is environmental in some sense, and the more the reader engages with it, the deeper is the experience of pre-conceptual being that Freeman speaks about. Such an experience may be the most authentic compared to other pleasures of reading, such as astonishment, the joy of discovering and the ability to enjoy different emotions. However, it is also the most complicated for the reader to achieve and for the writer to establish in its material form. An assumption could be made that the experience of pre-conceptual being is essential for the purposes of ecocriticism too, as it provokes deep emotions, possibly resulting in conceptual considerations. The landscape depicted, specific location described, or particular urban environment inscribed in the narrative could serve as a trigger for a burst of associations, so the relationship between the realised imagination of the writer and the associations of the reader could be considered an ecology of reading.

Since one of the goals of literature is to provoke an experience of pre-conceptual reality, it can be assumed that fictional texts focusing on eco-catastrophes will be effective in cultivating readers' sensitivity to the environment. Having experienced an eco-catastrophe and its consequences at a virtual level of associations, readers are likely to become more conscious when thinking about ways to address environmental issues such as the climate crisis and the sustainability of consumption. Looking at the Lithuanian literary works discussed in this article, it seems that the representations of crises in social life and the consequences of eco-catastrophes differ in their impact.

In Lithuanian literature, ecological catastrophes are not particularly common, but there are quite a few texts that focus on unique micro-ecological catastrophes affecting individual characters, such as the loss of a beloved domestic animal, a cherished tree or an old homestead (the short stories *Skerdžius* (*The Herdsman*) by Vincas Krėvė – Mickevičius (1882–1954), *Gyvenimas po klevu* (*Life under the Maple*) by Romualdas Granauskas (1939–2014), *Mėsą galima valgyti* (*Meat Can Be Eaten*) by Vytautas Martinkus (1943) could be mentioned here). Such micro-ecological catastrophes emphasise the importance of relationships in general and represent a personal catastrophe that, for a particular character, equates to a global disaster, often ending with the character's death due to the unbearable pain of separation.

This article will analyse two literary texts that can be interpreted as social and eco-dystopias. An assumption could be made that, for a country that has not yet experienced major ecological catastrophes (in this case, Lithuania), the literary representation of social dystopia may be more impactful than the meanings encoded in eco-dystopia.

Lithuanian Rural Fiction as a Hideaway

Natural landscapes and rural environments were characteristic of most Lithuanian texts of fiction, starting with the first Lithuanian masterpiece written in the eighteenth century by Kristijonas Donelaitis (1714–1780) [his poem, *Metai* (*The Seasons*), was written approximately between 1760 and 1775 and published in 1818]. Speaking about the situation of Lithuanian literature of the twentieth century, Lithuanian literary researcher Justina Petrulionytė stated:

> In literature (especially in emigration), the village is sacralized, created as an image of the lost paradise. This literary trend was undoubtedly reinforced by Soviet censorship. Criticism in this literature perceived a parallel between the (vanishing, disappearing) village, nature, and Lithuanian identity. . . . Works depicting rural life became a kind of resistance expression against Soviet power structures, and in culture – a contrast to the Sovietized city.
>
> (Petrulionytė 12–13)

Lithuanian literature experienced the extremely severe test of censorship during the Soviet period (1945–1989); therefore, the authors at that time were seeking particular topics and forms of writing, which could mask the messages they were trying to convey. The extract from the interview by Lithuanian prose writer Mykolas Sluckis (1928–2013), given to the poet of younger generation and literary researcher Giedrė Kazlauskaitė, reveals the significance of the rural landscapes inscribed in Lithuanian fiction:

> The development of literature was also served by the theme of rural life, often protected by the authorities. True, this homogenized the

> landscape of prose, but for Lithuanian prose writers, pig farming or flax cultivation (compared to the universally praised cotton cultivation in Uzbek literature) never became a leitmotif, let alone an ideal. A much more humane motivation emerged – the farmer's concern to survive, to still feed the family, relatives, the city in extreme conditions. On the other hand, rural fiction vividly reminded of what began to erode and fade – ancient customs, traditional way of life, peculiarities of character (e.g., diligence, honesty), and other traits still included in the conception of national identity today.
>
> (Sluckis 6)

Environmental/rural narratives, enriched by existential ideas and cultural intertexts, preserved the vitality of Lithuanian literature and helped the readers to hold on during the first decades of Soviet occupation. The dystopia of Soviet regime in factual world was somewhat mitigated by escapist works of fiction. Contemporary Lithuanian fictional narratives, which will be analysed here, reflect a dystopian past or possible future, but the events in those narratives unfold in urban settings.

The main aim of this article is to analyse the ways in which dystopian situations are revealed in the novels by two contemporary Lithuanian writers and to consider the potential for dystopian narratives to establish a particular state of being – the effect of pre-conceptual experience. In order to reach this aim, it would be useful to have a look at two Lithuanian novels – *Dorandobongas* (2015) by Jurgis Volandas (Juozas Glinskis (1933)) and *Ėko* (2021) by Valdas Papievis (1962) – paying attention at the strategies of writing, used by the authors.

Volandas' and Papievis' Modes of Writing: Towards Dystopia

Lithuanian writer and playwright Marius Ivaškevičius interprets dystopia emphasising its specific educational function:

> The goal of dystopia as a genre is to serve as a warning. It's an attempt to depict a terrifying reality through literature, theater, and cinema. All of this is aimed at dissuading people from trying it out in reality, in their lives. Hope lies in resistance to that finality, which occurs in a dystopian work. Perhaps because of this, the ending cannot be hopeful or bright, because otherwise, you'd remain calm, and in this case, it's important to evoke that unease.[1]

Obviously, Ivaškevičius refers to the main characteristic of fiction: the ability to put the reader closer to the essence of being, rather than merely depicting and reflecting the factual world. When discussing contemporary Lithuanian dystopian novels, there should be mentioned books by Gasparas Aleksa (*Ėriukėlis stiklo ragais* or, *The Glass-horned Lamb*, 2003), Marijus Gailius (*Oro* or, *Thirsting for Air*, 2018) and Unė Kaunaitė (*2084*, 2023). These fantasy novels depict a future world emphasising the mutation of today's reality caused by climate changes or humanity's engagement with high technologies and AI. Though the writers' intention may be to warn the readers, the effect on the reader after reading such texts is usually not fear or worry, leading to serious considerations, but rather astonishment evoked by the images of transformed human beings or virtuality which cannot be distinguished from reality. It seems that the effect of pre-conceptual experience is best achieved through forms of writing that more closely resemble the actual world or evoke associations based on cultural intertexts or allusions. Therefore, it would be useful to discuss more specific and subtle ways of conveying dystopian atmosphere, as Jurgis Volandas and Papievis do.

The "informal novel" *Dorandobongas* was published by Juozas Glinskis under the pseudonym Jurgis Volandas. Glinskis is a Lithuanian playwright who started his career as a writer with a book of short stories, *Verdenės* (*The Rivulets*), published in 1967. Glinskis is laconic when speaking about himself:

> I wrote my first play *Grasos namai* (*The house of Threat*) at the age of thirty-seven (1970). By 1990, I had published ten plays of various genres – from lyrical drama and comedy to tragic farce and dystopia. We were working under conditions of ideological censorship at that time. To avoid getting into trouble, you had to employ an "internal censor" who controlled the behavior and dialogues of the characters.[2]

Speaking about his play written in 2006, Glinskis remarks on the shift in the way he writes:

> *Vieno tėvo vaikai* (*Children of One Father*) is my first play in which the will of the characters essentially dominates. I tried my best not to interfere with the voices emerging from the depths. I am still exploring the meaning of the characters' "blundering out" and paradoxes of their behaviour.[3]

His last creative work, *Dorandobongas*, is written without any internal censor and illustrates his position, revealed in the appendix of the book (the appendix is entitled "How to Write a Good Novel"):

> Cross out everything that has ever been stated anywhere. It is impossible to foresee what and when will appear from anywhere. All the more to explain. It appeared and that's it. You put it down and you see: there it is. . . . The understanding and explanation will come later, after returning to the analysis of the beginning. It may come or it may not. (Volandas 222)

Entangled storylines and various writing techniques reflect Glinskis' decision to succumb to spontaneous flashes of imagination.

Lithuanian writer Valdas Papievis currently lives in Paris, France. His first book of short stories, *Užmaršties slėnis* (*The Valley of Oblivion*), was published in 1989, the same year his first novel, *Ruduo provincijoje* (*The Fall in the Provinces*), appeared. The writer describes his way of writing like this: "I write very spontaneously. I only need the first paragraph or word. When I start writing, I never know how everything will develop or how it will end – I don't have a genre description in mind."[4] The book "Ėko" has no subtitle indicating the genre, but Papievis defines it as a novel in the form of notes. The writer does not refer to it as a dystopia, although he emphasises dystopian pandemic situation of 2020, during which he started writing the second part of the text; the first part was begun in 2016. Nevertheless, the urban landscape revealed in the novel clearly depicts dystopian decay and melancholy following the mysterious eco-disaster.

Post-Soviet Dystopia in Volandas' *Dorandobongas*

Volandas' novel *Dorandobongas* evokes the disquiet, focusing on the ecology of society through an extremely specific form, an extracynical language, and by creating ridiculous, extreme situations and a complicated text structure. The author defines the genre of the book as an "informal novel", as if to cover himself or to warn the reader of possible unconventional content. The plot of the novel is put in several sentences by the author:

> The main focus of the novel *Dorandobongas* is children, boys and girls, caught in a web of sexual exploitation. This business is controlled by the porn mafia boss Gazolis. His wealth is amassed through bribed law enforcement officials and doctors. The children are saved by the novel's humane characters: the eccentric prince Hubert, his illegitimate children Gailutis and Koreckytė, the zoopsychologist Dr. Tramtamtinas, a dog that understands human language, and other animals.
>
> (Volandas 226)

It is quite difficult to extract the plot for the reader, because the text is too dense, full of unexpected twists and includes episodes that are not developed later and appear more like insertions or inclusions. Instead of a single story with multiple plot branches, the reader encounters several separate stories across different genres, seamlessly transitioning between these new registers of reading.

Dystopic atmosphere is created from the very first passages, using mock-journalistic style: "Experts in parapsychology claim that places where humans (and not only humans) have been tortured and murdered retain traces of the aura, evoking acute sensations. This could be true" (Volandas 5). The location where the main action is centred is clearly indicated: it is the children's ward of the psychiatric hospital in Naujoji Vilnia. As the events depicted in the novel unfold, the locations of the action change: the reader is guided through various districts of Vilnius, taken to the outskirts of the city, then transported to a hotel in the Alps, to virtual space and to the king's castle, where the novel concludes. Changing locations indicate an interchange of genres: at first, the reader emerges in a thriller, and later finds himself or herself reading a tale, a mock-romance and a script of a Hollywood movie. The image of a psychiatric hospital encompasses several focuses: it should be a place where patients can heal and a space where countless possible worlds flourish, because the patients in such hospitals often have unique imaginations and unusual consciousness. At the beginning of the novel, one of the main characters, Gailutis, comes to the hospital as a resident doctor, unveils a crime committed by the hospital authorities and eventually becomes a patient in the same hospital. The reader is free to choose his/her own imaginative space: to either appear in the textual factual world, where magical things happen, or to enter the imagination of one of the characters, Duke Hubert, and ascribe all the miracles to his dreams or visions. The narrator presents Duke Hubert in laconic sentences:

> François Hubert, who titled himself the Duke of Burgundy, was not a mental patient. He was ill with virtual trips and hunting. He needed no compasses or maps, no trains or plains; upon hearing a voice calling his name, even in the middle of the night, he would take his haversack, coat, and hat,. . . . and leave.
>
> (Volandas 116)

The emphasis on the character's mental health and the assurance that Hubert is not mentally ill makes the reader question the narrator's reliability and sincerity, as the solutions to various situations in the novel most often oppose similar scenarios activated in the reader's mental space. On the other hand, "virtual trips" (Volandas 116) explain the semantic shift – Hubert's affair with a female wolf. Different forms of the text, such as third person narration, the dialogue of the narrator as a writer with his characters (". . . Gailutis, the author does need you more than you need yourself: if you don't exist, I will not exist too. Wait a bit. I am to write a good piece, till I'll get at you, though I am not quite sure, how to do that. Salvinija is up to something, but she doesn't tell me. She doesn't associate with me at all and behaves in such a way that I, the author, must run after her caprices like a paparazzo with my tongue out, in order not to lose the thread of the story" (Volandas 86)), and reconstructed intertexts of ancient sagas, create an effect of dissipation. On the other hand, the reader is capable of grasping the entirety of the vast network of provoked associations and immersing themselves in the atmosphere of a dystopian post-Soviet society, as indicated by the time of the narration: "The events I want to tell you about started approximately three decades ago" (Volandas 6). The dystopic atmosphere in the novel is created using the image of put-upon children. The children are depicted as caricatures of mental patients or neglected homeless teenagers and most often are presented through dialogues with perverted adults, as in this example:

"Richard:
'Hellow, Uncle Parksas.'
'Hellow, little Elly. One of the girls said that you are a boy.'
'I am definitely a boy.'
'Such boys as you usually are not boys. They only look like boys.'
'But I am a boy.'
'Some of our girls thought they were boys as well. But now they do not think so anymore.'
'What have you done to them?'
'To the girls? Nothing. We have done something to the boys.'
'Surgery?'
'Professor Blauberg's clinic does such things.'
'But if the boys do not want it, you do the surgery nevertheless?'
'There are no boys who do not secretly imagine that they are girls . . .'"

(Volandas 183)

The form of a play enlivens the flow of the narrative and intensifies the pace of the events. The seemingly light-hearted chat and grotesquely comic situation describing the supposed misapprehension of sexual abuser uncle Parksas evoke a feeling that is quite difficult to define in words: it could be a mixture of horror, surprise, resentment, disgust and worry. The narrator disappears in this episode, leaving the reader with nobody to blame for the cynicism and feeling as if left alone to ponder the *unimaginable* – a concept discussed by Jennifer Wenzel in her book *The Disposition of Nature: Environmental Crisis and World Literature* (2019). Later on the mock-journalistic style gradually turns into a mock screenplay for an imaginary movie, eventually as if immersing the reader in the movie itself.

The postmodern collage of episodes in different mock-styles is infused with humorous cynicism, which is likely more meaningful to the Lithuanian reader who is familiar with post-Soviet realities and capable of recognising particular codes in the names of characters or various places (side characters of the book are named after non-Lithuanian names – Gigašvili, Marat Hadžijev, Chabibulin, Bžegovoj). According to Paul Smethurst,

> postmodernism signals a radical loss of differentiation between the real world as historical and geographical referent, and representations of the real world. . . . this loss of the real occurs not in the age of machines that were visible and palpable, but in the age of information, where so much is invisible and falsifiable.
>
> (Smethurst 3)

Volandas is capable of doing the opposite: by creating unexpected characters and placing them in quite simple scenarios; he evokes the effect of pre-conceptual experience, the most genuine reality of the real. For example, the dialogue about Gailutis' situation between the nurse Nasturcija and a hamster named Karambolinas in one of the psychiatric hospital wards reproduces a simple pattern of polite conversation, while foregrounding the simplicity of sincere communication and immediacy:

> How are you, nurse Nasturcija?
> The old lady looked at him once, looked at him twice, and said:
> – Go, wash yourself, then we will have a talk.
> The hamster recoiled from her, rubbed his paws on the fur, stroked his whiskers, and recoiled again.
> – Washed, nurse Nasturcija.
> – He is bad, Karambolinai. He does no eat. Does not take any medication. As soon as I leave him, they will take him out . . .
>
> (Volandas 102)

Such intimate episodes, like the horrific ones discussed a few passages before, are interspersed with pseudoscientific descriptions that mock the hierarchy of Soviet authorities, criminals or gang fights as in this example:

> Sergeant Žebrauskas was not shot enough times to say farewell to this world. Nevertheless, it should be mentioned that one bullet ruptured the riot gear and intercostal cartilage, reached the area of the heart, but did not cause fatal correction, because it lodged itself between the pericardium and pulmonary artery.
>
> (Volandas 157–158)

The kaleidoscopic change of references to different genres engages the reader and helps establish an emotional relationship with the text.

The novel by Volandas is environmental in a specific sense: it offers the reader an opportunity to observe the autopsy of post-Soviet Lithuania and to experience a unique gestalt by emotionally linking various imagined spaces and reproducing particular moments from personal experiences of living or reading. Culture and nature are forged in this novel as in pagan belief: Gailutis is rescued by his dog, who can speak; he communicates with a hamster named Karambolinas; Hubert falls in love with a female wolf, who gives birth to his child; and a stone-man named Dorandobongas, presented as mythological creature living in the heart of the mountain, helps Gailutis, Hubert and children from psychiatric clinic to get rid of the gangsters, sexual abusers and transplant surgeons. Happy-end occurs in the king's castle, which might be the product of Hubert's virtual trips, leaving no hope that the dystopian society of the textual factual world can change.

The intentions to abuse the most vulnerable members of society – mentally ill children – along with the exaggerated cynicism of both the abusers and ordinary members of society ("It was necessary to kill those children before their birth. They are incurable. . . . Of course they are not to blame. Nobody blames them. They are just unnecessary. It is a burden for the society and purposeless expenses" (Volandas 162)), create dystopian disaster and highlight the crisis of humanity. Therefore, insights of Wenzel seem extremely accurate when she states that

> the urgent task, then, is not to look to literature as a "solution" but to understand its role in calculating what counts as "nature," "environment," "crisis," or even "human": the social dynamics and cultural logics that not only *cause* crises but also inflect how crises are experienced and recognized as such, by whom. This means recognizing the work that literature and cultural imagining do *all the time* in naturalizing ideas about nature and shaping constituencies of caring and regimes of visibility, as well as their exclusions and occlusions.
> (Wenzel 16)

Dystopian aspect of Volandas' novel, first of all, turns the reader's attention to the question of what counts as "human", because nature seems quite stable and faithful to its eternal laws. Wenzel's considerations on the opposition of imagining and *unimagining* could be extremely suitable for defining the effect Volandas' novel has on the reader. According to Wenzel,

> to label something *unimaginable* is to contain it: to draw a comfort line of distance and difference around it, to pull back from the work of engagement and understanding, of disentangling and finding oneself entangled, that might implicate a person in the network of relations and processes that produced the situation deemed *unimaginable*.
> (Wenzel 16)

Volandas trespasses the "comfort line" (Wenzel 16) and uses forms of writing that could be imagined as unimaginable even in fiction (the narrator of the novel presents Gailutis' patients in an extremely cynical way:

> Little doctor got the ward of little debils. The little debils were as expected – some of them were quite fine, children as children, you would not suspect anything, if not for their medical record; but there were also some that you don't really feel like looking at.
> (Volandas 9))

The reader is enabled to immerse themselves in "the work of engagement and understanding, of disentangling and finding oneself entangled" (Wenzel 16), and to consider the circumstances for *unimaginable* to appear. Most likely, the conclusion will be unanimous: the catastrophic Soviet regime gave rise to a dystopian society.

The Dystopian Return to Nature in Papievis' *Ėko*

Papievis' novel *Ėko* is quite different from Volandas' *Dorandobongas*. The plot of the novel is disclosed in several sentences by Lithuanian literary researcher Violeta Kelertienė:

> Paris is already sinking apocalyptically. Nature is reclaiming the city spaces, with ponds and fountains appearing blind in parks. There are almost no people left. The narrator asks himself: who am I in this background of destruction? Lonely and withdrawn, he discovers Echo – a sentient dog, a companion for future adventures in the deserted city.
> (Kelertienė 140)

The Lithuanian migrant author chooses to reveal the urban landscape of Paris, his second city. Paris could be interpreted as a particular code for culture, arts and architecture, imbuing the depicted place with symbolic meaning. The first passage of the novel could be understood as a particular key to the interpretation of the text:

> So there it is: it seems that the laziness of summer, the summer itself will never end. The leaves of the trees are still green, not yet turning brown; the grass is also green, warm summer rains refresh it sometimes. The bliss of laziness, when the rhythm of blood slows down along with the slowing rhythm of the city, when after the ebb of the avalanche of work, you gaze obliviously at the mobile cranes stiff above *La Samaritaine*, bound by the scaffold, or in the embrace of the willow, you comb with your eyes its branches, eventually sinking into their thickness while stretched out on the embankment of the Seine, towards the spire of the *Cité*. There is nothing left to do, everything has been done already, everything has been created, everything told; if you do anything, say anything, you will only disturb. Yourself, the others, those around you.
> (Papievis 7)

When everything is said and done, there is only one option left: to rethink and reuse. It seems that after the first passage, the "imagination loop" (Zittoun et al. 143) unfolds, and the narrator starts imagining, enabling all three dimensions of the loop – temporal, that of a distance and of "the degree to which the actual content or outcome of imagination is socially validated in a given sociocultural context and time" (Zittoun et al. 143). It seems that the narrator feels like a representative of a society that has fulfilled its tasks and expired; he finds himself on the eve of apocalypse, but the suspicion of inevitable end does not disturb him at all. The narrator draws a conclusion that seems intended or suspected from the very beginning of humanity's efforts to urbanise nature:

> And indeed, the fewer people were left in the city, the more it seemed that the city was returning to nature, or, to be more precise, nature was coming back to it. Listening to the sough of the plane trees, looking at the ripples of the Seine, I used to surrender to the play of imagination: the creepers had already rioted before my eyes; tangling and clinging to the walls of the buildings, they were gradually disrupting them; herbage was penetrating through the tiles of the pavement and choking them inch by inch.
>
> (Papievis 9)

Such a position of the narrator reflects the admission that the realm of culture is at nature's disposal and does not depend on human will.

Papievis chooses long passages of descriptions and reflections, making the flow of the narrator's thoughts slow and sometimes monotonous, melancholic and nostalgic; the decay of the city, the tents of migrant people in the centre of it, the transformed everyday life give the narrator the possibility to succumb to blissful boredom and to contemplate the essence of a human being as a species:

> Believe it or not, but this life after life, the life only of inertia, was so blissfull, as blissfull can be the sky of a fall, when suddenly the sun blazes up after hazy days and deceptively promises the return of summer. When the light, penetrating through the leaves falling but not yet fallen, throw the deepest shadows on the pavements and the paths, when the conkers pop under your feet like mines of overripe feelings and their pops echo in our deaf and dumb temples in vague echoes. That was the life of the nomads: we lived in the city as in the steppes.
>
> (Papievis 26–27)

The fusion of two seemingly unforgeable spaces – the city and the steppe – saturates the imagination with a sense of completeness. However, to reach such a state of being, one needs to lose the comfort of civilisation or be on the edge of such a loss. As Wenzel states,

> nature becomes known to us in large part through narrative and other patterns of imagining. That is to say, particular literary genres, aesthetic modes, and narrative templates provide the forms through which human understandings of nonhuman nature and its dispositions are forged. Paradoxically, these cultural forms shape our sense of what is natural, or just: these human constructs naturalize nature and its relation to the social.
>
> (Wenzel 13–14)

Papievis' text illustrates such ideas, emphasising not a pastoral escapistic aspect but a dystopian one. There is no reason given for the apocalyptic end in the novel; the narrator makes his own assumptions:

> May be everything started – I tell myself in my mind – because we no longer wanted to acknowledge the self-contained beauty of our lives, were unable to face up to everything that was given to us, because we wanted to outdo ourselves, were striving for more and more, and in our striving, were trying to run away from ourselves and ran past ourselves, because we became diffident about sincerity, wanted to escape eternal truths and virtues, as they are like rocks and pebbles, polished by the sea for so many centuries to a disgraceful banality.
>
> (Papievis 60–61)

Existential considerations are presented in a long stream of consciousness without any codes or subtexts, as if summing up the conclusions of the modernist epoch of the twentieth century. Papievis' narrator refers to characters from the masterpieces of fiction of the previous century or to characters from his own previous books. The atmosphere of the novel and the main characters – the narrator and his companion Emi, whom he met on the bank of the Seine – could be easily compared to Oliveira and La Maga from Julio Cortazár's novel *Hopscotch* (1963).

Using the same technique as Volandas, Papievis enables his narrator to communicate with the characters of the novel, making the reader construct his identity as a writer:

"– Melanie!
You freeze.
– Melanie . . . We haven't seen each other for so long.
– Yes, quite a while.
. . .
– I didn't think that we would ever meet anywhere.
– I didnt't think so either.
. . .
She added after a pause:
– Sorry, I didn't say farewell that summer. I thought it would have been too much.
– Would you like me to rewrite?"

(Papievis 62–63)

Such a technique reminds the reader that the events unfolding in the text are to be treated as the narrator's imagination, referential but not part of the textual factual world. The temporal dimension of the "imagination loop" (Zittoun et al. 143) is realised by referring to both the future and the past. The dystopian future encompasses the past via fictional characters from books written by talented authors and warns about the elemental force of nature, though this threat is not presented as revenge or punishment for humanity's ignorance: "it [nature – I. Ž.] was neither angry with us nor asked for any help; it simply detected that the time had come to say goodbye to us, I would put it that way" (Papievis 61). The dimension of distance that the "imagination loop" (Zittoun et al. 143) encompasses could be characterised specifically: the narrator ponders serious existential problems but rethinks them through allusions to fiction, supposedly enjoying the experience of pre-conceptual experience literature suggests. The dimension of degree could be evaluated effortlessly: the postmodern sociocultural context stipulates dystopian visions even in everyday life. Thus, the narrator's imagination simply makes this prospect come through.

The structure of Papievis' text parallels the construction of an "imagination loop" (Zittoun et al. 143). Summer boredom provokes a burst of imagination, allowing the narrator to envision the cycle of humanity's existence, concentrated in a multicoloured cultural bubble and treasured in the realm of fiction. The fire at Notre Dame Cathedral concludes this daydreaming. The narrator interprets the flames engulfing the temple as purifying: "As if the Cathedral has redeemed the city, I see it revive from the ruins of my imagination" (Papievis 133). The "imagination loop" (Zittoun et al. 143) encompasses reconsiderations of a human being's position in the world and reveals a possible catastrophic future as the inevitable defeat of culture against nature: the city of art and architecture will eventually be swallowed by creepers, tree roots and grass. Capable of creating divine cathedrals, human beings are too narrow-minded to fully grasp the essence of existence. In Papievis' novel, the ecological catastrophe seems chronic – nature gradually destroys, swallows and suffocates – yet humans no longer see the point in resisting, as if subconsciously punishing themselves. People no longer control the situation, seemingly from fatigue or overindulgence in culture and consumption. Papievis' eco-dystopia is highly specific – elegant and aesthetic – but this aesthetic provokes much deeper reflections than those evoked by Marijus Gailius' novel *Oro* (2018), which portrays life in a technologically saturated consumer society after a flood caused by climate warming.

Conclusions

A brief analysis of the novels by Volandas and Papievis reveals two different ways of writing: Volandas employs a postmodern collage of different styles and is not scrupulous about the structure of the narrative, allowing the reader to deviate from the main storyline and immerse in different modes of reading. In contrast, Papievis is more restrained and modernistic, presenting existential considerations and melancholic scepticism. Speaking about the ability of fictional text to evoke the experience of pre-conceptual being, Volandas' strategy seems more effective: flashes of imagination are materialised in different forms – mock-journalistic, pseudoscientific descriptions, dialogues and the form of a play – as if constructing many different "imagination loops" (Zittoun et al. 143). Each cycle of such an imagined loop establishes a unique experience of a pre-conceptual state of being and enables the reader to feel a unique sense of authenticity, while simultaneously developing an awareness of the dystopian social environment.

Papievis' novel is written in the form of a loop, which could be interpreted as "imagination loop" (Zittoun et al. 143) defined by experts in sociocultural imagination. Using long descriptions, simple dialogues and first-person narration, Papievis creates particular intimacy, allowing the reader to read his text as a poem. Urban landscapes revealed in great detail draw the reader's attention and make them admire the vibrant atmosphere of decaying Paris. However, the phenomenon of pre-conceptual being may not occur due to intellectual considerations presented in very long sentences: when reading the text in Lithuanian the reader can hear a specific inner rhythm, which orientates towards a monotonous flow of thoughts and a specific way of reading, such as following allusions to other texts of fiction and frequent references to the realm of literature and its significance in general. The narrator is too straightforward in conveying his aesthetic experience and insights, so the reader can only agree or disagree, but not immerse themselves in their own imagination, seeking the landmarks of pre-conceptual states of being. Nevertheless, reading Papievis' text involves a specific melancholic mood, leading the reader to ponder the threats posed by nature while also evaluating the contribution of humans to the world they inhabit.

It may be too daring to draw a definitive conclusion, but an assumption could be made that specific dystopias like those of Volandas' and Papievis' are more effective in impacting the reader than fantasy novels referring to dystopian futures. The evoked effect of a specific state of being – where it is difficult to find words to define the experienced emotions – is the most effective way to understand the world objectively. Such understanding should be the first step towards considering a human being's role in the world.

Thus, the ecology of reading, understood as a relationship between the reader and the text, is as valuable as any other relationship in nature. If such an ecology were established, all the intentions of ecocriticism experts to make people think about their environment, their place and other forms of life and nature would be realised.

Notes

1 See www.bernardinai.lt/m-ivaskevicius-distopijos-tikslas-perspejimas-tai-baisios-realybes-pabandymas-literatura/.
2 See www.juozasglinskis.com/.
3 See www.juozasglinskis.com/.
4 Papievis, www.15min.lt/kultura/naujiena/literatura/nauja-knyga-pristates-valdas-papievis-man-patiko-aprasyti-griuvanti-paryziu-286–1647052.

Works Cited

Freeman, Margaret. *Cognition in Language: Volume in Honor of Professor Elzbieta Tabakowska*. Edited by Wladyslaw Chlopicki, Andrzej Pawelec and Agnieszka Pokojska, 2009, www.academia.edu/57337358/Poetic_iconicity.

Ivaškevičius, Marius. "Distopijos tikslas – perspėjimas, tai baisios realybės pabandymas Literatūra." 2021, www.bernardinai.lt/m-ivaskevicius-distopijos-tikslas-perspejimas-tai-baisios-realybes-pabandymas-literatura/.

Kelertienė, Violeta. "Anotacija." *Ėko*, edited by Valdas Papievis. Odilė, 2021.

Papievis, Valdas. *Ėko*. Odilė, 2021.

———. "Man patiko aprašyti griūvantį Paryžių." 2022, www.15min.lt/kultura/naujiena/literatura/nauja-knyga-pristates-valdas-papievis-man-patiko-aprasyti-griuvanti-paryziu-286-1647052.

Petrulionytė, Justina. "'Nematomi miestai' lietuvių literatūros, vartų sergėtojų 'diskurse'." *Literatūra*, 2015, etalpykla.lituanistika.lt/fedora/objects/LT-LDB-0001:J.04~2015~1489156979643/datastreams/DS.002.0.01.ARTIC/content.

Phillips, Dana. "Ecocriticism, Literary Theory, and the Truth of Ecology." *New Literary History,* vol. 30, no. 3. The Johns Hopkins University Press, 1999, www.jstor.org/stable/20057556.

Purves, Ross S., Olga Koblet, and Benjamin Adams. *Unlocking Environmental Narratives: Towards Understanding Human Environment Interactions through Computational Text Analysis*. Ubiquity Press, 2022.

Sluckis, Mykolas. "Apie pradžią, išlirkamumą ir paraleles." *Metai*, 2006, www.tekstai.lt/zurnalas-metai/123-2006-nr-89-rugpjtisrugsjis/397-apie-pradi-iliekamum-ir-paraleles-raytojas-mykolas-sluckis-atsako-giedrs-kazlauskaits-klausimus.

Smethurst, Paul. "The Postmodern Chronotope." *Reading Space and Time in Contemporary Fiction*. Rodopi, 2000.

Volandas, Jurgis. *Dorandobongas*. Juozas Glinskis, 2015.

Wenzel, Jennifer. *The Disposition of Nature*. Fordham University Press, 2019.

Zittoun, Tania, Vlad Glăveanu, and Hana Hawlina. "A Sociocultural Perspective on Imagination." *The Cambridge Handbook of the Imagination*, edited by Anna Abraham. Cambridge University Press, 2020, www.jsdp.jp/contents/workshop-symposium/workshop2019_2_05.pdf.

Part V

Climate Change and Environmental Disaster in Cinema

15 *Don't Look Up*

Political Satire Crashes into the Contemporary Disaster Film

Georgios Dimoglou

Introduction

Disaster films have long been connected to man-made environmental threats, such as toxic waste, nuclear disaster and climate change. Alexa Weik von Mossner mentions how eco-disaster films have serious messages to deliver, even during the 1950s and later, in the case of comedies (166). This continues to hold true in the twenty-first century, with films such as Pixar's *WALL-E* (2008, directed by Andrew Stanton, which von Mossner also uses as a case study) and *The Day After Tomorrow* (2004, directed by Roland Emmerich). According to Douglas Kellner, "*The Day After Tomorrow* uses the conventions of the disaster film to dramatize the dangers of climate change and global warming, ignored by the Bush-Cheney administration" (81–82). More recently, Adam McKay's *Don't Look Up* (2021) was released, similarly charged with an environmental and political identity in a comedic/satirical context.

The goal of this research is to reveal the ways *Don't Look Up* plays with the disaster film formula in order to satirise the contemporary landscape of political and social indifference towards climate change. Even if characters in the film do not deal with the issue of climate change, the director is rather overt when discussing his vision. As Cara Buckley writes (2022), McKay sent an email to *The New York Times*, in which he clearly outlined the film's connection to the issue. In his own words, he explains, "I'm under no illusions that one film will be the cure to the climate crisis . . . but if it inspires conversation, critical thinking, and makes people less tolerant of inaction from their leaders, then I'd say we accomplished our goal." In that same article, Buckley also notes how numerous scientists came to the film's defence as a call to action against global warming (2022). Thus, this article argues that *Don't Look Up* should be understood in light of the threat of climate change, specifically as an allegory of the potential consequences that may come with political inaction against such timely environmental threats. The film is placed against previous definitions of disaster films and eco-disaster comedies (e.g. Maurice Yacowar, Robin L. Murray and Joseph K. Heumann) and is analysed primarily according to approaches on the disguised political identity of Hollywood films (Betty Kaklamanidou).

DOI: 10.4324/9781003602750-21

Don't Look Up belongs in a specific category of disaster films that are based on the premise of a comet, asteroid or similar celestial object heading towards Earth, threatening life on the planet with an extinction event. This research will bring attention to previous renderings of this concept on the big screen, arguing that such texts actually reflect (Western) society's viewpoint on environmental issues and global-scale threats. Based on the idea that films "reflect the times in which we live" and that disaster films in particular "are said to be borne out of times of crisis" (Stephen Keane 5), this essay uses *Don't Look Up* to demonstrate how society has seemingly become less optimistic and more apathetic as humanity moves forward without taking substantive action against man-made climate change.

A Comet Made of Rock, Politics and Climate Change

McKay is known for delivering satirical films with rather sharp edges, usually at the expense of financial and political systems (e.g. *The Big Short,* 2015; *Vice,* 2018), and *Don't Look Up* is not an exception. The film begins with astronomy PhD candidate Kate Dibiasky (played by Jennifer Lawrence), who discovers what seems to be a comet. Soon, her team and professor Randall Mindy (Leonardo DiCaprio) realise that it is actually headed directly towards Earth and that they only have six months to stop it. However, US President Orlean (Meryl Streep) ignores their warnings until a sex scandal emerges, and she decides to use comet "Dibiasky" to divert people's attention. Meanwhile, Dibiasky and Dr Mindy have already leaked the news on *The Daily Rip*, a popular morning talk show; once again, they are not taken seriously. In fact, Dibiasky becomes a target of online memes, sexism and mockery because of her ranting about the threat. Eventually, when the time comes to strike the comet with nuclear missiles in order to alter its course, President Orlean withdraws the mission in a last-minute attempt to preserve the comet, as it carries rare, immeasurably valuable materials. Peter Isherwell (Mark Rylance), one of the richest people on Earth, has made this discovery and proposes a plan to fragment the comet and gather the materials once they fall into the Pacific Ocean, demonstrating his immense influence on the government. Dibiasky and Dr Mindy end up forming the "Just Look Up!" movement, while President Orlean calls for the world to "Don't Look Up!" Unfortunately, Isherwell's plan fails, as well as other countries' attempts to stop the comet. As a result, it strikes and causes an extinction event while Isherwell, Orlean and a few selected others escape in a spacecraft, in cryo-sleep.[1]

The concept of the "killer comet," inspired by the impact that famously led to the annihilation of dinosaurs during the Cretaceous–Paleogene extinction event,[2] has reached our screens more than once. Such films usually receive mixed responses, with *Meteor* (1979, directed by Ronald Neame), *Armageddon* (1998, directed by Michael Bay) and *Deep Impact* (1998, directed by Mimi Leder) being notable examples. However, they tend to succeed in the box office; both *Armageddon* and *Deep Impact* were among the highest-grossing films of 1998, with *Armageddon* being number one. On the same note, *Don't Look Up* proved to be highly popular, breaking viewing records on the Netflix streaming platform through which it was released (K.J. Yossman), although, again, critics are more divided than audiences.[3]

Generally, the disaster film genre is chiefly built around a specific moment in the film, when mass destruction is triggered, making it more difficult to define than other genres. Maurice Yacowar explains that "the basic imagery of the disaster film would be disaster, a general, spectacular destruction, but usually this imagery occurs only at the end, though often with brief and promising samples along the way" (320). Discussing 1970s disaster films, Nick Roddick locates six main attributes regarding the disaster event, which must be:

> diegetically central; factually possible; largely indiscriminate (in that it could happen to all sections of the population). . . unexpected (though not necessarily unpredicted); all-encompassing, in the sense that potential victims cannot simply opt out of it; and finally, ahistorical, in the sense of not requiring a specific conjuncture of political and economic forces to bring it about.
>
> (246)

Most of these attributes seem to remain relevant to this day, with small variations. Usually, the main narrative goal is to stop the disaster from happening, minimise the damage and/or protect as many people as possible. Also, the threat is theoretically realistic, even if parts of it or the world's reaction are not (e.g. because of current technological limitations). As Emily E. Auger points out, "the disaster film often engages both the discourse of science and the genre of science fiction" (49). Such films tend to begin with the discovery of this threat which takes the world by surprise and by defining the timeframe humanity has in order to counteract: if and when the disaster takes place, there will be no escape for most people. Finally, the disaster itself has no motive other than being a natural event or man-made accident.

The coming disaster usually works as a drive for a specific individual or small group of people to become heroes given the chance to save the world. According to Jule Selbo, "in a disaster film an ordinary person has the opportunity to do something extraordinary" and "the major plot points . . . should revolve around decisions or actions concerning the approach or presence of a disaster" (259–260). This is the case with both *Armageddon* and *Deep Impact*, for example, where a group of people are selected to face the comet based on specific characteristics and one of them is explicitly the head of the mission.

Based on Yacowar's typology of disaster films, killer comet films primarily belong in the first category of "natural attack" disaster films (313) and may overlap with other categories as well, such as "survival" and "the comic" (317–318). An instance of a survival film with a comet as its narrative object is *Greenland* (2020, directed by Ric Roman Waugh), a film that focuses on a family's attempt to reach a safe location in Greenland, designated by the US government, in a world where there is no escape from the comet (it was falsely believed that most fragments of the comet would never reach Earth, or that they would fall in safe locations). On the other hand, *Don't Look Up* is a natural attack – comedy film, parodic and satirical in nature. Robin L. Murray and Joseph K. Heumann observe that the disaster genre reached a point of saturation by the 1980s, as audiences "didn't need to be warned or taught about environmental problems," as there were already "institutions in place that took the issue seriously" (114). Additionally, according to John G. Cawelti, "when the generic patterns have become so well known that people become tired of their predictability . . . parodic and satiric treatments proliferate" (296). This essay claims that, similarly, audiences have become quite familiar with the killer comet narrative, allowing for films such as *Don't Look Up* to emerge. In turn, this film proves that contemporary society no longer believes that there are credible "institutions" that actively deal with climate change.

Discussing *Armageddon* and *Deep Impact*, Keane notices how the latter is relatively more serious than the first, viewing *Deep Impact* as "more passive and survivalist" and *Armageddon* as "more urgent and determined, but, with the fate of the world in the hands of a squabbling bunch of drillers, ultimately tongue-in-cheek" (81). Both films, however, retain their melodramatic, epic and serious attitude until the very end. They are more or less described by Yacowar's many conventions of the disaster genre: among others, which are less genre-specific (e.g. a romantic subplot), he notes the feeling of immediacy and contemporary significance, the representation of the entire society in the cast, the idea of isolation, savagery hiding under the rug of civilisation, the specialist figure and the failure, corruption, or absence of systems such as the government and the church (320–328). Interestingly, Yacowar adds that, despite people's actions and the outcome itself, optimism permeates these films (324). This is not only true in the case of *Deep Impact* and *Armageddon*, where an extinction event is ultimately avoided, but also in *Greenland*, where the world has been destroyed but the survivors return to the surface after a certain period, ready to rebuild.

Don't Look Up, as well as *Armageddon* and *Deep Impact*, can be considered political films, according to Betty Kaklamanidou's definition (*Greenland* is excluded, as it adopts a ground-level perspective, following a family and the trials it faces in such conditions). Kaklamanidou considers such films to be political "in disguise,"[4] as Hollywood avoids explicitly presenting them as such. Elaborating further, she explains that

> The political film is a narrative whose story can take place in the past or the present, follow the dramatic or comedic mode, originate from real or fictional events – or a combination of both – with a subject/protagonist whose narrative object is situated in the political structure

> of a state and its attainment will affect society as a whole in a positive or negative way.
>
> (21)

Almost immediately after its appearance, the comet attains a political dimension in all three films. The US President is always the one coordinating and supervising the mission to deflect it. In addition, the comet provides the opportunity to criticise government efficiency and structure in various ways. *Armageddon* portrays government officials as unqualified to deal with the situation; in an early scene, a scientist (Jason Isaacs) dismisses the government's plan to detonate bombs on the comet's surface and proposes to drill and cause an explosion from within the comet. In *Deep Impact*, the US Secretary of Treasury has to resign after the discovery of the comet, which remains classified information until the President (Morgan Freeman) is ready to present countermeasures to the public. This event also leads to the President's decision to impose martial law when it becomes apparent that the comet will strike. Finally, in *Don't Look Up*, President Orlean initially hides the news from the public, so that her party will win the midterm elections. Then, she launches a campaign to save the world, simply to divert people's attention from a scandal, while she also fires staff who have to take the blame for the delayed response. The protagonists do not have to be within the political structure (they are either specialists in drilling, space exploration and astronomy, following the genre's conventions) and, clearly, if they achieve their goal of destroying the comet, the entire world will be preserved. Kaklamanidou's conventions, such as iconography related to Washington DC, the Oval Office, the President, and media spectacle (40–44) are all present in these films as well.

And yet, such films do not generally require "a specific conjunction of political and economic forces to bring [the disaster] about," as Roddick states (246). We witness governments trying to cooperate and save the planet (the US government is always shown to be the protagonist in such efforts). Even if they make wrong decisions, they do not come from a place of ill intentions but fear, pressure and uncertainty. *Armageddon* makes a slight comment early on about the limited government budget on protection against such events and generally questions the skills of those in power: for example, the President (Stanley Anderson) decides to detonate the bomb from a distance, believing that the crew has failed, even though he is informed that an explosion on the comet surface will change little to nothing (luckily for the team and the world, this never happens). Nevertheless, his ethics are never questioned. In *Deep Impact*, the President is shown to be cold and calculating, trying to avoid disclosing the news to the public. However, under his leadership, a number of plans have been developed to tackle the issue. Furthermore, if everything else fails, underground shelters have been created, containing space for animals, plants, 200,000 selected specialists from various fields and 800,000 randomly selected American citizens (other countries have also taken similar measures). Freeman's President is even more active, if distant, addressing the public frequently. Again, throughout the film, no character questions him, the validity of his words or the upcoming disaster. A similar "Noah's Ark" plan is developed in *Greenland* as well.

McKay's disaster film seems to conform to previous definitions of the genre. Its narrative object is the disaster itself (comet "Dibiasky"), a theoretically possible event that only reveals itself six months before it hits Earth, giving little time for humanity to prepare themselves. Moreover, the destruction it is about to bring will most probably lead to the end of humans and most other species. Still, as the film progresses, the viewer realises that humans also play a role in bringing about their own doom. *Don't Look Up* does not share the kind of hopefulness that is found in other killer comet films; it subverts our expectations by actively holding "political and economic forces" responsible for the world's destruction. When Dibiasky and Dr Mindy present their findings to President Orlean, their 5–10 kilometre-wide discovery is met with ignorance and indifference. The President is more concerned about the midterm elections in three weeks than a possible extinction event in six months. Government incompetence and corruption become even more evident when billionaire Isherwell stops the mission to deflect the comet in order to secure its valuable materials. It remains unknown if the missiles that were to strike the comet would change its route or not, but it is a possibility. At that moment, the film asserts that it is no longer nature's unpredictability but humanity's greediness and idiocy that will eventually lead to their own unmaking.

The world's reaction to such news is another point of difference between these films. In *Armageddon* and *Deep Impact*, people all around the planet are shown to be calm and collected for the most part. Many go on with their lives as usual and continue to learn the news from their television screens, while others are shown to pray peacefully and in unison. The films mention instances of rioting and there may be moments where some people act desperately; however, there is no general breakdown of society. In *Greenland*, a much more recent film, societal collapse begins as soon as the US government implements its plan of moving randomly selected citizens to a secure location. The world immediately becomes a hostile environment, as those who are not chosen will not hesitate to steal, kill and perform other heinous crimes in order to survive. While such violence is not as widespread in *Don't Look Up*, the film follows a different scenario: taking place in a world where everyone has access to and is familiar with the internet, it uses the comet as an opportunity to raise issues of fake and alternative news, conspiracy theories and outright dismissal of scientific evidence, among others. Fake news always existed, as Melissa Zimdars and Kembrew Mcleod point out; however, the issue is exacerbated because there is a "much larger communication problem in the digital age" (3–4). This social climate comes in deep contrast with *Armageddon* and *Deep Impact*, where people accept everything, they hear as facts. Violence outbreak takes place in *Don't Look Up* only when comet Dibiasky is finally too close and people realise its existence to be true. This research supports that, in their own way, the terrifying settings of *Greenland* and *Don't Look Up* capture current sociopolitical problems of distrust against science, the government and other people in general, reflecting our own, much more pessimistic viewpoint, compared to 20 years ago.

Kaklamanidou claims that "there is a link between the release of these political films and their time of production and consumption" (2) and *Don't Look Up* affirms this statement. In his interview with Mark Olsen of the *LA Times*, McKay himself described President Orlean as a "mixture" of real US Presidents in variable amounts, from Donald Trump to Ronald Reagan and everyone in between. It can also be argued that the character brings other known politicians to mind as well, such as Hillary Clinton (Streep's President is seen at one moment hugging a photo of Bill Clinton with her). In the same interview, McKay also hints towards celebrity billionaires such as Mark Zuckerberg and Elon Musk. Through Isherwell's character, the film satirises the relationship of corruption that exists between the upper-class elite and government officials. Michael Ryan and Douglas Kellner explain that, on one hand, disaster films are not "unidimensionally conservative" and "they warn about the dangers of unrestrained corporate capitalism" (52). Nevertheless, the films do "advocate corporatist solutions" and even if

> the vision of hope offered by the disaster films is a response to troubled times . . . the need for those conservative solutions is made to seem all the more pressing by the increasingly hopeless literality of the disaster metaphors.
>
> (52–55)

Don't Look Up differentiates itself from previous disaster films, as it does not consider corporatist intervention to be the solution; rather, McKay presents it as the prime factor leading humanity to its doom.

The film is not only a satire but also a parody of the genre. Colonel Benedict Drask (Ron Perlman), an offensive, unfiltered war veteran who is from a "different generation," as others describe him in the film, is a character that functions as a parodic version of the hero that responds to the call. He evokes both the protagonists in *Armageddon* and *Deep Impact* as someone who is sent into space to inspire the world because of his status (such as Robert Duvall's character in *Deep Impact*) and who is rough around the edges and with no particular experience in space exploration (such as Bruce Willis' character in *Armageddon*). *Don't Look Up* also makes fun of the concept of drilling the comet in order to destroy it; in this case, Isherwell plans to drill the comet for profit, making it explode while also preserving its rare materials. Through being a parody of the killer comet narrative, *Don't Look Up* avoids being entirely direct in its criticism. Discussing social implications in Hollywood genre, Jean-Loup Bourget states the following:

> The freedom of Hollywood directors is not measured by what they can openly do within the Hollywood system, but rather by what they can

> imply about American society in general . . . if they describe the breakdown of a social system, they must somehow end on a hopeful note and show that both order and happiness are eventually restored.
>
> (76)

Indeed, *Don't Look Up* disguises its political identity in its treatment of the genre, although it certainly does not end on a hopeful note; the film is not afraid to show the end of humanity (and potentially, the entire world), unlike other comet killer films before it. Lawrence Buell predicted such texts, stating that "as ecocatastrophe becomes an increasingly greater possibility, so will the occasions for environmental apocalyptic expression and the likelihood that it will suffuse essay, fiction, film" (308). The creation of *Don't Look Up* signals this negative development, and its pessimistic viewpoint loudly reverberates the common fear of many: that, at this point, it might be too late to reverse climate change and save our planet.

Conclusion

Adam McKay's *Don't Look Up* is a film that parodies the popular comet killer narrative of disaster films and satirises social and political negligence in regard to climate change. Previous comet killer films show the people's trust in their own government and portray leaders as flawed yet moral individuals who try to deal with the issue to the best of their abilities. In sharp contrast, *Don't Look Up* reflects the growing distrust against governments and corporations, as they seem to ignore, downplay and/or flat-out deny the existence of climate change and other global-scale threats. Furthermore, the internet is shown to be a platform that can be used for leaking government-suppressed news and for people to express their opinion in a cyber-democratic environment, although it also plays a role in the spreading of fake news and misinformation. Theories on disaster and political Hollywood films, as well as on the connection between cinema and contemporary society were applied to the film's analysis.

The film might not have received positive reviews from professional critics; however, it has been lauded by the scientific community and by activists as well, as Donna Lu of *The Guardian* notes. Writing for *Rolling Stone*, David Fear considers the film to be "so caught up in its own hysterical shrieking that it drowns out any laughs, or sense of poignancy, or points it might be trying to make." This essay believes that such reactions could very well be part of the movie itself, which laments media for having to keep things "light," as the presenters of *The Daily Rip* state to Dr Mindy in the film. Kate Dibiasky, or "the yelling lady," as she becomes known in the film's world, is actually *Don't Look Up* incarnate; the only reason she is not taken seriously is because she takes the threat of the comet "too seriously." Nevertheless, it is probably better to judge a very urgent message based on its validity, not the messenger that bears it.

Notes

1 In a post-credits scene, we see the spacecraft reaching an Earth-like planet thousands of years later. President Orlean gets eaten by a creature called a "Bronteroc" while similar creatures approach Isherwell and the rest of the group.
2 *Armageddon*, one of the most well-known examples of comet disaster films, begins with a prologue on dinosaurs going extinct more than 60 million years ago because of a comet's collision with Earth.
3 On the *Rotten Tomatoes* website, which gathers data from multiple film reviews, *Don't Look Up* has a 55% score of approval with critics and a 78% score with audiences (according to data as they were formed on 24 April 2024. For more information: www.rottentomatoes.com/m/dont_look_up_2021).
4 Kaklamanidou points out that there is not a clear definition of the term "politics," since it is used both inclusively and exclusively to describe actions from governments and institutions to individuals and social groups (5). This research follows Sheldon S. Wolin's definition that politics are "institutionalized processes and settled procedures regularly used for handling public matters" (7).

Works Cited

Armageddon. Directed by Michael Bay, performances by Bruce Willis, Liv Tyler, Ben Affleck. Buena Vista Pictures, 1998.
Auger, Emily E. *Tech-Noir Film: A Theory of the Development of Popular Genres*. Intellect, 2011.
Bourget, Jean-Loup. "Social Implications in the Hollywood Genres." *Film Genre Reader IV*, edited by Barry Keith Grant. University of Texas Press, 2012, pp. 69–77.
Buckley, Cara. "Don't Just Watch: Team behind 'Don't Look Up' Urges Climate Action." *The New York Times*, 11 January 2022, www.nytimes.com/2022/01/11/climate/dont-look-up-climate.html.
Buell, Lawrence. *The Environmental Imagination: Thoreau, Nature Writing, and the Formation of American Culture*. Harvard University Press, 1995.
Cawelti, John G. "*Chinatown* and Generic Transformation in Recent American Films." *Film Genre Reader IV*, edited by Barry Keith Grant. University of Texas Press, 2012, pp. 279–297.
Deep Impact. Directed by Mimi Leder, performances by Robert Duvall, Téa Leoni, Elijah Wood. DreamWorks, 1998.
Don't Look Up. Directed by Adam McKay, performances by Jennifer Lawrence, Leonardo DiCaprio, Meryl Streep. Netflix, 2021.
Fear, David. "'Don't Look Up'... or You Might See One Bomb of a Movie Hurtling Right Toward You." *Rolling Stone*, 24 December 2021, www.rollingstone.com/tv-movies/tv-movie-reviews/dont-look-up-review-leonardo-dicaprio-jennifer-lawrence-1268779/.
Greenland. Directed by Ric Roman Waugh, performances by Gerard Butler and Morena Baccarin. STXfilms, 2020.
Kaklamanidou, Betty. *The "Disguised" Political Film in Contemporary Hollywood: A Genre's Construction*. Bloomsbury, 2016.
Keane, Stephen. *Disaster Movies: The Cinema of Catastrophe*. 2nd ed. Wallflower, 2006.
Kellner, Douglas. *Cinema Wars: Hollywood Film and Politics in the Bush-Cheney Era*. Wiley-Blackwell, 2010.
Lu, Donna. "'It Parodies Our Inaction': Don't Look Up, an Allegory of the Climate Crisis, Lauded by Activists." *The Guardian*, 30 December 2021, www.theguardian.com/environment/2021/dec/30/it-parodies-our-inaction-dont-look-up-an-allegory-of-the-climate-crisis-lauded-by-activists.

Murray, Robin L. and Joseph K. Heumann. *Ecology and Popular Film: Cinema on the Edge*. Suny Press, 2009.
Olsen, Mark. "Why Adam McKay's Apocalyptic Comedy 'Don't Look Up' Is a Disaster Movie for a Disastrous Moment." *Los Angeles Times*, 10 November 2021, www.latimes.com/entertainment-arts/movies/story/2021-11-10/dont-look-up-explained-netflix-adam-mckay.
Roddick, Nick. "Only the Stars Survive: Disaster Movies in the Seventies." *Performance and Politics in Popular Drama: Aspects of Popular Entertainment in Theatre, Film, and television 1800–1976*, edited by David Brady, Louis James and Bernard Sharratt. Cambridge University Press, 1980, pp. 243–269.
Ryan, Michael and Douglas Kellner. *Camera Politica: The Politics and Ideology of Contemporary Hollywood Film*. Indiana University Press, 1990.
Selbo, Jule. *Film Genre for the Screenwriter*. Routledge, 2015.
Von Mossner, Alexa Weik. "Love in the Times of Ecocide: Environmental Trauma and Comic Relief in Andrew Stanton's *WALL-E*." *Eco-Trauma Cinema*, edited by Anil Narine. Routledge, 2015, pp. 164–177.
Wolin, Sheldon S. *Politics and Vision: Continuity and Innovation in Western Political Thought*. Princeton University Press, 2016.
Yacowar, Maurice. "The Bug in the Rug: Notes on the Disaster Genre." *Film Genre Reader IV*, edited by Barry Keith Grant. University of Texas Press, 2012, pp. 313–331.
Yossman, K. J. "Adam McKay's 'Don't Look Up' Smashes Netflix Viewing Records with over 150 Million Hours Viewed." *Variety*, 6 January 2022, variety.com/2022/film/news/dont-look-up-netflix-weekly-viewing-records-1235147910/.
Zimdars, Mellisa and Kembrew McLeod, editors. *Fake News: Understanding Media and Misinformation in the Digital Age*. The MIT Press, 2020.

16 The City That (Never) Dies

Film Noir Imagines the Urban Disorder, Disease and Disaster

Dimitris Papacharalampous

Film Noir, a genre and cinematic style that emerged in the 1940s, is distinguished by its pervasive atmosphere of cynicism, moral ambiguity and visual chiaroscuro. Rooted in post-World War II anxieties and disillusionment, it reflects the complexities of urban life and the darker aspects of human nature, often portraying themes of crime, corruption and existential despair. The term "noir," derived from the French word for "black," encapsulates the genre's visual style, which is characterised by low-key lighting, high-contrast imagery and shadow-laden compositions, drawing influence from German Expressionism. Noir movies typically feature protagonists who are morally compromised anti-heroes, grappling with issues of fate, guilt and treachery, or femme fatales who use their allure to manipulate and entrap male characters. While the classic period of Film Noir extends to the late 1950s, its thematic and stylistic legacy persists, thanks to the development of neo-Noir movies from the 1970s to the present day.

In both classic and neo-Noir, catastrophe operates as a central thematic element, though its manifestation evolves in response to the historical and cultural contexts of each period. Classic Noir's emphasis on social decay and moral ambiguity resonates with the anxieties of its post-war audience, while neo-Noir's exploration of environmental and societal collapse reflects modern concerns about sustainability, global instability and the fragility of human civilisation. What remains constant is the sense that catastrophe – whether personal, social or environmental – is an inevitable consequence of human hubris and the flawed nature of human institutions.

The natural environment, while less explicitly addressed in classic Noir, is often metaphorically represented by the urban jungle – an environment that is as dangerous and unpredictable as any natural disaster. The focus on human fallibility and vulnerability serves as a parallel to the precariousness of life in a world where order seems extremely fragile. This sense of impending disaster, however, is primarily focused on the moral and social dimensions rather than the literal environment. Neo-Noir builds upon these themes but also reflects more contemporary anxieties, especially those related to environmental and systemic crises. It expands the Noir scope to include

DOI: 10.4324/9781003602750-22

larger-scale catastrophes, such as environmental degradation, the rupture in governance and the consequences of unchecked capitalism.

More specifically, the imagination of city disorder, disease and disaster becomes a canvas upon which filmmakers paint vivid and haunting portraits of urban decay. Through evocative visual imagery and narrative storytelling, directors harness the collective anxieties of their audiences to create dystopian landscapes that mirror the tumultuous realities of the modern world. The city, once a symbol of progress and promise, transforms into a nightmarish labyrinth plagued by moral ambiguity, corruption and existential despair. Crumbling infrastructure, dilapidated buildings and shadowy alleyways serve as metaphors for the decay of social institutions and the erosion of moral values, while the spectre of disease and epidemic underscores the fragility of human existence in the face of unseen threats. In this cinematic landscape, characters navigate a world teetering on the brink of collapse, grappling with their own inner demons as they confront the harsh realities of urban life. Through the imagination of city disorder, disease and disaster, Film Noir invites audiences to confront uncomfortable truths about the human condition and to ponder the precariousness of civilisation in an uncertain and ever-changing world.

Chinatown (Roman Polanski, 1974) is the quintessential urban-environmental (Neo) Noir. Combining mythic elements with a portrayal of urban corruption, moral ambiguity and ecological disruption serves as a paradigmatic lens for analysing the symbolic role of the city and nature within the Noir universe. The mythic undertones underscore the timeless struggle between humanity and the forces of nature, resonating with similar themes found in many (Neo)Noir films where cities serve as battlegrounds for existential crises and ecological degradation.

Los Angeles in Chinatown seems to be the node where many cultural, generational lines of representations of the city, disease and disaster meet and culminate. First, the contradictory myth and anti-myth of the L.A. which initially emerged as the New Eden, the last great city born of the idea of the West, and eventually transformed into a symbolic nightmare of the end of American Dream. Richard Lehan notes that probably no city has a more negative image (Lehan 29–30) and according to Mike Davis this was due to the overwhelming frustration of the expectations of immigrants, the working class and especially the intellectuals who sought refuge and inspiration at the West Coast (Davis 20). The disillusionment and despair generated among the lower and middle classes by the economic crisis of the 1930s merged with the existential and artistic dead ends of American or European refugees, writers, directors, visual artists, musicians, who were chased by fascism, and who were largely attracted to the city by the illusory glamour of the gigantic artistic and economic edifice of Hollywood. However, artists, like workers, experience the alienation, the dissipation of their talent and the crushing of their dreams by powerful economic mechanisms. In this way, the anti-myth of Los Angeles is gradually shaped mainly by hard-boiled detective novel writers and Film Noir filmmakers, who discover the most suitable city, on the one hand, to accommodate corruption, crime, the dangerous and morally precarious action of the heroes, and on the other hand, to receive their social and political criticism.

We could, however, further observe that *Chinatown* is also shrouded in the dark shadow of the wider, symbolic spectrum of the *sick* or *sin city*. The sick city, due to an inappropriate social and political reality, is a dominant metaphor in art and literature from the Iliad to the present day. In classic Noir period the *sick city* as both metaphor and literal condition is depicted in films like *Panic in the Streets* (Elia Kazan, 1950) and *The Killer that Stalked New York* (Earl McEvoy, 1950) where the spread of diseases – pneumonic plague and smallpox – leads to panic, violence and breakdown of social order. This kind of metaphor is related to the tendency to attribute the terms of birth and evolution of life or the terms of the anatomy of the human body to the natural, terrestrial or celestial environment. Thus, the city as a geographical and political term is presented as living, moving and suffering like a human organism, while the unjust acts of men, the hubris and excess, the unfair government seem to provoke an epidemic condition.

Los Angeles in *Chinatown* is, of course, a dying city because of the depraved governor. The drought corresponds to the tyranny of Cross and the corruption and injustice that plague every joint and function of the urban body. Cross is the ultimate personification of the ruthless and all-powerful criminal who goes unpunished for his sins. But perhaps his greatest crime is the fact that he has subjugated all the institutions of the city to his morbid authority. If incest with his daughter entails the symbolic dishonour of all citizens, his nefarious imposition on every aspect of life causes the moral decay of the entire society.

The association of city with incest and disease and its presentation as a sick organism and a suffering body perhaps alludes to an organismic conception of the city and culture. The organismic view identifies the evolution of political, social and cultural systems with the predetermined stages of evolution of living organisms: birth, growth, maturation, decay and death. Yet, the body of Los Angeles, torn by life and death at the same time, seems to contradict this simplistic, linear conception. The pathology of the city is not the stage that follows maturation but the basis on which it was founded. Disease may precede birth, and death is not necessarily the end. The fatality of urban illness and decadence is not identified with death but with the perpetuation of the confusion between health and disease.

Moreover, following biblical tradition, the city is presented from its origin as a source of death and corruption, since the first city is founded by Cain, while the image of the evil city plagued by moral depravity and sexual perversion has been attributed to many other towns, such as Babylon or Sodom and Gomorrah. Besides, the film's connection to the tragedy of Oedipus King (Belton; Linderman) allows us to treat the City of Angels as analogous to Ancient Thebes, which was struck, too, by plague, infertility, death, and also in the way Froma Zeitlin suggests that the tragic city serves as the opposite of the idealised urban environment, as it struggles to create a stable system, swinging between strict inclusion and exclusion policies and chaotic confusion of differences (Zeitlin 134).

Furthermore, according to Lee Horsley, Noir is one of the most powerful popularised expressions of the modernist pessimism of Eliot's *Waste Land* (Horsley 1). John Cawelti, more specifically, states that Polanski's version of 1930s Los Angeles reveals the transcendent mythic world of the sterile kingdom, the dying king, the Wasteland and the cyclical myths of traditional cultures (Cawelti 189). On the one hand, these themes emerge from the reservoir of primitive fears and beliefs that have been mixed with numerous cultures, and, on the other hand, they reappear persistently within a new social context and cultural correlations, while maintaining an undiminished intensity and similar function.

Of course, the city's morbid image has, also, deep roots in the soil of hard-boiled detective fiction and film. *Chinatown* alludes to Dashiell Hammett's *Poisonville*, Raymond Chandler's *Bay City* or *The Naked City*, *The City that Never Sleeps*, and *The Captive City* of Film Noir. A city that is a realm of death; cradle of crime and vice; a dark maze of concrete, metal and glass, of deserted or crowded *mean streets*, dangerous alleys or chaotic highways, tall menacing buildings; a place full of shadows or blindingly hallucinatory lights, horns and gunshots that are eerie projections of the troubled souls of the heroes. The city of Noir shares with its heroes the fatal perspective. Since they are heading towards an inevitable tragic end, the fate of the city is also ruled by the same death dynamic. Adding to Lewis Mumford's statement that in contemporary urban civilisation, there are internal forces that have the potential to destroy all evidence of its existence (Mumford 636), Nicholas Christopher argues that such a belief is inherent in the dynamics of the Noir city, which resembles the protagonist who carries within himself the seed of his death and destruction (Christopher 86).

Given the literary and mythic depth of Los Angeles's representation, we will examine how the disruption of the natural environment in *Chinatown* takes on symbolic dimensions reflecting a marginal psychological crisis, moral erosion, social and political disintegration.

At the scene of the City Council, one of the speakers exposes the seriousness of the circumstances, emphasising the precarious balance between abundance and scarcity in Los Angeles. He states that while they could easily reach the Pacific Ocean, its waters were rendered undrinkable due to the city's status as a desert community. Despite the proximity to the ocean, water scarcity loomed large, threatening not only individual consumption but also the sustenance of essential agricultural practices. The speaker's words paint a vivid picture of the city's reliance on water resources, the menacing presence of desert under the streets and around the city, the urgent need for sustainable management to avert potential disaster.

The image of a city facing the ultimate threat appears also in the alarmist newspaper headlines – that pleads officers to save the city, since Los Angeles is dying from thirst – and it is accompanied by the decimation of the animals and plants. The farmer in the orange grove attacks detective Gittes full of rage at the destruction of the harvest, the stiff trees and the poisoned wells, while the indignant cattle farmer, who invades the City Council chamber with his herd, passionately accuses engineer Mulwray, whom he holds responsible for the calamity, that he stole the water from the valley, ruined the grazing and starved his livestock.

In *Chinatown*, the ocean appears to surround the city along with the desert, reminding us of its constant dependence on and opposition to nature. These are natural entities that, while seemingly temporarily under control, on the other hand, they remain untamed, constantly alert and ready to launch an attack and take over the city. The constant presence of water is charged with a sense of an irresistible power, both life-giving and lethal. The person who controls and directs the dual power of water is Cross, who is symbolically given the name of Noah, the biblical master of the waters.

The threat to Los Angeles by the Pacific Ocean and the desert presents a precarious delineation of the city in relation to its eastern and western borders. However, evil seems to emanate mainly from the East, as Chinatown was seen as the main embodiment of Eastern fear. In the case of Los Angeles, the negative symbolism of *Chinatown* is consistent with historical reality, the racist treatment of the district and its residents by police and corrupt politicians. The Oriental Problem and the persecution of Asian immigrants intensified at the turn of the century, when yellow fever and plague afflicted all of California, and Chinatown was portrayed as the source of the epidemic. It should also be mentioned that Los Angeles' geographic location on the extreme western border means that the eastern threat emanates from the West. It is a place where the cultural, moral and spiritual compass goes mad, paralysed. East and West are confused, as are outside and inside, the tame and the wild, order and chaos, nature and civilisation.

Natural disaster seems to be the inevitable conclusion of the confusion, taking on the character of punishment, but at the same time of redemption. In the case of Los Angeles, literature and cinema take this tendency to the limits of excess. Davis counts 138 books and films that have as their central theme the destruction of the City of Angels, and he adds that what is remarkable is not so much the frequency of fictional destruction, as the pleasure that such apocalyptic stories evoke in readers and audiences. He presents the city as being defined by the dialectic of ordinary disaster. He points out that, since Los Angeles is built in an area plagued by recurrent natural hazards (earthquakes, floods, droughts, fires, plagues) and attacks by wild animals or supernatural monsters, city's life is identified with managing that everyday fear and defending itself against a permanent danger (Davis 274–277).

In *Chinatown*, the relationship between the city and the countryside is crucial to the plot development. The city is presented in constant contrast to the countryside and the calamity that befalls it is directly related to what happens beyond the urban periphery. The film insists on demonstrating L.A.'s connection to the diversity of California's landscape but also shows the city's stark juxtaposition with its natural environment. Nature's invasion is a response to man's fierce assault on it. Urban expansion – as Cross implements it – goes hand in hand with the encroachment and destruction of valleys, plains, rivers and lakes, and the city is presented as a ravenous monster that devours everything around it. The existence of Los Angeles attests to the violent superiority of man over the untamed elements of nature and the phenomenal predominance of civilisation over the Wild West. The film shows that not only is the echo of this conflict with the ruthless environment still evident but that the conflict itself is still in progress.

Detective fiction and film narratives are akin to Western fiction and in a way revise and re-evaluate the epic of the West. The action may no longer take place in the vast expanses, but in the shrunken and entrenched urban realm, yet the struggle to tame savagery and impose order continues in almost the same terms. For, the city ultimately failed to affirm itself as a realm of justice and it was transformed into an *asphalt jungle*. Thus, West's lone horseman is called upon to pass the baton of defending legality and battling evil to the – equally lonely – private detective. However, the two heroes seem to be trapped in their contradictory roles, so that they alternately identify sometimes with the hunter and sometimes with the prey. Although the deeds of West's hero solemnly affirm individual independence and celebrate the open horizon and the vast, untamed landscape, his heroic actions contribute to the eventual enslavement of the land. The detective comes to continue this heavy mission, and the detective story shows us the consequences of moving from the desire to control the earth to the desire to control people.

Returning to the film, we can, moreover, point out that Cross who rapes his daughter and the land to satisfy his sexual lust and greed for power and money might have had much to learn from the general and oilman Sternwood of Chandler's, *The Big Sleep*. His drilling rigs brutally disrupted the land, but in the end, he comes to understand that oil and water, wind and air are unruly elements that define us without our being able to manipulate them, and that death, the big sleep, is the force by which nature imposes its final dominion over proud man and his false conviction that he can control the world (Chandler 230).

Of course, it should, also, be noted that Noir from its birth is haunted by the image of the bombings of World War II and the atomic danger. Paul Arthur mentions that the secret names of the Noir city, its historic doppelgangers, are Dresden, Tokyo, London and Hiroshima (Arthur 170). The image of the ruined city functions as the source of fear, the tragedy to be averted, but also as the promise of a redemptive end. The roots of Noir lie at the end of World War II and the dawn of the atomic age, between an end that is avoided and an end that may come. Many Film Noir movies of the first period seem to be dominated by nuclear anxiety. Uranium and iridium serve as the main narrative keys in *Notorious* (Alfred Hitchcock, 1946) and *D.O.A.* (Rudolph Mate, 1949). The psychotic lawyer in *The Lady from Shanghai* (Orson Welles, 1947) seeks to run away from the city – any city – before atomic bombs start dropping. *Atomic City* (Jerry Hopper, 1952), *Split Second* (Dick Powell, 1953) and *City of Fear* (Irving Lerner, 1959) present the dangers of the American atomic program, while detective Hammer in *Kiss Me Deadly* (Robert Aldrich, 1955) fails to prevent a nuclear explosion. We should, also, note that the atomic bomb, with its potential for mass annihilation, looms as a metaphorical backdrop, casting a pall of doom over the characters' lives. The threat of nuclear war intensifies the characters' sense of alienation and hopelessness, as they grapple with the realisation of their own mortality in a world teetering on the brink of oblivion.

Moving to modern era and the neo-Noir cinema, the representation of city/nature disorder or destruction takes on new dimensions, reflecting contemporary concerns about environmental degradation, technological advancements and societal collapse, intertwined with themes such as surveillance, control, or the erosion of personal freedom. During the opening shots of *The Cruising* (W. Friedkin, 1980), the policemen on patrol predict that the city would one day blow up, and their words are spoken in a tone that is more wishful than threatening. The wish has been fulfilled in films like *Blade Runner* (Ridley Scott, 1982) and *The Element of Crime* (Lars V. Triers, 1984). The end has come, but it has not brought redemption. Everything is different, yet nothing has changed. The city, though deeply wounded, still lives the only way it knows how to live: by guarding and expelling, by hunting down killers and committing crimes itself, by being scared and terrorising.

If in *Chinatown* we witness the beginnings of L.A.'s damned mythology, the shaky foundations of the California miracle, in *Blade Runner* we are presented with its deplorable conclusion. The film is presumably set in a post-nuclear age, in a city repulsive, steeped in acid rain and polluted fog. A dark, terrifying maze, where explosions hurt the sickly sky and huge skyscrapers overwhelm the people, this abject huddle that crowds the filthy streets. In this ecological hell there is no longer natural environment; the animals have disappeared, and the only images of nature can be found in Deckard's dream, where the mythical unicorn runs through a fairytale forest. The extinction of flora and fauna casts doubts on the reproductive capacity of humans, feeding the threat of annihilation of the human species, too. In *The Element of Crime* a similar picture of devastation prevails, this time based on the Old Continent. Europe has become a swamp of rubbish and ruins. Poisonous water has flooded the streets, the houses, the souls and memories. Nature has been irreparably disrupted. As one character states, *there is water everywhere but not a drop to drink; there are no seasons, summers are not summers*. Like in *Blade Runner*, everything is shrouded in eternal darkness, eternal night.

We saw that in *Chinatown*, natural disaster was associated with incestuous perversion along with lust for money and power. The destruction of natural habitats, the depletion of resources and the encroachment of industrialisation are themes explored also in other films of the genre such as – the *Chinatown's* sequel – *Two Jakes* (Jack Nicholson, 1990), *L.A. Confidential* (Curtis Hanson, 1997), where the ruthless exploitation of the land by oil tycoons serves as a metaphor for the destructive forces of capitalism and greed. In *Blade Runner*, it is implied that the same greed has led to war and world destruction, arrogant violation of the environment, exploitation of natural resources and uncontrolled technological development. Tyrell, the businessman and ruler of the globalised capitalist totalitarianism imposed on Earth – who emphasises that his primary goal is the profit – can be seen as a descendant of Noah Cross. In *The Element of Crime* there is no justification for the disaster, which seems inexplicable or rather as the normal, inevitable outcome of human history and nature.

In *Blade Runner*, and specifically in its second version, all evidence of Deckard and Rachel's escape of the megacity to the supposedly untainted North has been removed. That seems consistent with the metaphor of the *prison city* that runs through the Film Noir corpus. The motif of the prison city is deeply intertwined with the theme of fate. The city serves as a metaphorical prison, where characters find themselves trapped by circumstances beyond their control, ultimately leading to their inevitable downfall. Noir heroes are imprisoned in a predetermined existence, where their choices seem futile in the face of a relentless and indifferent universe. The urban landscape becomes a symbol of this existential entrapment, with characters struggling to find meaning and purpose in a world governed by chaos and uncertainty. Whether they are trapped by their own desires and vices or by the machinations of fate itself, the characters in Film Noir often find themselves caught in a cycle of existential despair, grappling with questions of free will and determinism.

The tragic refutation of the heroes' longing to free themselves from urban bondage is a common element of many films. In *Detour* (Edgar J. Ulmer, 1945) the protagonist hitchhikes his way across the country, only to find himself stranded in a bleak and desolate cityscape. As he becomes embroiled in a series of unfortunate events, the city becomes an inescapable web of death. In *Out of the Past* (Jacques Tourneur, 1947) Jeff Bailey attempts to break free of his troubled past by retreating to a small town and adopting a new identity. But the past/fate, the city and the crime seem to form a triangular cell from which there is no escape, with tragic results. In *Asphalt Jungle* (John Huston, 1950) and *The Killing* (Stanley Cubrick, 1956) the characters' attempt to run away from the city is embodied by their ambitious, but, ultimately, disastrously failed heist plan.

The Noir theme of the city as a prison is culminated during the neo-Noir era in *Escape from New York* (John Carpenter, 1981), by Manhattan's transformation into a literal, lawless penitentiary, where society's outcasts are confined. Similarly, but in a more metaphorical sense, in *Se7en* (David Fincher, 1995), there is no way out of the city. Detective Somerset's statement that we should all go and live in a hut, which reflects his desire to live in the quiet countryside, is never fulfilled. The final scene of *Seven*, though set on the sunny plain, offers no redemption away from the polluted rain and sinful city streets. Instead, beneath the ominous figures of the power poles, symbols of expanding urban domination and transmitters of negative energy, serial killer's criminal plan comes to a horrific conclusion. The urban setting of the city, depicted as a grim, oppressive environment characterised by crime, corruption and decay, symbolises the darker aspects of human soul and the consequences of uncontrolled urbanisation. The absence of redemption outside the city underscores the film's overarching theme of moral ambiguity and the pervasive reach of evil, suggesting that no place is immune to its influence and that even the most remote areas are tainted by the horrors of humanity.

That's because the whole world is now urbanised. The city has invaded the countryside; as we saw in *Chinatown* it has swallowed rivers and lakes, has encroached on plains, and has flattened mountains. The countryside has been corrupted by the expansive urbanisation; the small towns have been infected by the sick city. In films such as *Fargo* (Joel & Ethan Cohen, 1996) *Lone Star* (John Sayles, 1996), *Affliction* (Paul Schrader, 1997), *Insomnia* (Christopher Nolan, 2002), *Winter's Bone* (Debra Granik, 2010) local communities are plagued by alcoholism, drugs, widespread aggression, juvenile delinquency and organised crime. As Frank Krutnik notes, although Film Noir treats the modern city as a threat to American society, it refuses to accept the small town and the country as a redemptive alternative (Krutnik 88). In the films mentioned earlier, the danger of death, the social disintegration and the moral depravity seems to emanate similarly from the harsh, hostile natural environment. The physical landscape functions as a counterpoint or expression of the psychological landscape: fog-lack of insight, midnight sun versus dark soul, snow versus intense passions. In *Insomnia* the rugged terrain of rural Alaska serves as a metaphor for the protagonist's inner turmoil and moral conflict. In *Affliction*, the snow-covered landscapes of New Hampshire mirror the protagonist's descent into madness and violence. In *Fargo*, the barren plains of North Dakota become a backdrop for a series of macabre and absurd events, highlighting the stark contrast between the beauty of the natural world and the brutality of human nature. In *Winter's Bone*, the dialogue between the wild savagery of Ozark Mountains with the human and social savagery within the rural community serves as a reminder of the harsh realities and primal instincts that govern both the natural world and human society. Finally, in *Lone Star*, the vast expanse of the Texas desert becomes a symbol of the characters' search for identity and redemption in a land haunted by the ghosts of the past. So, the heavy shroud of snow, the deadly cold, the mountain peaks that overwhelm vision and hope, the blinding and tormenting light or the immense desert define the geography of the soul and shape a climate of emotional turmoil, anxiety, threat and death.

Nature, instead of soothing and balancing, seems to stimulate wild passions, the propensity for crime and fatal rivalries. Even in the *provincial* early Film Noir *So Dark the Night* (Joseph Lewis, 1946) detective Cassin feels his murderous instincts awakened in the seemingly idyllic landscape of the French countryside. In the picturesque villages, dirt roads, bridges and fields lurk the same dark forces that have terrorised the alleys of Paris; the peaceful villagers prove ready to be torn apart by the murderous passions – common to all – for love, power and profit. *On Dangerous Ground* (Nicholas Ray, 1951) is set against a stark, wintry backdrop, using nature to enhance its Noir narrative. The film's setting – snow-covered mountains and treacherous roads – functions almost as a character itself, creating a harsh, unforgiving environment that reflects the protagonist's emotional turmoil. In an analogue way, *Key Largo* (John Huston, 1948), set in a Florida hotel where a group of people becomes trapped by both a hurricane and a gangster, suggests that, like urban corruption and personal moral failings, nature itself can be an unpredictable and destructive force.

In the iconic rural Noir *The Night of the Hunter* (Charles Lawton, 1955), nature is portrayed as both gorgeous and threatening, reflecting the duality of the protagonist's character and the themes of innocence versus corruption. The landscapes of West Virginia, with its slow-moving rivers, dense forests and rolling hills, initially evoke a sense of safe tranquillity. However, as the narrative unfolds, the environment becomes increasingly menacing, expressing its predatory nature. Finally, the nostalgic image of the serene, healthy countryside, as opposed to the image of the degenerate city, seems like a deceptive urban fantasy, designed to exorcise nightmares that we think are modern, but they are in fact age-old. City and countryside seem eternally attuned to the same violent rhythm, carrying on a constant dialogue of destruction.

Resorting to René Girard, we may finally accept that the problem is always the same: the violence, that neither individuals nor societies can avoid. In the face of the constitutive violence of the world, the generative violence of life, prohibitions seem to be the only way of resistance; however, prohibitions also end up contributing to the violence they are called upon to suppress, because – according to Girard – they are ultimately the same thing, since violence is both the disease and the cure (Girard 232) The city, as an assemblage of prohibitions and laws, is a form of violence. Its conflict with nature, then, is not something that leads to its separation from savagery, but it is an act of definitive fusion with it and resonance with the primordial, violent rhythm that governs everything. For although the city seems to symbolise the rival awe of nature, every city exists by nature and is directed by the instinct of life which is identified with the instinct of violence. Violence, as Girard points out, often appears in the form of contagious diseases or droughts and floods. Consequently, the extreme natural phenomena and the disease signify the overt and catalytic presence of violence and the world's alignment with it.

The reign of inevitable and universal violence inserts the city, countryside and nature into the cycle of perpetual Heraclitean warfare, but without any trace of balance and invisible harmony. Perhaps human existence is incompatible with harmony, or at least that is what the disharmonious world of Film Noir seems to suggest. Since the Garden of Eden, the history of humanity has been in tandem with the radical disruption of the natural and artificial environment that begets us and we beget it, owns us and we own it, threatens us and we threaten it, fights us and we fight it. We appear doomed to cause or participate in the disruption of nature and of society while enduring its destructive consequences. In a world like this, the dialectic of everyday disaster coincides with the dialectic of life, while the ecology of fear is elevated to an ontology of terror.

Works Cited

Arthur, Paul. "Murder' s Tongue: Identity, Death and the City in Film Noir." *Violence and American Cinema*, edited by D. Slocum. Routledge, 2000, pp. 153–174.

Belton, John. "Language, Oedipus and Chinatown." *MLN*, vol. 106, no. 9, 1991, pp. 933–950.

Cawelti, John G. "Chinatown and Generic Transformation in Recent American Films." *Film Genre Reader*, edited by B. K. Grant. University of Texas Press, 1986, pp. 183–191.

Chandler, Raymond. *The Big Sleep*. Vintage Books, 1992.

Christopher, Nicholas. *Somewhere in the Night. Film Noir and the American City*. Henry Holt Company, 1997.

Davis, Mike. *City of Quartz*. Pimlico, 1998.

———. *Ecology of Fear*. Picador, 1999.

Girard, Rene. *The Violence and the Sacred*. Translated by P. Gregory. Continuum, 2005.

Horsley, Lee. *The Noir Thriller*. Palgrave Macmillan, 2001.

Krutnik, Frank. "Something More Than Night: Tales of the Noir City." *The Cinematic City*, edited by D. Clarke. Routledge, 1997, pp. 83–109.

Lehan, Richard. "Los Angeles Novel and the Idea of West." *Los Angeles in Fiction*, edited by D. Fine. University of New Mexico Press, 1995, pp. 29–42.

Linderman, Deborah. "Oedipus in Chinatown." *Enclitic*, vol. 5, no. 6, 1981–1982, pp. 190–203.

Mumford, Lewis. *The City in the History*. Brace & World, 1961.

Zeitlin, Froma. "Thebes: Theatre of Self and Society in Athenian Drama." *Nothing to Do with Dionysos? Athenian Drama in Its Social Context*, edited by J. Winkler and F. Zeitlin. Princeton University Press, 1990, pp. 130–167.

17 "Toward an Otherwise"

Decolonising Epistemology and Ecology in *The Last Wave*

Will Underland and Matthew Spencer

There is a tendency for critics to depoliticise Peter Weir's psychological drama, *The Last Wave* (1977), a tendency perhaps borne of the ease with which the adjective "psychological" comes to mind to describe the film. Because epistemology is so clearly a theme of the film, it has been difficult for critics to not see it as strictly about how its protagonist David (Richard Chamberlain) perceives the world. Steven F. Walker, for instance, takes a Jungian approach and reads the film in a strictly psychological sense, concluding of David that "it is he who is transformed, not the world" (5). For Walker, as is typical for Jungians, the psyche has autonomy, a nature and essence independent of the external world. Thus, Walker's reading posits a psyche impervious to the external environment, which of course includes political reality. Whether this philosophical position is true or not, it is not the reality which Weir depicts in *The Last Wave*.

Even a less dogmatically psychological reading of the film by Luke Buckmaster in *The Guardian* shies away from the fullness of the film's political content. While Buckmaster does mention that "Current concerns about climate change give the film's prophetic ideas more potency," he also suggests that "*The Last Wave* pits human beings against awesome natural phenomena and makes a point about our helplessness in the scheme of things." Weir, though, goes to great lengths to demonstrate that the catastrophe of David's premonitions is man-made, is in fact a return of the repressed – a repression inherent to the colonial mindset which pervades and defines the modern city of Sydney. Only once we understand the inseparability of psyche and environment, their co-constitutive relationship, can we discern the full force of the film, the complexity of which far exceeds Walker's limiting individualised interpretation and Buckmaster's lukewarm comments about the climate crisis.

Part of understanding the relation between inner and outer worlds means taking place seriously, and the setting of *The Last Wave* is rife with historical connotations. As Francesca Merlan puts it in the introduction to her book *Dynamics of Difference: Indigenous Past and Present in a Settler Country*, "The colonization of Australia and its aftermath shaped broadly similar, continent-wide structural patterns of domination, expropriation, and

DOI: 10.4324/9781003602750-23

subsequent state interventions" (1–2). A film set in Australia, therefore, and which dramatises a clash between indigenous and nonindigenous cultures, is an inherently political one and a far cry from Walker's argument that the film is about an individual's transformation. Yes, the nonindigenous and indigenous characters are both embodiments of ways of perceiving, representations of vastly differing epistemologies, *but embodiments that are innate to the political realities of the respective characters.* That is to say, there is a reason why these characters *work* as psychological metaphors.

Merlan alludes to the long history of the belief in "the incommensurability between indigenous and nonindigenous ontologies; ways of being that are irreconcilably different, which nonindigenous institutions and agents seek to police and make commensurate with their own terms" (2). At the beginning of the film, the protagonist, David Burton, is firmly situated in such a "nonindigenous institution"; he is a corporate tax lawyer in Sydney. But when he takes as clients five indigenous men accused of murdering another indigenous man, David's worldview slowly unravels. David comes to see that, as Boaventura De Sousa Santos argues in *The End of the Cognitive Empire: The Coming of Age of the Epistemologies of the South*,

> The dominant criteria of valid knowledge in Western modernity, by failing to acknowledge as valid kinds of knowledge other than those produced by modern science, brought about a massive epistemicide, that is to say, the destruction of an immense variety of ways of knowing. (8)

Through this unravelling Weir highlights the pervasive colonial mindset of modern-day (in 1977) Sydney while showing that the major institutions of Western life – the legal system and Christianity, especially – are not equipped for ecological catastrophe. In fact, Weir shows that the colonial mindset as expressed in these institutions is the root cause of imminent ecological collapse. *The Last Wave*, that is, imagines apocalyptic eco-catastrophe as neither god's wrath nor a problem solvable by science, technology or legislation in courts, steeped as they are in the colonial ideology causing the ecological crisis. Rather, the film imagines apocalypse as the inevitable result of the white world's unwillingness to shed its supposed supremacy and its ideological instantiations in institutions. The denial of alternative worlds and ways, the film says, the apocalypse the white world inflicted on indigenous Aboriginal people, ultimately results in a cataclysmic return of the repressed. That denial of the other turns out to be a disavowal of the self and life itself.

To clue us into the world from which David will seek to escape, early on, Weir establishes the city of Sydney as repressed and repressive. The city streets are congested with cars and crowded with busy people. As David sits in a traffic jam he sees a logo on a public bus, an image of a leopard. Moments later he sees an advertisement for the Taronga Zoo with an image of an orangutan. His line of vision in the advertisement is briefly interrupted by a pedestrian carrying a potted plant. In an impressively short amount of screen time, Weir shows us several images of domesticated nature and thereby establishes the city as repressive of wilderness. In Sydney, the only nature available is the vestige of wilderness – images of nature that have been totally appropriated and now exist only as commodities.

Adding to this is the emphasis on the illusion of control of water in the city. There are shots of a child drinking from a hose, of a man drinking from a water fountain, of sprinklers on an immaculate lawn, and of an aquarium inside David's home. These images demonstrate the human domination of nature and the hubris of such domination. Moreover, several scenes take place at a toxic waste site, implying that the modernity of the city is afforded by environmental degradation. Catherine E. Walsh and Walter D. Mignolo go so far as to coin the "compound expression: modernity/coloniality," given that "Coloniality is constitutive, not derivative of modernity" (4). Though, crucially, in *The Last Wave*, modernity/coloniality's domesticated nature hides in plain sight and its environmental degradation is well out of view of the people of Sydney.

Such repression is also the *modus operandi* for many people in Sydney in terms of their relationship with and treatment of Aboriginal people. When David's wife, Annie (Olivia Hamnett), learns that David has invited one of the aboriginal men accused of murder, Chris (Gulpilil), to their house for dinner, she says, "I'm a fourth generation Australian and I've never met an Aboriginal before." Accordingly, she treats them with, at first scepticism and then outright fear. David's fellow lawyer, Michael Zeadler (Peter Carroll), is also representative of the mainstream dynamic between the indigenous and nonindigenous in Sydney. He says that, despite the Romantic mythology surrounding Aboriginals, and despite their history as victims of colonial violence, Aboriginals today are no different than "depressed whites." This view is an expression of what Merlan, quoting Stanner, calls "a cult of forgetfulness practiced on a national scale" (9). While Zeadler's perspective goes on to be discredited, early in the film as he expresses these views, we are meant to understand them as practical-minded, as indicative of the average city-dweller at the time. The irony, of course, is that the residual impacts of colonialism – coloniality – inform the very belief that colonialism is a thing of the past in its claim over the historical narrative of colonised people.

In terms of the film's narrative, the images of repressed nature and history are used to establish the worldview from which David will try to escape. This escape is dramatised largely through spatial metaphors, metaphors for the decolonising of David's mind. To emphasise the metaphors as about, as Weir puts it, "systems of perception" (Weir), windows play a major role in the film. In the opening scene, falling hail breaks the window of an elementary school classroom. Characters (including the first scene with David) are constantly looking out windows and seeing things which frighten them. When David first meets with the Aboriginal men in a legal aid office, on the window behind him someone has written "Do not open." As Gaston Bachelard puts it in his classic *The Poetics of Space*, "The window is the house's eye" (34). Thus, Weir emphasises windows to emphasise perception and the scary "otherness," which exists just beyond the limits of his characters' worldviews. David's suburban home itself, in accordance with Bachelard's corollary argument that "[t]he house, even more than the landscape, is a 'psychic state'" (72), comes to stand for David's mind. Later, when Annie sees Charlie (Nandjiwarra Amagula) outside the house, she becomes terrified and whispers to herself several times, "Go away." She turns from the window and locks the door. "A lock," says Bachelard, "is a psychological threshold" (81).

As David lets the "otherness" in, his house, once populated by his wife and children and other markers of normalcy and success for, in Val Plumwood's term, "Rational Economic Man" (32), is increasingly invaded by nature. Early in the film, just as it was raining torrentially during the traffic jam, David's house starts to flood due to a clogged drain in the bathtub; the cause of the clog is his children's toy cars. By the end of the film, as the weather has become downright apocalyptic, David's house is effectively destroyed; trees smash through the windows and the lights go out. The house, like David's mind, is reclaimed by nature. Or, perhaps "excavated" is a better word for the process of David's experiences. The previously mentioned toxic waste site, which leads to the hidden sacred site of the Aboriginal tribe, displays a sign that reads, "An Excavation Area." That which was buried in David is indeed brought to the surface.

For all of the film's subtle use of space to convey meaning, time is an even more prominent theme – one whose relevance is made explicit in David's consciousness of it. As David investigates the murder and his clients remain hesitant to share the true story of how their fellow Aboriginal man died, he visits an anthropologist to learn more about Aboriginal culture. The anthropologist explains,

> Aborigines believe in two forms of time – two parallel streams of activity. One is the daily objective activity to which you and I are confined. The other is an infinite spiritual cycle called the Dreamtime, more real than reality itself. Whatever happens in the Dreamtime establishes the values, symbols, and laws of Aboriginal society.

Such a distinction between concepts of time is an expression of what Walsh and Mignolo, 41 years after Weir's film was released, include in their explanation of their term "the pluriversal." As they explain in the introduction to their book *On Decoloniality: Concepts, Analytics, Praxis*, "[The pluriversal] opens up coexisting temporalities kept hostage by the Western idea of time and the belief that there is one single temporality: Western-imagined fictional temporality" (3). However, just as the anthropologist in the film says that the Aboriginal perspective allows for "two forms of time," Walsh and Mignolo allow for the non-hegemonic co-existence of forms of time. Again, of the pluriversal, they write, "Such perspective does not mean a rejection or negation of Western thought; in fact, Western thought is part of the pluriversal" (3). The undoing of David's worldview by way of his encounter with Aboriginal culture is, in Walsh and Mignolo's terms, an "undoing" of "the singularity and linearity of the West" (3).

Far from informing David's worldview alone, Weir depicts "the singularity and linearity of the West" as inscribed into the major institutions of Sydney. Early in the film, we learn that David's stepfather is a Christian minister. Immediately, Weir sets this character's worldview in contrast to the Dreamtime epistemology encroaching on David's mind. When David's stepfather learns that he has not been sleeping well, he reminds David that as a child he (David) had bizarre dreams and would often wake up tired. The contrast between the Western-Christian perspective (*epistle*mology?) and Dreamtime hinted at here is later made explicit. Late in the film, when David has come to believe in and experience Dreamtime, he is confronted by his stepfather for sending his family away from the city. This confrontation leads to David all but screaming, "Why didn't you tell me there were mysteries?" His stepfather replies, "David, my whole life has been about a mystery." "No!" David says. "You stood in that church and explained them away!" Weir suggests that Christianity values dogma over mystery and further emphasises the distinction between such a view and Aboriginal spirituality with the recurring image of a snake – the Christian symbol associated with the knowledge of good and evil. Near the end of the film, as Chris leads David to the sacred site beneath the city, he points to the image of a snake carved into the stone above the dark doorway to the sacred site. Chris says, "Down there, past the snake, you will find what you are looking for." What David is "looking for" is of course his true identity, an identity which exists beyond the rationalistic, egoic Western world. Such an identity, Weir suggests, is only obtainable by going "past the snake," past, that is, an understanding of the Western form of knowledge as the only way of knowing.

Insofar as Christianity is at odds with the Dreamtime spirituality of the Aboriginal men, the legal system is equally at odds with it, for Weir shows the legal system to be ideologically consistent with and inflected by a Christian worldview. The Aboriginal men accused of murder are all too aware of this problem. The "murder," it should be noted, by this point in the film, has been shown to be a result of the dead man's breaking of a tribal law – stealing sacred objects and attempting to sell them, a commodification that should remind us of those early images of domesticated and commodified nature. The tribal (and later, global) punishment for this crime is death. The men on trial, though, did not physically attack him in any way. The tribal leader, Charlie, merely points a particular bone in the man's direction and his heart stops beating. That this "murder" is perpetrated by means beyond Western rationalism presents major problems both for the accused and their lawyer, David. From the beginning, the pragmatic way to approach the case is embodied by Zeadler, who argues for pleading guilty in hopes of getting the men a shorter prison sentence. Zeadler's view is one which operates in total deference to the dominant culture, one which sees the way of knowing implicit to the legal system as inscrutable and immutable. As a result of their differing views, David eventually removes Zeadler from his defence team. Zeadler is irritated but relieved. He says, "I don't want to make a fool of myself. Or of them." Zeadler epitomises Merlan's "nonindigenous agents" seeking "to police and make [indigenous culture] commensurate with their own terms" (2).

David, though, inundated as he is in Aboriginal epistemology, enters a plea of not guilty. He believes that if the accused men tell the truth in court and admit that they are members of a tribe, the judge will dismiss the case. The absurdity that Zeadler sees in this defence is predicated on the belief that no Aboriginal tribe members live in the city, a belief that the audience by now knows is false. When Chris takes the stand, he begins by answering David's questions as he was instructed, telling the truth about how Billy died and admitting that he and the other men are part of a tribe that resides in the city. But Chris starts to feel the truth as a forsaking of his tribe. Despite being in an official court of law, his allegiance is to tribal law. Earlier in the film, Charlie reveals to David the Aboriginal men's orientation to tribal law, saying, "We're nothing but the law. We learn from our forefathers." "But surely," David protests, "men are more important than the laws." "No," says Charlie. "The law is more important than just man."[1] By the time Chris takes the stand it has become clear that the two men were talking about two very different ideas of the law, and Chris ultimately decides to lie in the court in order to uphold tribal law. He eventually tells David and the rest of the court, "No more stones. No more sacred things that our fathers did. You think they still happen? No. You got me mixed up. It was long time ago. We got drunk, had a fight, and that's all."

While Weir teases us with the notion that David will win the case by depicting the judge as open-minded to the tribal defence angle, he has already suggested that Chris's decision to tell the court what they already believe is a foregone conclusion. Before Chris's just quoted speech, as he is being sworn in to take the stand, we hear the familiar, "Do you swear that the evidence you are about to give will be the truth, the whole truth, and nothing but the truth?" Then the same voice says, "Say 'So help me God.'" Chris obeys and says, "So help me God." Through a close-up of Chris's hand on the Bible as he is sworn in, Weir invites reflection on what is happening ideologically, which is that Chris is entering a Western, Christianity-inflected rationalistic epistemology that renders him not so much unwilling but *unable* to tell the truth. When he looks around the courtroom, he sees his all-white jury and judge, and we understand that there is no way for him to remain faithful to both ideas of the law. And while this does conform to the oft-depicted belief in the incommensurability of indigenous and nonindigenous cultures (Merlan 2), Weir makes it clear that that incommensurability is the result of the rigidly rationalist orientation of the nonindigenous institution.

If the two cultures are to be commensurate, Weir seems to say, the non-indigenous world must be the one to change, for theirs is the dominant one and the one creating the conditions for the looming ecological collapse. By the time the trial happens, the threat of a catastrophic event has become ominous. During a recess, mid-trial, David becomes frustrated with Chris for not telling him the truth, and David realises that Chris's reticence to speak has to do with a secret about water (rather arbitrarily, for this realisation comes as David happens to be standing by water leaking down a wall next to him). David says, "The secret is linked to water. Black rain, the weather. Is that what it is? Tell me. It affects us all doesn't it? Doesn't it!. . . My dreams are about water. . . Am I dreaming your secret?" David here refers to the haunting correspondence between the content of his dreams and the worsening weather. One of his dreams features water spilling from his car stereo; in another, the city of Sydney appears to be underwater. "Black rain" – incidentally, the title under which the film was originally released in the United States – is a reference to the earlier event of petroleum raining down from the sky, the event which Buckmaster aptly describes as "disturbingly literal." As if we needed any more evidence that *The Last Wave* is a political film, not simply the psychological journey of an individual, Weir depicts the impending catastrophe as a direct result of the petro-economy and even has David unsubtly announce, "This affects us all." The eco-catastrophe, therefore, is shown to be inseparable from the coloniality that passes for normalcy in Sydney. What could be both more normal and more catastrophic than petroleum?

Such political signifiers lead the Australian film critic, David Stratton, to straightforwardly call *The Last Wave* "a climate change film" ("David Stratton"). Supporting this reading is how clearly David's premonitions fit the description of what E. Ann Kaplan, in her book *Climate Trauma: Foreseeing the Future in Dystopian Film and Fiction*, calls "pretrauma," or "Pretraumatic Stress Syndrome" or "PreTss" (1–2). PreTss, Kaplan explains, relates to "disturbing future-oriented cognitions and imaginations as measured in direct temporal reversal of conceptualizations of past-directed cognitions in the PTSD diagnosis" (2). While Pretrauma can refer to any number of future-oriented fears, Kaplan restricts her study to "fears about the total collapse of natural and social environments" (2), from which David clearly suffers and which environmental films often depict. Weir's film, though, shows the spiritual roots of such problems, the ideological foundations which lead to Pretrauma. The film's ubiquitous theme of epistemology is one that suggests the visceral connection between worldview and the material circumstances of the world, without shortchanging either aspect. As critic Diane Jacobs puts it, *The Last Wave* insists "on the tangible power of spiritual life."

That tangible power is only fully displayed in the film's final scene. As the storm grows stronger and David's house is destroyed (his mind totally excavated), Chris shows up having changed his mind about his secret and leads David to the sacred site beneath the city. He directs David through the abandoned industrial plant, the toxic waste site and then "past the snake." What David finds is a cave with colourful images on the walls. Among the images are a calendar which seems to have reached its end and a large wave. David also finds a mask which fits and mirrors his face perfectly. We realise that David has found his true identity as Mulkurul, a powerful spirit which can take human form. David, it turns out, is just a vessel for Mulkurul. Despite David's narrative arc, his long epistemological and spiritual journey from corporate tax lawyer to Mulkurul, *when confronted with his true identity, he rejects it.* In the cave, Charlie appears suddenly and in full shaman attire and face paint. David screams with fear both at Charlie and at the fate depicted on the wall. He picks up a sacred stone and kills Charlie with it. David flees the cave and only after several wrong turns in the Sydney sewers discovers an exit which leads him onto the beach. There he drops to his knees and looks out at the ocean where an apocalyptic wave gathers force. David closes his eyes and awaits his fate. Walker's argument that the film's conclusion depicts an "intrapsychic fantasy" (2) does not take into account that *David's journey is a failed one, that he rejects his identity as Mulkurul.* If, as Walker says of the scene on the beach, David's perception has been sharpened, it is not to suggest his psychical revolution; rather, David's clearer perception only helps him to see the eminent catastrophe on the horizon, which, again, "affects us all." Furthermore, contrary to Walker's claim that the wave is "beautiful, but deliberately unrealistic" (5), the image of the wave is in fact grittily realistic. The images are grainy (they look as if they could have been taken by a particularly daring surfer), and their documentary-like quality calls attention to Weir's literalism – his ambitious and audacious attempt to show ecological catastrophe as a result of colonial epistemology in the latter's repression of nature and indigenous ways of knowing.

According to Indian novelist and critic Amitav Ghosh, such alternative ways of knowing may be the last hope in mobilising effective action against the climate crisis. In his treatise on climate catastrophe and its fictional representation (or lack thereof), *The Great Derangement: Climate Change and the Unthinkable*, Ghosh argues that "it is increasingly clear to me that the formal political structures of our time are incapable of confronting this crisis on their own" (214). Ghosh goes on to claim that "climate change represents, in its very nature, an unresolvable problem for modern nations in terms of their biopolitical mission and the practices of governance that are associated with it" (214). Just as Weir depicts the institutions of Sydney as constitutionally incapable of understanding Aboriginal tribal spirituality and thus its ecological warnings, Ghosh argues that the wider framework of modern nation states in which such institutions operate is not equipped to address a problem as foundational as climate change. This is why, Ghosh says, he sees as a sign of hope, in fact "the most promising development," the willingness of some Hindu, Muslim and Buddhist groups to voice their climate concerns (213). The reason why this is "the most promising development" is that it addresses the ideological roots of the climate crisis not just its material consequences. "Religious worldviews," Ghosh explains,

> are not subject to the limitations that have made climate change such a challenge for our existing institutions of governance: they transcend nation states, and they all acknowledge intergenerational, long-term responsibilities; they do not partake of economistic ways of thinking and are therefore capable of imagining nonlinear change – catastrophe, in other words – in ways that are perhaps closed to the forms of reason deployed by contemporary nation states.
>
> (215)

While "religious worldviews" is a broad category, and one may balk at the claim that "all" religious worldviews do not think economistically – Christianity is noticeably absent from Ghosh's list of religions voicing their climate concerns – the essence of Ghosh's point is valid. Only a worldview that can transcend rationalism (not necessarily one that is *irr*ational), both Ghosh's passage and Weir's film suggest, can address the climate crisis, which is as much a material problem as it is a spiritual or epistemological one. This is why Weir's narrative, if we zoom out and look at the general arc, is one of the Western subjects – from the profane modern/colonial city of Sydney to the sacred cave of the Aboriginal tribe. This is also why Ghosh boldly claims that "it is impossible to see any way out of this crisis without an acceptance of limits and limitations, and this in turn is, I think, intimately related to the idea of the sacred, however one may wish to conceive of it" (215). Insofar as *The Last Wave* is consistent with Ghosh's ideas, it is equally bold. Insofar as we in the real world reject these alternative ways of knowing and acting, we are, like David in his rejection of Mulkurul, subject to apocalypse.

The Last Wave is unique in its intersectionality, its prescient ability to see and dramatise colonialism as the residual and unconscious worldview informing present-day issues of racism and ecology, that the powers which oppress and suppress formerly colonised people are the very same powers oppressing the environment and exacerbating the climate crisis. *The Last Wave* is an emphatic challenge to coloniality and its rationalistic epistemology, especially in the film's insistence that the voice of true reason (i.e. sustainable reason) are the voices of indigenous cultures. As Edward Said says in *Culture and Imperialism*, "There is incorporation; there is inclusion; there is direct rule; there is coercion. But there is only infrequently an acknowledgment that the colonized people should be heard from, and their ideas known" (50). Weir tells this story from the Western perspective simply as an entry point for Western viewers, as a sort of invitation to self-scrutiny that might lead the dominant culture further away from a colonial mindset and, in Walsh and Mignolo's words, "toward an otherwise" (4).

Note

1 In an interview for the Criterion Collection in 2001, Weir explains that this line was a direct statement made by Nandjiwarra Amagula in pre-production, and the theme of the importance of the law was a prerequisite for Amagula's participation in the film.

Works Cited

Bachelard, Gaston. *The Poetics of Space*. Beacon Press, 1994.

Buckmaster, Luke. "The Last Wave Rewatched – Mysticism, Prophecy and the End of Times." *The Guardian*, 15 August 2014, www.theguardian.com/film/australia-culture-blog/2014/aug/15/the-last-wave-rewatched-mysticism-prophecy-and-the-end-of-times.

"David Stratton on *The Last Wave*." *YouTube*, uploaded by ACMI, 12 August 2010, www.youtube.com/watch?v=qYRLpB2NhIo.

Ghosh, Amitav. "Part III: Politics." *The Great Derangement: Climate Change and the Unthinkable*. Penguin Books, 2016, pp. 157–217.

Jacobs, Diane. "The Last Wave." *Criterion.com*. The Criterion Collection, 26 November 2001, www.criterion.com/current/posts/168-the-last-wave.

Kaplan, E. Ann. "Introduction: Pretrauma Imaginaries." *Climate Trauma: Foreseeing the Future in Dystopian Film and Fiction*. Rutgers University Press, 2016, pp. 1–22.

Merlan, Francesca. "Introduction: Persistent Difference." *Dynamics of Difference in Australia: Indigenous Past and Present in a Settler Country*. University of Pennsylvania Press, 2018, pp. 1–17.

"Peter Weir on *The Last Wave* (1977)." Directed by Peter Weir. The Criterion Collection: no. 142, 2001.

Plumwood, Val. *Environmental Culture: The Ecological Crisis of Reason*. Routledge, 2002.

Said, Edward W. "V: Connecting Empire to Secular Interpretation." *Culture and Imperialism*. Vintage, 1994, pp. 43–61.

Santos, Boaventura de Sousa. "Introduction: Why the Epistemologies of the South? Artisanal Paths for Artisanal Futures." *The End of the Cognitive Empire: The Coming of Age of Epistemologies of the South*. Duke University Press, 2018, pp. 1–16.

Walker, Steven F. "Chapter One: Apocalypse, Transformation, and Scapegoating: Moving Myth into the Twenty-first Century." *Myth, Literature, and the Unconscious*, edited by Leon Burnett, Sanja Bahun, and Roderick Main. Taylor and Francis Group, 2013, pp. 3–26.

Walsh, Catherine E. and Walter D. Mignolo. "Introduction." *On Decoloniality: Concepts, Analytics, Praxis*. Duke University Press, 2018, pp. 1–32.

Weir, Peter. "Interview with Judith M. Kass." 8 January 1979, www.peterweircave.com/articles/articlei.html.

18 Scorched Earth and Precarious Existence

Representation of the Anthropocene in Bollywood Films *Kadvi Hawa* and *Jal*

Amit Mandal

Introduction

Planet Earth's systems or spheres, since life first emerged, have been providing the desired stability for all the biophysical elements to survive and flourish on the surface of the Earth (Harding 44). There are five major Earth's systems which include geosphere, biosphere, hydrosphere, cryosphere and atmosphere. All of these systems combine and interact with each other to create and maintain a habitable environment for all species to thrive on this planet; furthermore, these systems or spheres are all interconnected, and one can significantly affect another (Lenton 25–30). For example, all the water in solid form, found in polar ice caps, glaciers or on mountains, form the part of the Earth called cryosphere, and unfreezing of the particular Earth system will wreak havoc on other systems. The rapid rise of sea levels across the globe is proof of the following fact. For James Lovelock, all these systems are parts of a bigger whole, a gestalt system that characterises the life-world. This idea was the basis for his "*Gaia hypothesis.*" Gaia theory, named after the Greek goddess of Earth, is a wholistic concept, which posits the planet as a

> complex entity involving the Earth's biosphere, atmosphere, oceans, and soil; the totality constituting a feedback or cybernetic system which seeks an optimal physical and chemical environment for life on this planet. The maintenance of relatively constant conditions by active control may be conveniently described by the term homeostasis.
>
> (Lovelock 10)

The homeostasis that Lovelock talks about paved the way for humans alongside all natural beings to emerge and evolve (Lovelock 255). No singular species in particular, in the entire duration of Earth's history, had been able to acquire the immense ability to rule over any of the Earth systems, and destabilise the balance created by Gaia. Ever since, Homo Sapiens arrived and began to terraform the face of the whole planet, their arrival also marked a new paradigmatic transition for planet Earth.

DOI: 10.4324/9781003602750-24

The present human epoch, which is referred to as the "Anthropocene," might be taken as a rupture in Earth's history as it came right after the Holocene stability. It's hard to pinpoint when the Anthropocene as a geological period started. Researchers have claimed the onset of the Industrial Revolution in the eighteenth century as the decisive moment in history. The entire history of the Anthropocene, from the beginning to the present, is an incessant assault on the planet Earth and its resources. The impact of these anthropogenic forces on planet Earth is so profound, as Lewis and Maslin argue, that they "will probably be observable in the geological stratigraphic record for millions of years into the future, which suggests that a new epoch has begun" (1). Humankind, in this period, has emerged as a globe-trotting force that has changed whole landscapes and ecoscapes, erased millions of species and terraformed the face of Earth in such a manner that it's now called the Age of Man or the Anthropocene (Crutzen 23). The effects and magnitude of Anthropocene are so far-reaching that many researchers and posthumanist scholars like Donna Haraway, Jason Moore and Ana Tsing have coined various terms to grasp the enormity of the situation, such as Cthulucene (Harraway), Capitalocene (Moore), Plantationcene (Haraway), Anthrobscene (Jussi Parikka), Neganthropocene (Bernard Stiegler) and so on.

India is one of the fastest-growing economies in the world, and anthropogenic climate change is supposed to be wreaking havoc on its economy. Panda stresses the gravity of the situation by emphasising that,

> a country's vulnerability to climate change is decided by the presence of appropriate mitigation and adaptation options . . . in developing countries ecological environments are fragile, the susceptibility to economic systems to risks is high- and low-income levels of most citizens constrain their ability to cope.
>
> (105)

Hence, India's socio-economic position, its population of over a billion people, and its geographic location on map are some major factors that indicate the region is going to be one of worst hit places, when the climate crisis reaches its full potential in the decades to follow (Bisht 1). With a detailed analysis of the characters and their lived realities, the present essay tries to bring out the harsh realities of climate change in two severely impacted regions of India through the prism of two Bollywood films. The rationale of the essay is to deal with the uncertainties and vulnerabilities that the people of these regions face in their day-to-day lives, caused primarily due to the climatic shift, coupled with socio-political dynamics of the regions.

Kadvi Hawa

The film *Kadvi Hawa* (translated as "Dark Wind or Bitter Wind") by Nila Madhab Panda is set against the backdrop of a harsh drought situation in an

Indian region called Bundelkhand. Bundelkhand is primarily classified as a harsh semi-arid to completely arid region. The narrative focuses on the precarious lives of farmers and marginalised people in a village called Mahua. At the centre of the narrative lies the issue of ecological degradation and human's relation to Nature and weather-cycle systems. It primarily tells a horrifying tale of farmer suicides, in conjunction with crop failures and debt traps, and tries to emphasise how climate change chooses its victims. It demonstrates how the very existence of the farmers' lives is being affected by the changing climate, unpredictable weather patterns and incessant drought. In an agrarian economy like India, weather systems play a significant role for farmers because farmers mostly rely on rains for cultivation and a decent yield. Farmers' lives are invariably tied to the weather patterns, and in a fast-changing climate, agriculture has become a life-threatening risk which they depended upon throughout generations for their sustenance. Harshita Bisht's report on Bundelkhand suggests that

> agriculture forms the backbone of the rural economy in the region with almost 75% people dependent on agriculture and animal husbandry for their livelihood. Farmers face a crisis due to increased agriculture inputs when compared to agricultural outputs and benefits.
>
> (3)

Kadvi Hawa centres around a blind old farmer Hedu (Sanjay Mishra) who is always paranoid that his son Mukund (Bhupesh Singh) might commit suicide, because of the pressure of loans taken from bank. Farmers in Bundelkhand usually take loans from banks in hopes that they will be able to pay back after a decent earning from the crops. Unfortunately, when the crops fail due to lack of precipitation and farmers cannot pay back the loans, the following condition generates such extreme psychological distress among farmers that they are forced to take their own lives. Debt distress is quite common among villagers in Mahua; Hedu exclaims, "When a child is born here, we read his palm not to tell his fortune but how much debt he has" (Translation mine). In order to save his son, from the same depression and possible suicide due to his large sum of debt, Hedu forms an evil alliance with the loan recovery agent of Mahua, Gunnu Babu (Ranvir Shorey). In exchange for clearance of his son's debt that has been engulfing him lately, the deal demands accurate information about the wages of farmers in his village, thereby helping Gunnu Babu recover money at ease. Through the images of scorching heat, unyielding lands and events of farmer suicides, the film talks about the helplessness of poor and marginalised farmers who are at the forefront of this massive climatic shift and have little to no capacity to adapt to the crisis, or take required mitigation measures.

Jal

The film *Jal* (translated as "Water") by Girish Malik is another quintessential take on the climate crisis, portraying extreme examples of drought and ensuing water scarcity, in one of India's hardest drought-prone regions. Like *Kadvi Hawa*, Malik's eco-narrative is another effort by Bollywood towards the sensitisation of ecological issues amongst Indian masses which have so long been neglected. To put it simply, Bollywood films play a paramount importance on the Indian psyche when it comes to upholding major socio-political issues on screen, and in turn propound discourses of the same. In this respect, *Jal* is a timely endeavour to delve deeper into human consciousness to bring to the fore issues regarding ecological degradation, water scarcity and global warming, so that proper climate awareness can be formed, not only among urban but also rural masses, towards an urgent political attention. The lived realities of people, portrayed through the cinematography in their day-to-day scenarios, stand as a grave warning for worse things to come on this changing landscape.

Malik's eco-narrative, the *Jal*, is set against the backdrop of an enormous drought and freshwater scarcity in a region called "Rann (desert) of Kutch" in the western part of India, Gujarat. The name "Kutch" is also quite significant in this context, referring to a "tortoise," an amphibian species, having the ability to be both dry sometimes and sometimes wet. The Rann of Kutch is a shallow wetland that is submerged during monsoon season and becomes a salt flat when water recedes. It is quite harsh as a habitable environment, but people have adapted to the region and have been living there for generations. Climate change has exacerbated harshness of living conditions recently, "The analysis of the last 100 years (1891 to 1996) data on annual mean maximum and minimum temperature showed that the mean maximum temperature increased by 0.5°C, while the mean minimum temperature has declined by 0.5°C" (GEC 18). Both humans and nonhuman beings here now find it hard to adapt and be resilient in relation to the new warming and climatic trends.

The narrative focuses on the struggles of freshwater scarcity that Bakka (Purab Kohli) and the people living in his village have been facing due to lack of rainfall. It tells how Bakka, a water diviner (someone scouting for underground water), tries to find freshwater for his village and the flamingos who migrate here during every monsoon season. When a birdwatcher from Russia arrives in his village and finds that the place is drier than ever and migrating flamingos are dying due to lack of enough freshwater, she recruits Bakka to dig wells to find water for the flamingos. This account in the film can arguably be taken as a comment on the primacy of nonhuman beings for functional ecosystems in a changing world. Not only humans but other natural beings are also finding it hard to adapt to the changes. According to the analyses of the birdwatcher team, rising temperatures and drying up of freshwater lakes are what causing mass extinction of the bird species:

> There is a proper balance of freshwater and brackish water in the Kutch. It promotes the growth of the zooplanktons, algae and small fishes, and this is the diet of the flamingos. But now because of the increased salinity, it's killing these chicks.
>
> (Translation mine)

The countless bodies of dead flamingo chicks, lying dormant underneath shallow waters of Kutch, thus, bear instances of the destructive anthropogenic activities which are driving many nonhuman species to extinction.

The opening of the movie zooms in on the nightmarish and apocalyptic landscape, which presents visuals of scorching temperature, acute dryness and aridity, where Bakka roams around exclaiming, "I am a water-diviner and I will bring water" (Translation mine). The film also shows him desperately trying to dig a well as deep as he can into the scorched Earth, in search of water for his pregnant wife, while heavy dust winds blowing from all sides. His failure at not being able to find water for his unborn child is symbolic of the fact that the land can no longer sustain life, "Ran never gives water. Ran's heat takes lives on the other hand. It cannot sustain life" (Translation mine). The film tries to reiterate the fact that on this godforsaken parched landscape, life has a limited amount of resources and possibilities for survival. Regions like Gujarat's Kutch are also

> highly vulnerable to climate change because many are located in arid and semi-arid regions, and most derive their water resources from single point systems such as boreholes and isolated reservoirs. These systems, by their nature, are vulnerable because there is no redundancy in the system to provide resources, should the primary supply fail.
>
> (Watson et al. 5)

Moreover, people's mere efforts at finding underground water by digging wells inform us of bygone times when Earth stored plenty for the people of Kutch and nourished them on this harsh landscape. Nowadays, changing climate and disruption in the weather cycles have created this anthropogenic drought and water scarcity of epic proportions. The entirety of the narrative revolves around this very ecological lack and the resilience of people towards it.

Climate Change and Suicide Sickness in Bundelkhand Region

It is no surprise that communities depending heavily on climate for their sustenance and survival are more likely to be affected by changing climate and unpredictable weather patterns. A developing economy like India is always at the forefront of climate change and its effects because it relies heavily on climate-sensitive sectors like agriculture. According to the Economic Survey of 2017–18, more than 50% of the Indian total workforce is engaged in the agricultural sector (Sundar). According to the United Nations, India also surpassed China and became the largest population in the world in April 2023 at around 1.4 billion people (World Population Prospects). Hence, climate change in coming years can have a devastating effect on the Indian economy and its food security. Being an agrarian economy, it relies heavily on the whims of the climate. In a world of climate chaos, where all the natural systems are reeling under stress, the fate of any agrarian economy like India, hangs on a loose thread.

Massive spike in temperature in the recent decades, owing greatly to global warming, has created an epidemic in the form of farmer suicides in the Bundelkhand region and other Indian states like Maharashtra, Andhra Pradesh and Tamil Nadu and so on. Tamma A. Carlton in her study, based on the connection between temperature spikes and suicide rates in India, observes that

> fluctuations in climate, particularly temperature, significantly influence suicide rates. For temperatures above 20° C, a 1° C increase in a single day's temperature causes ~70 suicides, on average. This effect occurs only during India's agricultural growing season, when heat also lowers crop yields . . . find no evidence that acclimatization, rising incomes, or other unobserved drivers of adaptation are occurring. I estimate that warming over the last 30 years is responsible for 59,300 suicides in India, accounting for 6.8% of the total upward trend.
>
> (8746)

The hypothesis, that climate change and lack of rainfall are directly proportional to suicide rates, is further substantiated by the fact that with increasing rates of precipitation suicide statistics witness an exponential decrease. Unlike non-growing seasons, temperature variations in main growing season substantially influence agricultural outputs and in turn suicide rates (Carlton 8749). Hence, it can be argued that only a good amount of rainfall stands a chance against mitigating the suicide rates. Though, in a changing climatic paradigm, stability in the weather system is too optimistic to be practical.

Kadvi Hawa portrays the issue of this particular climate-related suicide epidemic but on a small scale. Mahua village represents Bundelkhand and provides us a glimpse into the broader mechanisms that connect global warming to farmer suicides. Degradation of climate and falling agricultural income are directly proportional to the rate at which farmers are taking their own lives. The film is a treatise on human callousness to our environment and nature's revenge upon us, by making the region dry, barren, unyielding and devoid of minimum natural resources that can sustain agriculture and in turn the region's population. The fact that the incidents of suicides are ever-increasing and ever-expanding sheds light on the gravity of the situation, making them nothing but closer to the definition of an epidemic. The protagonist Hedu calls these happenings a "*Rog*" (disease or sickness), which will soon engulf the whole village. His very paranoia in regard to the suicide epidemic is justifiable in the fact that the numbers of suicides are horrifying and ever-increasing, making him believe that sooner or later the epidemic will also consume his only son: "What if this sickness takes my son? The suicide sickness that took Janaki's father. It consumed Ramsharan. Slowly, it will kill everyone else" (Translation mine).

Major drivers like environmental degradation, depletion of resources and resulting agricultural crisis are not the only significant forces behind this suicide epidemic. The back-breaking debts have also played a major role in forcing farmers to induce self-harm upon themselves. Weather patterns are never stable in Bundelkhand, which has presented drought as a gruesome reality that farmers have no control over. Incessant drought and lack of adequate precipitation usually lead to crop losses and in turn shatter all hopes of good earnings for farmers, thereby creating uncertainty regarding their own livelihoods. The crop damages due to unpredictable weather patterns and subsequent hope for a better yield generally force farmers to take loans from banks or private landlords. Their hopes of better earnings are further dashed by bad weather and lack of rainfall, leading to the cause of acute psychological distress as it becomes clear they won't be able to pay the loans back on time. Their lives are further made worse when they are chased by loan recovery agents like Gunnu Babu. He is called Yamraj (God of Death) by villagers for no better reason than his ability to create circumstances to make people commit suicide: "Nobody escapes him. He was in Jalaun district before. He broke all recovery records. Wherever he goes one or two people hang themselves" (Translation mine). Besides, the lack of proper adaptive and preemptive measures from government to tackle the crisis and provide some relief to the farmers have also contributed to the suicide epidemic:

> Farmers' suicides in Bundelkhand are a result of several years of neglect of the agricultural sector and industrial backwardness. Neither the Uttar Pradesh nor the Madhya Pradesh government has made efforts to address the basic issues of ecological degradation, agricultural modernization and rural indebtedness.
>
> (Verma 1)

In 2009, the Government of India released a sum of approximately 7000 crores as Bundelkhand Drought Package to provide relief to the rain-starved region, but rampant corruption did not let it trickle down to the needy in time (Dhar). Policy interventions like this are meant to transform people's lives, but they are also motivated by electoral gains and "patronage-based implementation of welfare policies in drought hit areas" (Suthar 399). Hence, alongside ecological uncertainties, bad governance and corruption further contribute to the enormity of the crisis.

Climate Change and Ecological Degradation in Kutch

The Rann of Kutch, in Malik's *Jal,* was already an extreme environment for both humans and nonhuman species before the era of Anthropocene. It is generally considered an underdeveloped and backyard region of the prosperous state of Gujarat, a "punishment for bureaucrats" (Mehta 314). Nowadays, it is famously recognised as an ecologically fragile ecosystem that is struggling to cope with the recent anthropogenic changes. The climate is characterised by its dry and hot summers; the temperature from March to May usually soars above 40°C (GEC 18). Opening descriptions and visuals of Kutch's landscape prepare us beforehand for the hellish and uncanny environment we are about to step in. The panoramic view of the landscape presents the surface of Earth to be cracked and completely dry, while dust winds cover Baka's cottage as a blanket. Here, anthropogenic drought has substantially contributed to the worsening of living conditions by making people change their old habits and lifestyles on this parched landscape. Drought generally refers to disruption in the water cycles in a given location or some "deficit on water-related variables or water-dependent activities (e.g., Precipitation, soil moisture, surface and groundwater shortage, irrigation)" (Agha Kouchak et al. 1). It depicts a situation when all the natural variables are out of control, thereby creating a situation of uncertainty and precarity for all the people in Rann of Kutch. One can hardly ever hope for something better living in such a precarious condition as exemplified by an elder in Bakka's village: "These clouds also discriminate against us. It pours everywhere except this place. And whatever amount it gets, Ran further sucks it up" (Translation mine).

Uncertainty sometimes also creates more uncertainties. Socio-economic status of people living in Bakka's village further adds woe to their adaptive capacity. Most notably, their status as rural and underdeveloped communities further exacerbates their conditions. People mostly rely on agriculture (which depends heavily on rainfall) and livestock rearing for their livelihoods here. Though pastoralism has always been more favourable for earning livelihoods than agriculture; Kutch is always better recognised as a livestock-based economy than rain-dependent farming (Mehta et al. 1). Both agriculture and pastoralism are climate-sensitive sectors, and disruption in climatic variables has been exerting a substantial influence on the economic sector of the region. Srivastava et al. argue that

> they are struggling to cope due to combination of, on the one hand, changing rainfall patterns and intensities of droughts, and the other hand, compromised ability to tackle these extremes due to wider social and economic changes which have increased the precarity and vulnerability.
>
> (84)

Eco-tourism provides another alternative avenue for livelihood earning. Greater biodiversity of the Rann of Kutch attracts a lot of tourists each year. Aside from its climatic extremes, Kutch district is also a rich and biodiverse region, full of flora and fauna. The famous "flamingo city," based in the mudflats of the "Kutch Desert Wildlife Sanctuary," hosts thousands of flamingos each year. It appears to be a paradise for greater flamingos who come here to breed each year. The film depicts the arrival of the Russian eco-tourist Kim and her team studying the flamingos, and trying to find water for them after finding out they are dying due to the very lack of it. Their arrival creates opportunities for earning for Bakka and some villagers, when they get recruited to collect dead flamingos out of the shallow and muddy waters for the purpose of research. What is alarming is that owing to anthropogenic changes like scanty rainfall, the numbers of flamingos are dwindling each year, causing further heavy blows on their livelihood earning and adaptive capacity: "There was once a time when flamingoes used to come in large numbers but now they don't. It has affected the number of tourists and livelihoods of villagers" (Translation mine).

Hence, socio-economic concerns coupled with the struggles to cope up with the vulnerabilities of climate change have created an uncertain future for people of Kutch. The film also addresses the major issue of political and governmental negligence towards the region, which further aggravates the vulnerability of the region to the "pre-existing precarity that climate change finds in place" (Ribot 673). Bakka's failure at not being able to bring the drilling machine to his own village, leased out to the Russian team for drilling wells to quench the thirst of flamingos, showcases utter negligence and legislative failure at handling climate change. The fact that the helpless villagers ultimately opt to steal the government property, upon their petition to borrow the machine being unanswered and neglected for months, tries to stress the point that technology and good governance should always be ready at hand in a deepening crisis. Watson et al. further sum up:

> The impacts of climate change will depend on baseline condition of water supply system and the ability of the water resources managers to respond not only to climate change but also to population growth and changes in demands, technology, social, and legislative conditions.
>
> (5)

The argument that is being stressed here is that social, economic and ecological factors have created a form of "radical uncertainty" (Kay and King 38), where indigenous knowledge systems no longer help them determine the future course of action, and the future too looks grim, obscure and uncertain. Climatic variables and extreme weather events challenge their very notions of traditional weather systems on a day-to-day basis, thereby putting constraints on their adaptive mechanisms. All these uncertainties are driving people beyond their limits of coping, and in effect reducing their adaptive strength. The rapid rate, at which climatic changes are happening, is exerting substantial pressure on the natural world too, which has been highlighted by the deaths of countless migratory flamingos. Changing landscapes and natural habitat losses, due to scanty rainfall and extreme weather events, have been driving the bird species to slow death and mass extinction. In the Age of Man, animal populations and sometimes whole species, are being wiped out from the face of the Earth; "anthropocene defaunation crisis" has been one of the hallmark syndromes of the age, threatening the very existence of many species, still roaming on Earth (Finn et al. 2). The rapid rate at which species extinction is happening might be an indicator of the fact that Earth is presently entering into a new phase called "Sixth Mass Extinction" (Kolbert 233). The argument can be substantiated by the fact that "Globally, 48% (23–76%) of included animal species are undergoing population declines, 49% (23–76%) remain stable, and only 3% of species are undergoing population increases" (Finn et al. 4). The defaunation of multiple species including flamingos of the Rann of Kutch is just one chapter among countless others in Earth's sixth mass extinction process.

Surprisingly, the idea of the Anthropocene as an epoch, with all its monstrosities, is still a theory for many parts of the globe. Though climate change is recognised as an all-pervasive and global phenomenon, its effects are not felt or comprehended with the same intensity across the globe. It is recognised by all means that it is a totalising event, but the processes are not paced up along the same axes everywhere. Climate change as an idea encapsulates all totalities of existence, yet its immediate effects are more local than global. Localised impacts, born out of a global and totalising idea, thus, pave the way for the denial of the climate crisis. Countries like India in the Global South and other third-world nations are the worst affected by the climate crisis than the rest of the globe. Recent IPCC Sixth Assessment report in 2023 also speaks volumes of the fact:

> Approximately 3.3 to 3.6 billion people live in contexts that are highly vulnerable to climate change. Human and ecosystem vulnerability are interdependent. Regions and people with considerable development constraints have high vulnerability to climatic hazards. Increasing weather and climate extreme events have exposed millions of people to acute food insecurity and reduced water security, with the largest adverse impacts observed in many locations and/or communities in Africa, Asia, Central and South America, LDCs, Small Islands and the Arctic, and globally for Indigenous Peoples, small-scale food producers and low-income households.
>
> (5)

Therefore, the Indian population in the Global South, by comparison, should be more aware of the nature, scale and degrees of changes in overall climatic conditions. Experiential knowledge should hardly leave any space for climate scepticism or denialism. Bjoern Hagen's study on climate perspectivism in nine countries sheds some light on the same, "climate change was generally ranked as a low priority compared with other issues such as employment, crime, and education, and the levels of skepticism are significantly higher in the Netherlands, UK, and USA than in Brazil and Mexico" (Matthews 2). It can be argued thus that the climatic shift in the Global South is more immediate, practical, and urgent as a human-induced catastrophe than in the regions of the Global North. This is precisely where the local vs global debate on climate concerns in the Anthropocene perfectly fits in. Most importantly, the purpose of the essay is to highlight the vulnerabilities and uncertainties that lie ahead for India in the face of the current climate chaos.

According to the IPCC report, parameters such as increase in average temperature, sea level rise, loss of polar ice caps and glaciers, and displacement of millions of people are some of the unavoidable circumstances in relation to the present global warming trends. The Sixth Assessment Report also puts countries in the Global South, in particular, as the most vulnerable to the hazardous effects of climate catastrophes (IPCC 5–11). Most of the countries in this region including India are developing and poor countries, which make them more vulnerable for adaptation. Ludwig et al. opine that

> developing countries are affected more because of the economic importance of the climate sensitive sectors such as agriculture in combination with their low adaptive capacity. Many developing countries lack human and financial capacity to respond to the threats of climate change.
>
> (7)

Conclusion

In the Anthropocene, the idea of a humane and nurturing climate is quite non-existent; furthermore, the discourses about the climate generate much existential anxiety amongst the general populace. From films like *Waterworld* (1995), *The Day After Tomorrow* (2004) to Geostorm (2017) and *Don't Look UP* (2021), Hollywood has always tried to portray the following climate anxiety on the big screen through a plethora of films. Recently, Indian cinema has also understood the importance of environmental films and tried to produce significant numbers of them. *Kaun Kitney Paani Mein* (2015), *Sherni* (2021) and *All That Breathes* (2022) are some of the films which try to start a new discourse regarding climatic condition in India. In the essay, the discourse that both *Kadvi Hawa* and *Jal* attempt to touch upon is quite crucial as they both try to inform us that the climate we had in the pre-industrial era, which drove the weather systems, bringing rainfall for farmers, replenishing lakes and rivers, quenching the thirst of flamingos, was a natural cyclical process. The climate of the present, that we experience every day, is far from such "natural" courses. Climate in the Anthropocene is a hybrid one, a mix of both natural and artificial processes (Hulme 269). Therefore, it would be an understatement to characterise the climate of any particular region as a natural one, because artificial or anthropogenic forces are transforming them rapidly. Sometimes, changes are not so rapid and visible but regional climates are also part of the global climate system, and global warming is affecting everywhere. In this regard, Kutch and Bundelkhand are two extreme environments, where the combined effects of natural and artificial activities are more visible. Unlike other regions in India, they lack sufficient resources and technology to counter their hybrid climate system. This mere fact reduces their chances of adaptive capacity, making them more vulnerable to climate change. Their precarious existence foregrounds the idea that in the Anthropocene there are always losers and winners, based on the facts of resource distribution and geographical location.

Works Cited

Agha Kouchak, A. et al. "Anthropogenic Drought: Definition Challenges, Opportunities." *Review of Geophysics*, vol. 59, no. 2, 2019, p. 1, https://doi.org/10.1029/2019RG000683.

Bisht, H. *Climate Resilient Development in Bundelkhand Region of Madhya Pradesh, Synthesis Report*. Swiss Agency for Development and Cooperation, pp. 1–3.

Carleton, T. A. "Crop-damaging Temperatures Increase Suicide Rates in India." *Proceedings of the National Academy of Sciences. USA*, vol. 114, no. 33, 2017, pp. 8746–8751, https://doi.org/10.1073/pnas.1701354114.

Crutzen, P. "Geology of Mankind." *Nature*, vol. 415, 2002, p. 23, https://doi.org/10.1038/415023a.

Dhar, A. "Whither Bundelkhand Package." 12 February 2012, www.thehindu.com/news/national/whither-bundelkhand-package/article2883401.ece. Accessed 2 May 2024.

Finn, C., F. Grattarola, and D. Pincheira-Donoso. "More Losers Than Winners: Investigating Anthropocene Defaunation through the Diversity of Population Trends." *Biological Review*, Wiley Online Library, 2023, pp. 2–4, https://doi.org/10.1111/brv.12974.

Gujarat Ecology Commission. *Trends of Changing Climate and Effects of Eco-Environment of Kachchh District, Gujarat*. Gujarat Ecology Commission, 2011, p. 18.

Hagen, B. *Public Perception of Climate Change: Policy and Communication*. Routledge Studies in Environmental Communication and Media, Routledge, 2017.

Harding, S. *Animate Earth: Science, Intuition and Gaia*. Chelsea Green Publishing, 2006.

Hulme, M. "Cosmopolitan Climate: Hybridity, Foresight and Meaning." *Theory, Culture and Society*, SAGE, vol. 27, no. 2–3, 2010, pp. 267–276, 269, https://doi.org/10.1177/0263276409358730.

IPCC. "Summary for Policymakers." *Climate Change 2023: Synthesis Report*. Geneva, Switzerland: Intergovernmental Panel on Climate Change, pp. 5–11, https://doi.org/10.59327/IPCC/AR6-9789291691647.001.

Kay, J. and M. King. *Radical Uncertainty: Decision-Making for an Unknowable Future*. The Bridge Street Press, 2020.

Kolbert, Elizabeth. *Sixth Extinction: An Unnatural History*. Holt Paperbacks. 2014.

Lenton, T. *Earth System Science: A Very Short Introduction*. Oxford University Press, 2016.

Lewis, S. and M. Maslin. "Defining the Anthropocene." *Nature*, vol. 519, 2015, pp. 1, 171–180, https://doi.org/10.1038/nature14258.

Lovelock, James. *The Vanishing Face of Gaia*. Basic Books, 2009.

———. *Gaia: A New Look at Life on Earth*. Oxford University Press, 2016.

Ludwig, F. et al. "Climate Change Impacts on Developing Countries – EU Accountability." Policy Department: Economic and Scientific Policy, European Parliament, 2007, p. 7, www.europarl.europa.eu/thinktank/en/document.html?reference=IPOL-ENVI_ET(2007)393511.

Matthews, P. "Why Are People Skeptical about Climate Change? Some Insights from Blog Comments." *Environmental Communication*, vol. 9, no. 2, 2015, p. 2, https://doi.org/10.1080/17524032.2014.999694.

Mehta, L. *The Politics and Poetics of Water: Naturalising Scarcity in Western India*. Orient Longman, 2005.
Mehta, L., P. Joshi, and M.R. Bhatt. "How Pastoralists in Kutch Respond to Social and Environmental Uncertainty." 29 May 2020, steps-centre.org/blog/how-do-pastoralists-in-kutch-respond-to-social-and-environmental-uncertainty/#:~:text=Pastoralists%20here%20are%20attuned%20to,the%20uncertainties%20that%20they%20bring. Accessed 1 May 2024.
Panda, A. "Assessing Vulnerability to Climate Change in India." *Economic and Political Weekly*, vol. 44, no. 16, 2009, p. 106.
Ribot, J. "Cause and Response: Vulnerability and Climate in the Anthropocene." *Journal of Peasant Studies*, vol. 41, no. 5, 2014, pp. 667–705, https://doi.org/10.1080/03066150.2014.894911.
Srivastava, S. et al. "Between the Market and Climate Change: Uncertainty and Transformation in Kutch." *The Politics of Climate Change and Uncertainty in India*, edited by Lyla Mehta et al. Routledge, 2022.
Sundar, S. "Indian Economic Survey 2018: Farmers Gain as Agriculture Mechanisation Speeds Up, but More R&D Needed." *The Financial Express*, 29 January 2018, www.financialexpress.com/budget/india-economic-survey-2018-for-farmers-agriculture-gdp-msp-1034266/. Accessed 1 May 2024.
Suthar, S. K. "Bundelkhand Development Package, Drought and the Development Question in India." *Social Change*, SAGE, vol. 48, no. 3, 2018, pp. 398–410, 399, https://doi.org/10.1177/0049085718781660.
United Nations, Department of Economic and Social Affairs, Population Division. *World Population Prospects 2024: Summary of Results*. United Nations, 2024, https://population.un.org/wpp. Accessed 15 April 2025.
Verma, A, K. "Farmers' Suicides and Statehood Demand in Bundelkhand." Economic and Political Weekly, vol. 46, no. 28, 2011. p. 1, www.epw.in/journal/2011/28/states-columns/farmers-suicides-and-statehood-demand-bundelkhand.html.
Watson, R. T. et al. "The Regional Impacts of Climate Change: An Assessment of Vulnerability." A Special Report of IPCC Working Group II. Cambridge University Press, 1998, p. 5.

Part VI

Climate Change and Critical Thinking in Other Cultural Media

19 From One World to Another

Immersion in Digital Games and its Relevance for Climate Change

Laura Akers

Introduction

About two-thirds of Americans regularly play digital games (Entertainment Software Association). Millions of people of all ages and genders devote hours to their favourite games, and even those who don't identify as "gamers" often appreciate a momentary break with *Spider Solitaire* or *Candy Crush Saga*. As a medium with wide range and appeal, digital games may offer an opportunity to help communicate information on important issues, including climate change.

Scholars have suggested many potential objectives for games addressing climate change. They could "raise environmental awareness and educate players about the values of sustainability" (op de Beke et al. 9) and encourage critical thinking to "inform transformative civic action" (Germaine and Wake 489). More ambitiously, they could also promote "a more deep-seated adoption of system thinking" (Backe 110), "demonstrate the incommensurability between capitalism and the conditions of life on a finite planet" (op de Beke et al. 21), and even work to "destroy, challenge, fight with, and dismantle current destructive societal systems" (Vervoort et al. 192).

Two needs seem paramount: education and fostering resilience. As individuals, we need a fact-based understanding of how climate change works and what we can do about it: the skills and information to make informed and responsible choices as citizens and as consumers. We also need the kind of information that can support us in better coping with the ontological insecurity associated with climate change – the potential upheavals in our physical and cultural infrastructures that may arise from climate action or inaction.

This chapter will review the potential effects of video games on their players, focusing especially on broader insights about how the real world works and their own identities and aptitudes. It will then describe and critique a range of digital climate change games and describe recreational games from two genres unrelated to climate change that may also support players in developing personal resilience during systemic upheaval.

DOI: 10.4324/9781003602750-26

Immersion and Gameplay

The primary reason for playing games is emotional: games can be fun. Many gameplay experiences are positive – the satisfaction of achievement and skill mastery that come from completing challenging tasks, and the pleasures of exploration and discovery. In video games, such challenges often involve solving puzzles or achieving feats of timing and manual dexterity. Conversely, playing games can lead to frustration from setbacks and failures, but for those who persist, the reward of eventual victory can be all the sweeter.

Playing games creates immersion, a psychological state where one's attention is focused within the game world. Narrative transportation, in which a reader or viewer becomes "lost" in a story and temporarily loses some awareness of their own surroundings, is one form of immersion (Gerrig; Green and Brock). Digital games may include narratives, but there are also many other aspects of engagement: having or lacking awareness of one's physical interface with the game; sensory engagement, with a focus on what the player sees and hears; social engagement; and cognitive engagement, enabling strategic problem-solving (Bjørner and Schønau-Fog 134–136). With books and films, we can be *fully immersed*, a relatively kinetic experience where the viewer experiences a character's emotions as if happening to themselves, or *reflectively immersed*, a more cognitive experience in which our minds are making connections within the story world (as in a mystery) or between the story world and real life (Gholamain 27–28; Oatley 446). Reflection is possible with video games, too, typically when the game is encouraging it, for example, by creating a desire to connect the pieces of a story into a coherent whole. When playing a game, reflection would happen most often during relatively quiet points within the game (when one isn't actively making choices) or after a play session.

Immersion also makes possible other emotional aspects of gameplay, beyond the pleasure of playing itself. Many believe (perhaps falsely) that games featuring fighting, destruction or escape can provide an emotional catharsis (Gentile 491), evoking the fight-or-flight method of dealing with stress by eliminating it or putting ourselves beyond its reach. The game itself may generate such stress (with threats or challenges), and evidence shows that exposure to video game violence can spill over into real-world aggressiveness (Anderson et al. 4) but for some – such as veterans with post-traumatic stress disorder – playing such games may also help manage real-world stress (Hirst 495).

An alternative stress management approach allows us to remain engaged in the situation but to feel better about it: the "tend-and-befriend" strategy described by Taylor and colleagues (411). Just as in real life, we can find relief from stress through games featuring creativity and caring, practices that are also valuable in their own right.

Games and Learning

Learning is fundamental to playing games. Without some knowledge of the game's rules and procedures, it would not be possible to participate meaningfully. However, playing games – especially for recreation – doesn't necessarily lead to real-world behaviour change. Although exposure to violent video games may increase aggressiveness (Anderson et al. 4), the vast majority of those who play games that simulate combat do not commit real-world crimes (Markey et al. 1), and surely only a tiny fraction of the millions of players of the *Grand Theft Auto* games have stolen actual cars. It is probably relatively easy to compartmentalise one's real-world self from a character or role one plays in a game, if we see game-playing as a form of make-believe or acting.

The distinction between learning and transfer is important: "learning" refers to recognising and remembering information within a medium, and "transfer" involves being able to apply that information to other exemplars or contexts (Strouse et al. 2). Figure 19.1 illustrates the distinction between learning and transfer in the game-playing context. In the game world, we learn about the game world, which may include learning skills and facts, the game's stories, and how the game world works. We then have at least six main ways that our game-learning may translate into real-world learning. (1) The skills and facts learned in the game world may apply directly to the real world. (2) The empathy we have for game characters may lead to real-world empathy. (3) The game may illustrate new possibilities for how things can be done in the real world, and the systems we learn in the game

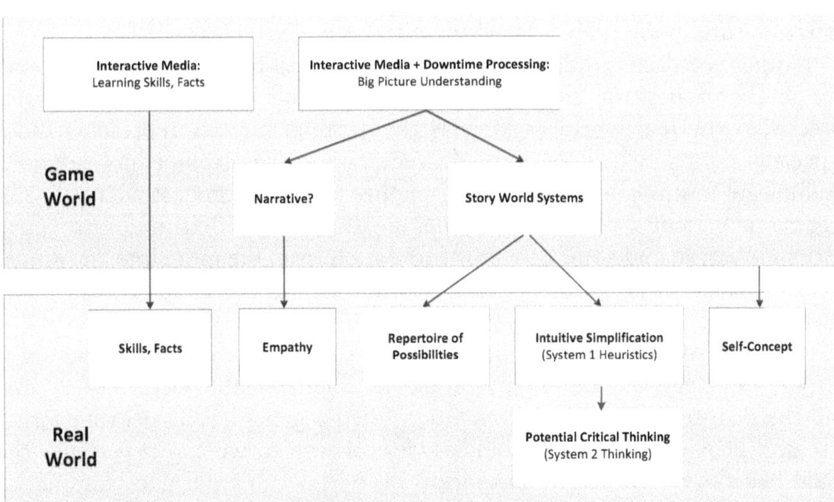

Figure 19.1 Learning and Transfer from a Game World to the Real World

may (4) reinforce or (5) encourage us to critique our real-world systems. Finally, (6) our experiences from playing games can influence how we think about ourselves.

The first key distinction between types of game-learning arises from the neuroscience of learning. When our attention is focused on an immediate task, we use a focused attention network in our brain (the Central Executive Network (Van der Linden et al. 954–955) or the Dorsal Attention Network (Anderson and Davidson 169)) to coordinate our actions and train our responses. This part of our brain is most active when we are using interactive media, like playing a game, and need fast, automated responses, or as Anderson and Davidson put it, "specific associations tied to concrete contexts presented within a narrow time frame" (176). We can use this network to learn skills, especially motor skills, and we can also use it to learn facts. When the game world represents the real world, we can gain factual knowledge about the real world; otherwise, we're gaining factual knowledge about how the game world works.

At other times, when our attention is not focused on an immediate task, we are more likely to be using a different brain network: the Default Mode Network, or DMN. This network allows us to think about ourselves in past, present or future situations (Van der Linden et al. 953). Whenever we need "broad situational awareness over time" (Anderson and Davidson 169), the DMN accesses observations and other memories or speculations and integrates them to create an overall interpretation or understanding. The DMN is thus "strongly implicated" in the processing and comprehension of narratives and their significance (Anderson and Davidson 170). It can also create mental models of how some aspect of the world works, using abstract representations and potentially connecting disparate ideas (Geary 29).

In other words, as Anderson and Davidson hypothesise, interactive media like games are best for learning skills and facts; this knowledge can transfer directly to the real world as long as the game accurately represents those aspects of the real world. Meanwhile, receptive (or passive) media like books or films are best for developing a big-picture understanding. If we're playing a game or watching an intense movie, creating that understanding requires both motivation and some downtime to switch from the more intense, immediate focus of an attention-specific network to the DMN so we can make sense of it all.

One type of big-picture understanding is created when we encounter a story – we spend mental energy on assembling a mental picture of events, and their causes and effects, over time. When a game has a narrative structure and provides us with a point-of-view character, we can experience the usual benefits of narrative transportation. As Lu and colleagues summarise it, transportation "promotes suspension of disbelief and reduction of counterarguments, enables the story experience as a personal experience, and creates the player's deep affection for the narrative protagonists" (199). They note that a well-crafted narrative about a desired health behaviour can give audience members the sense that they have performed that behaviour, and can allow them to experience the benefits of adopting that behaviour "more directly and vividly than from didactic instruction alone." Narrative transportation can show us what it's like to be someone different from ourselves, potentially increasing our empathy for others like our story's point-of-view character. Moreover, in a game, where a character's actions happen through one's own acts of agency, the experience may be even more compelling than what has already been determined for a character in a book. Narrative is not the only motivation for playing games, however; with many games (such as *Magic: The Gathering)* one can ignore the storyline altogether and still enjoy its other challenges.

In addition to potentially interacting with stories from the game world, we can also be influenced by its systems – the way things work there. Games can introduce us to new possibilities, expanding our repertoire of how things might happen and what we might be able to do, and perhaps reinforcing our conventional expectations and stereotypes or offering us alternatives. For example, in the real world we typically acquire items by buying them, receiving them as a gift, making them and so on. In games, these methods often exist, but other methods are also common, such as looting the corpses of newly slain monsters. Certainly the presence of shops in games reinforces our expectation of buying and selling objects, and although we are unlikely to start smashing pots found within others' homes and taking what was stored inside (as in many *Legend of Zelda* games), it is reasonable to expect that spending hours crafting with found items in *The Legend of Zelda: Tears of the Kingdom* might encourage us to think more often about real-world salvage and reclamation.

What about larger systems? Can games influence our understanding of how the world works? The answer is yes, but the information we learn about systems from playing games does not automatically lead to critical thinking. Decades of research by Daniel Kahneman, Amos Tversky, Shelly Chaiken, Paul Slovic and others have found that our default is not logical, analytical thought. Effortful thinking taxes our mental resources and is reserved for specific circumstances. Rather, our first, automatic response to information is to use our intuition, known as "System 1" thinking. On its own, System 1 leads us to create heuristics, simplified and often emotionally charged representations of situations (Slovic et al.), often with related decision rules (Chaiken 752).

A more rational and deliberate approach to thinking, System 2, is then available if we need it (Tversky and Kahneman; Kahneman). Before we can call on System 2, however, we need both motivation and sufficient mental capacity. Typical motives include a need for accuracy (if there's a high cost for being wrong), the enjoyment we might expect for thinking about this topic, or a sense of threat that we want to avert by shoring up our reasons for an existing position. Accuracy and pleasure are more likely to support open-minded, critical thinking, but having a preference for a specific outcome would typically lead to a focus on only the reasons that support our position, using what psychologists call "motivated reasoning" (Taber and Lodge 756). Even when one is truly open to where the facts may lead, however, System 2 also requires sufficient mental resources, which may not happen if we're experiencing stress, time pressure, fatigue, or other challenges. Further, results from System 2 thinking can become System 1 heuristics thereafter.

If we want games to support critical thinking, we first need to recognise that they are only likely to do so for players who are already motivated to spend time outside the game reviewing the information presented within the game for its usefulness. And, second, we need to expect that whatever players can learn from the game will already have been largely influenced by inferences from System 1.

As Bogost explains, games make claims about how the world works by the processes through which the game works, a "procedural rhetoric." He says, "Videogames simulate specific experiences that provide insights into the general relationships that drive those experiences," for a given topic (241). The game's underlying models are making claims about how the world works. If every encounter within a game is threatening, the player makes the simple inference that "encounters are threatening," an inference that may or may not apply to the real world too. Conversely, if the player's character can eventually succeed through persistence and resourcefulness, the player may learn that these are generally useful traits and that at least in this one context, they possess these traits. The underlying assumptions of how the game works and how we are intended to play it become messages ripe for use as System 1 heuristics; taking a further step and critiquing these assumptions would be an example of System 2 reasoning.

If the player believes that a game is intended to tell them something about the real world, these underlying game messages are our first, emotionally laden takeaways from the game-playing experience, available for modification by System 2, but potentially powerful on their own. Only if we have an incentive for more complex thought – the reflection or analysis that can happen during gameplay downtime or afterwards – are we likely to proceed to System 2 cognition. Thus, teachers often facilitate class discussions after students have read a work of fiction, to help them reach more considered and complex conclusions than are possible from intuition alone.

Returning to our list of potential real-world effects from game playing, our final category is self-concept. Our experiences in playing games can influence how we think about ourselves. If we do particularly well at a challenging game, we can experience the satisfaction of mastery and realise that we are capable of difficult things. If we learn to persevere over long periods of time, we experience ourselves as resilient. If we build unusual objects or solve difficult puzzles, we may come to think of ourselves as creative.

With all of this in mind, let's see what may be possible for climate change education with interactive media.

Climate Change Games

Two recent games illustrate the effects of varying climate-related public policies. Both games are very short – maybe 20 minutes – which can encourage the player to try again with different choices. In *Survive the Century*, the player is a senior editor with "the enviable power to set the news agenda, and thereby shift the zeitgeist" (SurvivetheCentury.net). For each decade, you choose among four options for your preferred response to that decade's key issue, leading down a branching path that concludes that decade with amusing yet pointed headlines. If you focus on reducing emissions while investing in jobs in education and health, you do survive the century – especially if you keep funding the military once national sovereignty (implausibly) becomes obsolete, because you'll need to defend the Earth against the billionaires who had emigrated to Mars. Conversely, if you trust the billionaires from the start, you "avoid the worst effects of climate change" but lose "any hope for democracy or fairness." The authors say, "We hope that this game helps you feel less hopeless and nihilistic about the future" (SurvivetheCentury.net), and someone playing this game could easily conclude that as long as the key decision-makers are reasonable, everything will be fine.

In *The Climate Game* you are a government minister charged with reaching net-zero carbon dioxide emissions by 2050. Specifically, you spend "effort points" each time you make a choice – it costs more to stop building coal plants, for example, than to trust the market. As 2050 approaches, your choices become more extreme and costly, and if you run out of points before you've made all of your choices, you get sacked.

A key underlying assumption, or "procedural rhetoric," of both games is that tackling climate change can be reduced to a simple task, requiring but a few minutes' attention. The players are left to hope that those in charge will be better-informed about the options than they are themselves – the fact that some 32% of *The Climate Game* players do keep warming below 1.5°C is heartening.

A far more elaborate game, *Fate of the World* focuses on optimal management of numerous resources. In this global strategy game spanning the next two centuries, players face more than 1000 possible impacts, including six tipping points, which they are to address through 117 potential policies, micromanaging the entire planet as a single (long-lived) government official. In order to succeed, however, the player must make multiple attempts, starting over each time the planet's biosphere fails. The underlying logic is far more opaque than it would be in real life, and several player reviews stress its flaws. One claims that the game pushes "a very specific narrative in which only cap-and-trade works to halt the eventual end of the world" ("DrSwordopolis"). Another says, "The moral lessons it tries to teach are questionable at best," because their only consistent path to winning was by targeting China, India and the Middle East with genocide ("Xenus Icelon").

The underlying messages in *Fate of the World* are at least threefold. First, the climate change problem is one of "soul-crushing difficulty"

(Makai 103), and the best that players can hope for is an "awe of understanding" from "watching the world burn time and again" (Makai 96). Second, "only policy-making bureaucrats have the necessary knowledge and authority to determine 'the fate of the world'" (Price 110), through "techno-fixes as opposed to other, more systemic solutions" (op de Beke 197), and "there is no solution to climate change that does not involve some form of violence" (Price 110). A more subtle point is made by op de Beke, who notes that game choices are limited by what we, today, believe is plausible, which "satisfies the desire to control the unknowability of the future by imbuing it with a sense of inevitability" (186).

The Climate Trail also takes a resource management approach to the topic, this time at an individual level, embedded in a story. You find yourself in the ruins of future Atlanta, after "The Burn" of massive wildfires everywhere and the ensuing Resource Wars, and decide to flee to Canada. Your task is the careful rationing of food and water throughout your journey. If you're travelling after 4°C of warming, you have a good chance of looting more food and water from abandoned stores and trucks; you can also collect rainwater, or you can drink pond water and risk dysentery. You decide how fast to travel and whether to take shelter from extreme weather events. If you go without food or water for too long, you die. Naturally, if you wait until 5°C or 6°C of warming, conditions are more extreme. If you make it to Canada, though, everything is green and pretty, although you're warned that acidification may soon destroy ocean life and lead to a massive extinction event anyway.

The Climate Trail is based on *The Oregon Trail*, but its takeaway messages are very different: "Here's what happened to these poor folks almost 200 years ago, but we're here today" may lead to empathy for the Oregon Trail pioneers and the satisfaction that at least our ancestors survived (if they did make the trek). "Here's what could happen to us and our children; we should relocate to a safe haven as soon as possible," by contrast, tells us that conditions are inevitably dire. As Julia Hoydis notes, the "bleak" narrative of *The Climate Trail* "tries to make the experience of climate change shockingly real but also shows the force of human agency and how each decision carries a consequence, some more immediate and some better or more fatal than others" (16). But without suggesting decisions that can meaningfully change the prospect of this future, *The Climate Trail* – like *Fate of the World* – can easily feed climate anxiety.

Climate Anxiety Games

A recent international survey found that most young people aged 16–25 are very or extremely worried about climate change (Hickman et al. e863). Many are experiencing climate anxiety: a sense of helplessness about an uncertain – or nonexistent – future. A research team led by Stefania Maggi is currently developing a game targeting these young people. *Kibou – The*

Guiding Light will explicitly tackle the challenges of climate anxiety, focusing on five tenets: "Mindful, Optimistic, Compassionate, Healing, and Innovations" in the context of adaptation and change. The intention is "to help youth navigate their emotions around climate change in a non-threatening way" through "solution-oriented communication" (Lynde-Smith).

Meanwhile, *Garden Story* already incorporates features promoting climate resilience. In this gentle recreational game, an ambulatory grape named Concord has been asked to help the four villages in The Grove by completing daily tasks, such as planting seeds, moving levers at a dam to reduce the risk of flooding, collecting shells and so on. The tasks were chosen, in part, to illustrate ways to create climate resilience, although it should be noted that The Grove is not dealing with climate change per se. If players do not realise the developers' intent, the information may not be obviously transferrable to the real world. The game does succeed at conveying the message that each of us is not alone and that no one individual bears the burden of saving others. If Concord cannot complete a day's tasks, someone else will.

Unlike the games about averting climate disaster, where the player works in heroic mode as the single person to decide the world's fate, climate anxiety games focus on building community for mutual support and scaling back the pressure on the story's point-of-view character. The main objective appears to be shifting the player's self-concept to foster resilience and modest optimism.

Recreational Games for Self-Care and Resilience

Most recreational games provide periods of respite. Rather than viewing game-playing as escapism, we can also consider it an active strategy for developing one's self-concept as someone who cares and perseveres. At least two categories of recreational games offer such opportunities: those focusing on building and caregiving, and those in the "roguelite" category.

Animal Crossing: New Horizons is a "cozy" simulation or "sim" game that epitomises the "tend and befriend" style of stress management. The player is given an entire island to modify – creating buildings, waterfalls, forests, gardens, and all the structures and decorations to go with them. As the land becomes more hospitable, anthropomorphic animals come to join the community, and the player can converse with them and do favours to make their lives better. Socialising with other players is possible, both those who share the same game console (typically family members and roommates) and those invited to visit, but the player can also treat the game as their personal haven – the undersea museum exhibits and the seashore at night are particularly restful.

The game rewards achievements in development and scenery with a maximum of five stars; stars unlock new milestones. Development encompasses the number of residents and the number and types of structures to accommodate them, while scenery refers to the placement of plants and crafted furniture. Points are removed if there is too much clutter lying about, or if the island is "too rural" – too many trees (*Animal Crossing* Wiki). In other words, the game favours an ordered aesthetic of development and habitation. Harris critiques the game's crafting system as bearing "a striking resemblance to the way colonizers will come into a space and then disrupt and construct the natural order into what they see fit" – "colonialism disguised as industrialization."

Conversely, Quirk (428) finds that the game promotes Shinto-style attention to the natural environment and mindfulness, "contrasting with the values promoted by capitalism and globalisation." She notes that the game takes place in real time (its clock and seasons correspond to the real world), and that all crafting and creation requires considerable patience and planning ahead. *Animal Crossing: New Horizons* is meant to be a long-term commitment and a place to spend time, not a game to "win"; thus, a second underlying theme is delayed gratification and appreciation of the present moment.

Another category of games supports a mentality of incremental growth and perseverance. Referring to the 1980 digital game *Rogue*, "roguelike" games are those in which one's character explores a "dungeon," fights monsters and collects treasure, until the character wins or dies. Character death is permanent (there are no "save spots"), and the player is expected to begin again fresh (RogueBasin). A subcategory of roguelikes, the "roguelites," permits the player to transfer experience, abilities or items from one attempt to the next. *Hades* is a roguelite that embeds its repeated failures in an overarching narrative: the story of Prince Zagreus's attempts to escape the underworld and to reconcile his parents, the god Hades who rules the underworld, and his wife Persephone. Zagreus repeatedly tries to fight his way through four regions of the underworld to reach Greece and find his mother, and each "death" provides dialogue opportunities that advance the larger story and give his trials a deeper significance.

Roguelite games are an excellent analogue for real-world living – each playthrough, like each new day, builds on the one before and gives us lessons that can inform future actions. In *Hades*, the player learns more of what it takes to succeed in the game world, but the interactions between Zagreus and other characters also model exemplary real-world relationship-building. Zagreus is an exceptional role model, navigating one-sided attraction and a power differential in one friendship, and apologising to Achilles and Orpheus for trying to repair their intimate relationships without consent.

The personal growth that can be supported by a roguelite game with a conscientiously scripted narrative is in stark contrast to the essentially roguelike playstyle encouraged by *Fate of the World*. In the former, each "life" corresponds to a day, or an episode, while in the latter, each "life" is an all-or-nothing endeavour with an entire planet at stake. Games like *Hades* can generate hope, while *Fate of the World* offers despair, or at best, a cultivated indifference to the outcome.

Conclusions

Numerous digital games have begun addressing climate change by educating players about the issues involved (Reckien and Eisenack 253). As this review shows, however, their explicit and underlying messages do not always match, and the underlying messages are not always psychologically helpful. Other games, both those intended to promote climate change resilience and others intended purely for recreation, are conveying messages that can support players in challenging times.

This short literature review has also revealed several potential research questions:

- Does critical/analytical thinking within a game (puzzle solving) reduce the likelihood of analytical thinking about the game world's premises as real-world analogues, because it uses up our critical thinking capacity for this context?
- Does repeat playing of short games lead to more critical thinking about their underlying algorithms?
- When does questioning of algorithms lead to overall real-world scepticism versus the mental substitution of potentially more valid assumptions for the algorithms?

In general, empirical work is needed to confirm the circumstances in which players will reach sophisticated real-world conclusions on their own, and which types of facilitation may help. Game designers should note that their players may be at any point along a spectrum of climate change attitudes: active denial, trust that others will take care of it, anxiety, anticipatory grief, grief around the experience of losses in the physical world or of one's cultural heritage, or even trauma from actual experiences, which may lead to depression, substance use and other problematic coping strategies. More attention to players' needs could go far towards helping people learn and practice emotionally healthy behaviours while also having fun.

Acknowledgements

The author would like to thank Jonathan Woodward, Ashley Alwin and David Akers for valuable discussions.

Works Cited

Anderson, Craig A., Douglas A. Gentile, and Katherine E. Buckley. *Violent Video Game Effects on Children and Adolescents: Theory, Research, and Public Policy.* Oxford University Press, 2007.

Anderson, Daniel R. and Matthew C. Davidson. "Receptive Versus Interactive Video Screens: A Role for the Brain's Default Mode Network in Learning from Media." *Computers in Human Behavior* 99, 2019, pp. 168–180.

Animal Crossing: New Horizons. Nintendo, 2020.
Animal Crossing Wiki. "Island Evaluation," animalcrossing.fandom/com/wiki/Island_evaluation.
Backe, Hans-Joachim. "Between the Lines: Using Differential Game Analysis to Develop Environmental Thinking." *Ecogames: Playful Perspectives on the Climate Crisis*, edited by Laura op de Beke et al. Amsterdam University Press, 2024.
Bjørner, Thomas and Henrik Schønau-Fog. "A Dynamic Engagement Model to Provide Ecological Awareness of the Climate Crisis through Video Games." *Ecogames: Playful Perspectives on the Climate Crisis*, edited by Laura op de Beke et al. Amsterdam University Press, 2024.
Bogost, Ian. *Persuasive Games: The Expressive Power of Videogames*. MIT Press, 2010.
Brock, Timothy C., Melanie C. Green, and Jeffrey J. Strange. "Narrative Impacts and Processes Within Individuals." *Narrative Impact: Social and Cognitive Foundations*. Psychology Press, 2002, p. 343.
Chaiken, Shelly. "Heuristic versus Systematic Information Processing and the Use of Source Versus Message Cues in Persuasion." *Journal of Personality and Social Psychology*, vol. 39, no. 5, 1980.
The Climate Game. The Financial Times and Infosys, 2022.
The Climate Trail. Deep State Games, 2019.
"DrSwordopolis." Reviews: *Fate of the World*, 2016, steamcommunity.com/profiles/76561197991359271/recommended/80200/.
Entertainment Software Association. "Video Games Remain America's Favorite Pastime with More Than 202 Million Americans Playing Regularly." 2023, www.theesa.com/video-games-remain-americas-favorite-pastime-with-more-than-212-million-americans-playing-regularly/.
Fate of the World. Red Redemption, 2011.
Geary, David C. *The Evolved Mind and Modern Education: Status of Evolutionary Educational Psychology*. Cambridge University Press, 2024.
Gentile, Douglas A. "Catharsis and Media Violence: A Conceptual Analysis." *Societies*, vol. 3, no. 4, 2013, pp. 491–510.
Germaine, Chloé and Paul Wake. "Imagining the Future: Game Hacking and Youth Climate Action." *Ecogames: Playful Perspectives on the Climate Crisis*, edited by Laura op de Beke et al. Amsterdam University Press, 2024.
Gerrig, Richard. *Experiencing Narrative Worlds: On the Psychological Activities of Reading*. New York: Yale University Press, 1993.
Gholamain, Mitra. *The Attachment and Personality Dynamics of Reader Response*. University of Toronto, 1998.
Green, Melanie C. and Timothy C. Brock. "The Role of Transportation in the Persuasiveness of Public Narratives." *Journal of Personality and Social Psychology*, vol. 79, no. 5, 2000, p. 701.
Hades, Supergiant Games, 2020.
Harris, Emily. "'Crafting' Your Reality: How *Animal Crossing: New Horizons* Uses Disarming Aesthetics and Game Play to Push a Colonialist Narrative." *The Classic Journal*, vol. 7, no. 2, 2022, theclassicjournal.uga.edu/index.php/2022/04/14/crafting-your-reality/.
Hickman, Caroline et al. "Climate Anxiety in Children and Young People and Their Beliefs about Government Responses to Climate Change: A Global Survey." *The Lancet Planetary Health*, vol. 5, no. 12, 2021, pp. e863–e873.
Hirst, Aggie. "'Videogames Saved My Life': Everyday Resistance and Ludic Recovery among US Military Veterans." *International Political Sociology*, vol. 15, no. 4, 2021, pp. 482–503.

Hoydis, Julia. "Dialogues with the Machine, or Ruins of Closure and Control in Interactive Digital Narratives." *Open Library of Humanities*, vol. 7, no. 2, 2021.
Kahneman, Daniel. *Thinking, Fast and Slow*. Macmillan, 2011.
The Legend of Zelda: Tears of the Kingdom, Nintendo, 2023.
Lu, Amy Shirong et al. "Story Immersion of Videogames for Youth Health Promotion: A Review of Literature." *Games for Health: Research, Development, and Clinical Applications*, vol. 1, no. 3, 2012, pp. 199–204.
Lynde-Smith, Jena. "Combatting Eco-Anxiety: Interactive Video Game Helps Youth Cope with Climate Change Fear." Carleton University, 2024, challenge.carleton.ca/kibou-video-game-combats-eco-anxiety/.
Makai, Péter Kristóf. "Do You Want to Set the World on Fire? Amplifying Player Agency to Demonstrate Alternatives to the Climate Crisis." *Ecogames: Playful Perspectives on the Climate Crisis*, edited by Laura op de Beke et al. Amsterdam University Press, 2024.
Markey, Patrick M., Charlotte N. Markey, and Juliana E. French. "Violent Video Games and Real-World Violence: Rhetoric Versus Data." *Psychology of Popular Media Culture*, vol. 4, no. 4, 2015, p. 277.
Oatley, Keith. "Meetings of Minds: Dialogue, Sympathy, and Identification, in Reading Fiction." *Poetics*, vol. 26, no. 5–6, 1999, pp. 439–454.
Op de Beke, Laura. "Premediating Climate Change in Videogames: Repetition, Mastery, and Failure." *Nordic Journal of Media Studies*, vol. 3, no. 1, 2021, pp. 184–199.
Op de Beke, Laura et al. "Ecogames: An Introduction." *Ecogames: Playful Perspectives on the Climate Crisis*, edited by Laura op de Beke et al. Amsterdam University Press, 2024.
Picogram. *Garden Story*. Rose City Games, 2021.
Price, Derek. "Damage Over Time: Structural Violence and Climate Change in Videogames." *Violence |Perception | Video Games: New Directions in Game Research*, edited by Federico Alvarez Igarzábal, Michael S. Debus, and Curtis L. Maughan, transcript, 2019, pp. 105–112.
Quirk, Monica Alice. "Island Time in Lockdown: *Animal Crossing: New Horizons*' Slow Culture and its Shinto Source." *Journal of Religion, Media and Digital Culture*, vol. 10, no. 3, 2021, pp. 428–444.
Rawitsch, Don, Bill Heinemann, and Pal Dillenberger. *The Oregon Trail*. MECC, 1971.
Reckien, Diana and Klaus Eisenack. "Climate Change Gaming on Board and Screen: A Review." *Simulation & Gaming*, vol. 44, no. 2–3, 2013, pp. 253–271.
Rogue. AI Design, 1980.
RogueBasin. "Welcome to RogueBasin," www.roguebasin.com/index.php/Main_Page.
Slovic, Paul, Melissa L. Finucane, Ellen Peters, and Donald G. MacGregor. "The Affect Heuristic." *European Journal of Operational Research*, vol. 177, 2007, pp. 1333–1352.
Strouse, Gabrielle A., Angela Nyhout, and Patricia A. Ganea. "The Role of Book Features in Young Children's Transfer of Information from Picture Books to Real-World Contexts." *Frontiers in Psychology*, vol. 9, 2018.
Survive the Century. Climate Interactive et al., 2021.
Taber, Charles S. and Milton Lodge. "Motivated Skepticism in the Evaluation of Political Beliefs." *American Journal of Political Science*, vol. 50, no. 3, 2006, pp. 755–769.
Taylor, Shelley E. et al. "Female Responses to Stress: Tend and Befriend, Not Fight or Flight." *Psychological Review*, vol. 107, no. 3, 2000, pp. 411–429.
Tversky, Amos, and Daniel Kahneman. "Judgment under Uncertainty: Heuristics and Biases: Biases in Judgments Reveal Some Heuristics of Thinking under Uncertainty." *Science*, vol. 185, no. 4157, 1974, pp. 1124–1131.

Van Der Linden, Dimitri, Mattie Tops, and Arnold B. Bakker. "Go with the Flow: A Neuroscientific View on Being Fully Engaged." *European Journal of Neuroscience*, vol. 53, no. 4, 2021, pp. 947–963.
Vervoort, Joost M. et al. "Games for Better Futures: The Art and Joy of Making and Unmaking Societies." *Ecogames: Playful Perspectives on the Climate Crisis*, edited by Laura op de Beke et al. Amsterdam University Press, 2024.
"Xenus Icelon." Reviews: *Fate of the World*, 2015,steamcommunity.com/profiles/76561198002567771/recommended/80200/.

20 Geopolitics of Climate Change Cartoons

Exploring Everyday Resistance through Visual Discourse

Shifana P. A. and Asha Susan Jacob

Introduction

The nexus of geopolitics and resistance presents a complex terrain where power dynamics and dissent intersect. Traditionally framed as an analysis of power relations among nations, geopolitics extends beyond hierarchical structures to encompass the everyday actions of individuals and communities. Conversely, resistance emerges as a counterforce to established power hierarchies and dominant narratives.

Resistance encompasses grassroots activism and cultural expression across various levels, including artistic forms such as cartoons. Within this context, cartoons offer a distinctive avenue for resistance within geopolitics. Through their visual medium, cartoons can challenge prevailing ideologies, critique authority figures and spotlight social injustices in a manner accessible to diverse audiences.

Cartoons disrupt established norms and amplify marginalised voices by subverting dominant narratives and presenting alternative viewpoints. They serve as platforms for dissent, facilitating the articulation of grievances and critiques against oppressive systems. In geopolitics, cartoons function as potent instruments of resistance, illuminating overlooked issues, exposing hypocrisy, and stimulating dialogue and debate.

The symbiotic relationship between geopolitics and resistance is evident, each shaping and influencing the other. As geopolitical landscapes evolve, so do the forms of resistance that emerge in response. Cartoons, with their ability to incite reflection, challenge conventions and mobilise action, assume a crucial role in this ongoing interplay between power and dissent.

This chapter contributes to the broader discourse on geopolitics and cartoons by examining the integral role of visuals in climate change communication. Additionally, it explores the nuances of visuality in the context of climate change discourse and debates. The chapter contends that our growing reliance on visual representations to understand and portray the world paradoxically complicates our comprehension of climate change – a concept inherently invisible, temporally complex and entangled with geopolitical narratives.

DOI: 10.4324/9781003602750-27

In contemporary climate change discourse, integrating cartoons into the narrative more broadly mirrors the interconnectedness observed in geopolitics and visual culture. The intricacies inherent in climate change make its visualisation both challenging and imperative. The crux of the matter lies not in the ability of visuals to faithfully depict the reality of climate change but in their effectiveness in portraying its geopolitical dimensions (Macdonald et al.).

This concern, coupled with the paradoxes associated with dominant visual mediums such as photographs, has prompted an inquiry into how political cartoons serve as vehicles for communicating geopolitical perspectives on climate change. While previous studies have explored climate change communication through avenues like fiction films and art exhibitions (Leiserowitz), the potential of cartoons in this realm remains largely untapped.

Despite their capacity to convey serious issues through humour and parody, cartoons have received limited attention in scholarly research on dissecting political representations (Dodds, "Popular Geopolitics and Cartoons"). Nevertheless, their ability to address substantial concerns lightheartedly does not diminish their significance for scholarly examination and critical analysis.

The chapter's empirical basis comprises cartoons on climate change selected from various websites through a purposive sampling technique. Each cartoon falls under the five themes chosen: fossil fuels, deforestation, desertification, glacier melt and unstable meteorological conditions. The subsequent sections are structured as follows: The first section reviews existing literature on geopolitics, visuality and political cartoons, focusing on elucidating "the communicative power of cartoons" (Kleeman). This review establishes an analytical framework for interpreting primary data discussed in part three. The second section aims to elucidate the workings of visuality within climate change discourses and underscore the geopolitical nature of these discourses. It contextualises discussions about climate change within broader inquiries into power dynamics and knowledge construction. This study delves into how political cartoons within critical geopolitics illuminate everyday resistance and challenge dominant narratives, contributing to discussions on power dynamics and visual culture.

Exploring Geopolitics, Visuality and Political Commentary

The scholarship on geopolitics and visuality underscores the significance of visual elements in contemporary geopolitical discourse, highlighting the reliance on visual evidence for geopolitical reasoning (Macdonald et al.). Moreover, it raises pertinent questions regarding the functioning and consequences of visuality, emphasising the impact and implications of visual representations (Campbell).

Building upon these foundational insights, this chapter ventures into the climate change communication and visualisation domain, exploring the theoretical underpinnings of cartoons as a visual medium. This theoretical journey serves as a precursor to an in-depth analysis of the role of cartoons in communicating geopolitical visions within the context of climate change.

Cartoons have been analysed through various lenses in the scholarly discourse, including popular geopolitics and geopolitical iconography (Dodds 2010 "Popular Geopolitics and Cartoons"; O'Tuathail and Dalby). They have been characterised as symbols (Diamond; El Refaie, "Multiliteracies"), conduits of visual news (Greenberg), modes of visual communication (Harrison; Ridanpaa), embodiments of cultural history (Clough), vehicles for political satire and social commentary (Kleeman) and forms of expression (El Refaie 2009). Despite the diversity of definitions, there is a consensus that cartoons serve as a means of visual expression and communication.

Exploring the distinctiveness of cartoons as an artistic form, it is pertinent to consider their distinguishing features compared to other visual mediums like paintings. For instance, Harrison's seminal work on "the cartoon code" delineates three key communication processes: levelling, sharpening and exaggeration (Harrison). Additionally, Dodds (2010) identifies further processes such as condensation, repetition, dramatisation and caricature. Cartoons also employ techniques like personification, objectification, analogy and familiarisation to convey their messages, as exemplified by the anthropomorphism of planets in certain cartoons.

The characterisation of cartoons as a "distinctive form of artistic expression" (Kleeman 144) prompts an inquiry into their distinguishing features compared to other visual art forms like paintings. A seminal investigation into "the cartoon code" by Harrison (11) elucidates three primary communication processes: levelling (or simplification), sharpening (of personal features) and exaggeration. Additionally, Dodds (16) identifies key elements such as condensation, repetition, dramatisation, exaggeration and caricature of prominent figures. Moreover, cartoons employ techniques like personification (of objects), anthropomorphism (e.g. representing planets as individuals), objectification, analogy and domestication or familiarisation.

Cartoons often employ symbols or visual metaphors, with "geopolitical iconography" denoting symbols crafted specifically "to represent international political events" (Dodds 17). Metaphors serve a broader purpose, symbolising abstract or invisible concepts (El Refaie 2009). Inventive symbols and metaphors in cartoons facilitate the visualisation of abstract concepts, such as climate change, which are inherently intangible. Moreover, cartoons can create fantastical scenarios and alternate realities by blending reality with the imaginary. This suggests that cartoons possess significant potential as geopolitical narratives, even when they do not offer direct visual evidence of climate change.

Cartoonists often utilise written text to convey their messages effectively. Harrison (1982, 2) describes "the thought balloon or the speech balloon" as central elements of cartoon symbolism. Additionally, Kleeman (2006, 150) highlights that cartoonists may employ captions to complement the non-verbal aspects of cartoons. However, as demonstrated in the subsequent section, cartoons can communicate geopolitical messages without textual elements, emphasising the significance of symbols, metaphors and the underlying cartoon code. Writing, therefore, does not solely define the essence of cartoons, which transcend mere textual expression to embody a form of discourse.

Political cartoons are typically associated with their publication in print media and their focus on political content. Historically, the term "cartoon" has been used to refer to humorous or satirical illustrations found in newspapers and magazines (Kleeman 144). Similarly, El Refaie (2009, 184–185) defines political cartoons as illustrations published in newspapers' editorial or comment sections, usually in a single panel. These cartoons address current political issues, events, social trends, or prominent personalities, often taking a definitive stance or presenting a specific viewpoint.

While humour, including parody, satire, or caricature, can serve to raise awareness and focus attention on "politically serious matters" (Ridanpaa 729–733), it is essential to note that political cartoons are not exclusively comedic. They encompass a spectrum of approaches, ranging from deliberate provocation or shock to subtle persuasion, all aimed at engaging viewers and prompting action.

The essence of political cartoons lies in their content, as El Refaie (2009) points out. They are characterised by their treatment of political issues, often conveying a message, taking a stance, or presenting a viewpoint. Political cartoons can satirise politicians, expose their hypocrisy (Ridanpaa), offer commentary and critique on political events (Dodds; El Refaie), and serve as a form of resistance against abuses and power imbalances (Hammett). Thus, the central themes of political cartoons revolve around exposure, commentary and critique of political matters.

In essence, political cartoons serve as a form of visual discourse and communication, transcending their physical location in newspapers or the presence of accompanying text. Each cartoon operates through its unique communicative codes, utilising imaginative symbols and visual metaphors to convey a particular perspective or commentary on topical issues, events, or trends. While political cartoons may not necessarily be confined to popular newspapers, they emerge within specific contexts, shaped by the subject matter and the viewers' capacity to comprehend and interpret their messages.

The Role of Imagery in Climate Change Communication and Everyday Forms of Resistance

Doyle (2011) emphasises three keyframes of climate change: environmental, neo-liberal consumerist and humanitarian. Dittmer et al. (2011) and Manzo (2010) highlight the prevalence of neo-realist and liberal discourses, contrasting them with a discourse emphasising global unity and vulnerability. These diverse frames rely on visual elements, such as maps and photographs, to convey the reality of climate change, drawing on the historical foundations of photography and Western scientific knowledge (Macdonald et al.).

In climate change discourse, the invisibility and latency of its effects make it challenging to deduce from a single photograph. As noted by photojournalist Nick Cobbing, a solitary image cannot convey whether a glacier is melting or receding, let alone indicate a problem. Comparative visualisations, such as paired photographs of the same glacier with dates showing different periods, are more effective in contextualising climate change as a real and pressing issue (Cobbing, qtd. in Smyth). Often associated with the catastrophe frame, these comparative images represent climate change's symptoms and purported effects, such as floods or droughts.

Textual explanations of climate change often accompany visual imagery to enhance understanding and engagement. For instance, newspaper headlines linking CO_2 emissions to economic recession are often paired with photos of smoking factory chimneys (Harvey). Visual representations of climate change include depicting potential solutions like solar panels and energy-efficient light bulbs, symbolising concepts such as renewable energy and climate change mitigation (Schmidt and Wolfe; Slocum). Computer-generated images, like the UK government's "future worlds images," also illustrate potential adaptation solutions for urban and rural environments (Defra).

Despite its paradoxical nature, this reliance on visual technologies and practices is evident across all six climate change discourses. While visual mass media, such as television and newspapers, are common sources of information about climate change, studies indicate low levels of understanding and engagement among the public (Whitmarsh). Whitmarsh's (2009) study distinguishes between factual knowledge of climate change and a deeper understanding of its integration with other environmental issues, moral considerations and cultural discourse. It attributes restricted public understanding to the media's limited capacity to communicate complex information effectively and public mistrust of the media.

Photographs, despite their prevalence in climate change communication, are not always effective in conveying the complexity of the issue, as they often oversimplify or fail to capture the temporal aspects of climate change. While they contribute to visualising climate science, they may not fully represent the intricate relationship between present actions and future consequences. Doyle (2011) highlights the challenge of communicating the reality of climate change without adequate visual evidence, yet the very concept of visual evidence reinforces scientific paradigms that struggle to convey the invisible and temporally complex nature of climate change.

A significant paradox addressed in this chapter is that certain visual representations can render climate change visible while obscuring its geopolitical dimensions. This challenge is underscored by calls to recognise climate change as a global issue and to acknowledge the power dynamics inherent in climate change discourse (Barnett; Doyle). Development and humanitarian NGOs like Oxfam and Christian Aid advocate for a shift in focus from environmental concerns to human impacts and social injustices associated with climate change (Manzo 2010a). This approach challenges dominant frames such as neo-realism, environmentalism, catastrophe narratives and neo-liberal perspectives, which often overlook or downplay geopolitical issues (Hulme).

Many academic studies examine the geopolitical implications of prevalent visual images and forms. These studies raise critical questions about the appropriateness of iconic images, their impact on viewers and the colonial visions they may perpetuate (Bravo; Slocum; O'Neill and Nicholson-Cole; Manzo; Doyle). Doyle (2011) offers a comprehensive critique, highlighting the diffusion of enlightenment conceptions of nature, vision and time from Western science into popular media and environmental movements. She argues that these conceptions, which separate nature from culture and frame it as an economic commodity, contribute to climate change. Moreover, she examines the pervasive sense of urgency in media coverage of climate change, linking it to enlightenment notions of time as a finite commodity (Doyle).

In short, climate change communication intersects with broader discussions on geopolitics and visuality, particularly regarding the efficacy and consequences of visual representations. This study explores how political cartoons comment on climate change geopolitics and their impacts, as climate change communication and everyday resistance employ diverse strategies to engage stakeholders and challenge dominant narratives.

Drawing from James C. Scott's theory of everyday resistance, this chapter explores how resistance manifests within climate change communication, from overt protests to subtle subversion. Visual representations, language choices and community mobilisation serve as avenues for resistance, fostering critical engagement and advocating alternative perspectives. By employing Scott's framework to analyse political cartoons, this study sheds light on how humour, irony and visual symbolism challenge mainstream climate change discourse, contributing to a deeper understanding of cultural influences on environmental advocacy.

Methodology of Analysis

This study adopts a systematic approach to analysing the visual and textual elements of the selected cartoon, focusing on symbols, metaphors, caricatures and spatial dimensions that convey geopolitical messages on climate change. By deconstructing these elements, the analysis reveals how visual discourse shapes perceptions of environmental crises, exposing power imbalances and contested narratives.

Particular attention is given to how the cartoon embeds hidden transcripts of resistance within its imagery, amplifying marginalised voices, critiquing power structures and engaging with broader social justice issues. Through close reading, the study identifies explicit and implicit acts of defiance, illustrating how visual satire challenges dominant narratives and reframes discussions on climate justice.

Situating the cartoon within broader climate change discourses allows for critically examining its alignment with or opposing prevailing ideological frames, including environmentalism, neoliberalism and humanitarianism. This contextualisation highlights how visual media reflects and reshapes public debates on ecological crises.

Finally, the study explores the geopolitical implications of the cartoon, interrogating its representation of global power dynamics, colonial legacies and cross-cultural dialogues on climate change. While the original cartoons analysed in this chapter cannot be reproduced due to copyright constraints, detailed descriptions and critical analysis will illustrate their visual and linguistic significance. This approach underscores how visual discourse functions as a site of resistance, shaping public perceptions and political debates on climate change. By examining how visual satire disrupts hegemonic narratives, this analysis contributes to a deeper understanding of the intersection between political cartoons, environmental discourse and the geopolitics of climate communication.

Analysis of Cartoons

The famous cartoon *The Tree of Life* presents a powerful commentary on environmental destruction and the fragility of ecosystems. In the centre of the image, a large, flourishing tree teeming with wildlife symbolises nature's resilience and the intricate interdependence of all living beings. A human figure, positioned at the base of the tree, aggressively cuts into its trunk with a chainsaw. The loud "SCRREEEEE" sound effect highlights the violent disruption caused by this action. This figure embodies unchecked human exploitation, whether through deforestation, industrial expansion or resource extraction, driven by short-term economic interests disregarding long-term environmental stability.

The Tree of Life critiques the prevailing narrative that industrial expansion and resource extraction are necessary for economic progress. The cartoon exposes the power imbalance between human intervention and ecological resilience by contrasting the thriving tree and wildlife with the destructive human figure. The exaggerated juxtaposition of destruction and survival forces viewers to question the justifications for environmental degradation and confront the harmful consequences of unchecked human actions.

Simultaneously, the cartoon reveals everyday resistance through the enduring presence of the tree and its wildlife. Despite the immediate threat posed by the chainsaw, the tree continues to stand tall, sheltering various life forms. This survival represents the persistence of marginalised communities and ecosystems, which resist domination through their mere existence. The image suggests that resistance is not always overt but often embedded in the natural world's ability to endure and adapt, even when confronted with systemic exploitation.

The cartoon unpacks the power dynamics within environmental discourse by employing visual satire and metaphor. The chainsaw and the accompanying sound effect emphasise the violence of human exploitation and serve as a disruptive force that demands attention. This visual intervention challenges viewers to reconsider the assumptions surrounding climate change and environmental degradation, shifting the conversation from abstract policy debates to the tangible realities of everyday resistance against ecological harm.

The *Tree of Life* contributes to critical geopolitics by challenging state-centric and corporate-driven environmental narratives through the interplay of destruction and defiance. It repositions ecological destruction as a profoundly political act and highlights how power struggles over climate change play out in official spaces, ecosystems, and visual culture. The cartoon, therefore, is more than just an artistic expression; it is a form of political critique, urging viewers to reconsider the consequences of human intervention and the resilience of nature.

Another cartoon, *Trust Your Mirrors*, critiques the environmental impact of excessive carbon emissions, particularly from automobiles in industrialised societies. The central image of a large SUV, labelled "Luxury Sport RX," likely refers to fuel-guzzling vehicles that contribute heavily to pollution. As the vehicle moves forward, it leaves behind a massive, dark footprint of smoke labelled "Carbon Footprint." This ominous shape reinforces the environmental consequences of excessive fossil fuel consumption.

A speech bubble from the driver reads: "Remember, 'Objects in mirror are closer than they appear,'" a common phrase on vehicle mirrors, repurposed here to suggest that climate change and its consequences – often seen as distant threats – are imminent. The irony in the cartoon highlights the dangers of ignoring carbon emissions and their accelerating impact on global warming. The exaggerated visual elements emphasise the urgency of addressing this issue.

Hidden transcripts within the cartoon offer insight into marginalised voices and social justice issues, critiquing power structures that perpetuate environmental exploitation. The "Blindspot EX" vehicle choice challenges attitudes towards environmental responsibility, while the smoke cloud forming a footprint subtly critiques the power structures that perpetuate environmental harm. The desolate road surrounding the vehicle amplifies the voices of environmental advocates and indigenous communities who are fighting against exploitation. These hidden transcripts challenge prevailing assumptions, highlight the disproportionate impacts on marginalised communities and critique the systems perpetuating environmental degradation.

Trust Your Mirrors challenges dominant frames in climate change discourse, including environmentalism, neoliberalism and humanitarianism. While it underscores the consequences of unsustainable behaviour and carbon emissions, it critiques neoliberal ideologies that prioritise economic growth over sustainability and indirectly critiques humanitarianism for focusing on short-term needs at the expense of long-term environmental preservation.

The cartoon also carries significant geopolitical implications, drawing attention to the disproportionate contribution of wealthy nations to carbon emissions and critiquing colonial legacies of environmental exploitation. Culturally, it challenges prevailing attitudes towards consumerism and environmental responsibility, amplifying the voices of affected communities. By fostering cross-cultural dialogue, *Trust Your Mirrors* encourages solidarity in addressing climate change and highlights the need for collective action to mitigate the urgent challenges posed by this global crisis.

The cartoon *Fish Inside an Oyster Shell* depicts a surreal and symbolic scene, with a fish trapped inside an open seashell, struggling to survive in its limited space. Surrounding it is a vast, cracked and barren landscape, indicating extreme drought and water scarcity. The contrast between the aquatic life inside the shell and the lifeless land outside emphasises themes of climate change, habitat loss and environmental degradation. The cartoon powerfully symbolises resilience in the face of water scarcity, directly addressing the challenges of climate change.

The image of a fish seeking refuge in an oyster shell amidst a parched landscape highlights the struggle for survival in environments affected by water scarcity due to climate change. This stark visual representation underscores the harsh reality faced by ecosystems and communities impacted by water shortages. As a symbol of resilience, the fish challenges perceptions of helplessness in environmental crises and inspires recognition of adaptation, even in adverse circumstances.

Additionally, the hidden transcript of the fish seeking refuge within the shell subtly resists the adverse effects of environmental degradation exacerbated by climate change. This resilience challenges the narratives of vulnerability often imposed on marginalised communities, particularly those disproportionately affected by climate change. The cartoon critiques dominant frames like neoliberalism and humanitarianism, questioning ideologies prioritising economic growth over sustainability and short-term humanitarian interventions over long-term environmental preservation.

Geopolitically, the cartoon reflects debates on resource management and global governance related to climate change. It highlights the disproportionate impact on vulnerable populations, many already experiencing the adverse effects of water scarcity. Culturally, the cartoon promotes environmental responsibility and collective action to address climate change. Amplifying marginalised voices and fostering cross-cultural dialogue encourages solidarity in confronting the complex geopolitical challenges posed by environmental degradation.

A popular cartoon titled *Polar Bears* depicts a polar bear reading that they are on the endangered species list, sitting on a melting ice fragment. The image shows the polar bear asking its sinking companion, "Are you listening?" oblivious to the fact that the ice is disappearing beneath them. The cartoon symbolically conveys the environmental impacts of climate change on Arctic ecosystems, where melting ice poses an existential threat to species like polar bears. It underscores the urgency of addressing climate change by showing the polar bears' precarious situation amidst the rapid decline of their habitat.

The melting ice and the sinking companion of the polar bear serve as a metaphor for broader environmental degradation, emphasising the interconnectedness of ecosystems and the necessity for collective action. The cartoon critiques environmentalism, neoliberalism and colonial legacies by addressing the political and social dimensions of climate change. It calls attention to the disproportionate impacts on vulnerable species, like the polar bear, while emphasising the importance of systemic change in response to environmental threats.

Through this, *Polar Bears* fosters cross-cultural dialogue by using a universal visual language to discuss the climate crisis. It amplifies the voices of marginalised species and ecosystems disproportionately affected by environmental degradation, urging viewers to confront the realities of climate change and take action. The hidden transcript within the cartoon also highlights social justice issues and critiques power structures contributing to environmental harm, reinforcing the need for a comprehensive response to climate change that addresses ecological and geopolitical concerns.

Another cartoon titled *Climate Forecast* features a polyptych depicting erratic weather patterns daily, showing extreme conditions such as floods, heatwaves, tornados, fires and droughts. The word "freak" runs across all the days except Saturday, where an ostrich with its head in the sand symbolises denial in response to the climate crisis. The cartoon critiques the normalisation of extreme weather events by repeatedly using the word "freak" and emphasises the urgency of addressing climate change. The ostrich image highlights the complacency and denial that often prevail among those benefiting from the status quo, urging viewers to confront the reality of the climate crisis.

The cartoon's erratic climate forecasts subvert conventional notions of weather patterns, symbolising resistance to complacency and the established norms surrounding climate change. Hidden transcripts within the cartoon reflect resistance to climate denial and complacency, urging action against the normalising of extreme weather. The ostrich represents denial prevalent in systems that perpetuate environmental harm, while the absence of a forecast on Saturday symbolises collective inaction, challenging marginalised communities to mobilise in response to climate injustice.

Geopolitically, *Climate Forecast* critiques the disproportionate impact of climate change on vulnerable regions and communities, calling attention to global power dynamics that contribute to environmental degradation. It indirectly challenges influential actors who contribute significantly to greenhouse gas emissions while reflecting diverse cultural perspectives on climate change. The cartoon emphasises the need for decolonisation and equitable resource management in response to climate injustice, fostering solidarity and collective action across communities facing the impacts of the climate crisis.

These cartoons intricately weave narratives of resistance against climate change, using visual satire and metaphor to confront and challenge dominant environmental discourses. They question the established norms surrounding environmental exploitation and shine a spotlight on marginalised voices, amplifying the often overlooked or silenced communities most affected by climate change. Critiquing the power structures that perpetuate environmental harm – whether through corporate greed, state-sponsored policies or neoliberal economic models – the cartoons forcefully highlight the imbalance of power that drives environmental degradation and social inequality.

Rather than simply depicting environmental destruction, these cartoons push for a deeper, critical examination of the systems and ideologies that allow such exploitation to persist. They expose the stark realities faced by vulnerable populations and ecosystems, urging viewers to reconsider the prevailing narratives that justify resource extraction, industrial expansion and unsustainable consumption. The juxtaposition of destruction and resistance in these cartoons serves as a call to action, imploring individuals and communities to challenge the status quo and engage in collective efforts to reverse the harm done to the planet.

Through their powerful visual storytelling, these cartoons illustrate the urgent environmental crises of our time and emphasise the need for solidarity, empathy and cross-cultural dialogue. They advocate for a collective global response transcending borders, encouraging a united front to address the environmental, social and political issues exacerbated by climate change. Ultimately, these cartoons serve as more than just artistic expressions – they are vehicles for political critique and instruments of change, demanding that we confront the realities of climate destruction and take bold, unified action to secure a sustainable future for all.

Conclusion

In conclusion, this research underscores the intricate relationship between geopolitics and resistance, wherein power dynamics and dissent intersect. Geopolitics extends beyond traditional analyses of nation-state power to encompass the everyday actions of individuals and communities, while resistance emerges as a counterforce to established power hierarchies and dominant narratives. Within this context, cartoons offer a unique avenue for resistance, challenging prevailing ideologies and spotlighting social injustices through accessible visual mediums.

Simultaneously, this study highlights the profound impact of cartoons as potent tools for effectively communicating the effects of climate change. Cartoons convey intricate geopolitical messages to raise awareness and inspire action for environmental sustainability by simplifying complexities, evoking emotions, and transcending linguistic and cultural barriers. Through a detailed analysis of selected political cartoons like "The Tree of Life," "Carbon Footprints," "Fish Inside an Oyster Shell," "Polar Bears" and "Climate Forecast," this research reveals how these visual narratives challenge dominant narratives, amplify marginalised voices and critique systems perpetuating environmental exploitation.

Ultimately, cartoons foster empathy, critical thinking, and advocacy for environmental sustainability among diverse audiences worldwide, making them invaluable assets in shaping public discourse and promoting collective action to address the urgent challenges posed by climate change. By shedding light on the intersection of geopolitics, resistance and climate change communication, this research contributes to a deeper understanding of how visual culture influences public perceptions and responses to pressing global issues.

Works Cited

Barnett, Jon. "The Geopolitics of Climate Change." *Geography Compass*, vol. 1, no. 6, 2007, pp. 1361–1375.

Bravo, Michael. "Voices from the Sea: The Reception of Climate Impact Narratives." *Journal of Historical Geography*, vol. 35, 2009, pp. 256–278.

Campbell, David. "Geopolitics and Visuality: Sighting the Darfur Conflict." *Political Geography*, vol. 26, 2007, pp. 357–382.

Clough, J. "Cartoons 'n Comics: Communication to the Quick." *Center for Media Literacy*, 1982, p. 20.

Defra (Department for Environment Food and Rural Affairs). "Future Worlds Images: How Do We Prepare for a Changing Climate?" 2011, archive.defra.gov.uk/environment/climate/documents/interim2/future-worlds.pdf. Accessed 22 January 2024.

Diamond, Matthew. "No Laughing Matter: Post-September 11 Political Cartoons in Arab/Muslim Newspapers." *Political Communication*, vol. 19, 2002, pp. 251–272.

Dodds, Klaus. "Popular Geopolitics and Cartoons: Representing Power Relations, Repetition and Resistance." *Critical African Studies*, vol. 4, 2010, pp. 1–19.

Doyle, James. *Mediating Climate Change*. Ashgate Publishing, 2011.

El Refaie, Elisabeth. "Multiliteracies: How Readers Interpret Political Cartoons." *Visual Communication*, vol. 2, 2009, pp. 75–95.

Greenberg, J. "Framing and Temporality in Political Cartoons: A Critical Analysis of Visual News Discourse." *Canadian Review of Sociology and Anthropology*, vol. 39, 2002, pp. 181–198.

Hammett, Daniel. "Zapiro and Zuma: A Symptom of an Emerging Constitutional Crisis in South Africa?" *Political Geography*, vol. 29, 2010, pp. 88–96.

Harrison, Roy. "How Cartoons Work: The Cartoon Code." *Center for Media Literacy*, 1982, www.medialit.org/reading_room/article154.html. Accessed 28 March 2022.

Harvey, Fiona. "Worst Ever CO2 Emissions Leave Climate on the Brink." *The Guardian*, 30 May 2011, pp. 1–2.

Hulme, Mike. "Why We Disagree about Climate Change in Art," www.interclimate.org/challenge/index.php?option=com_content&view=article&id=203&Itemid=221. Accessed 28 March 2022.

Khalili Mehr, Tayebe. "Fish Inside an Oyster Shell." *Cartoon*, Irancartoon Web Gallery, The Exhibition of Cartoon about Drought in Somalia, n.d. Irancartoon, www.irancartoon.ir/gallery/album515/Tayebe_Khalili_Mehr_Iran_2.

King, J. "Polar Bear." *Cartoon*. Originally published in EarthFirst.com. RealClimate.science, realclimate.science/2012/05/18/polar-bear-drowning-season-kicks-off/#gsc.tab=0. Accessed 18 May 2012.

Kleeman, Grant. "Not Just for Fun: Using Cartoons to Investigate Geographical Issues." *New Zealand Geographer*, vol. 62, 2006, pp. 144–151.
Leiserowitz, Anthony A. "Day after Tomorrow: Study of Climate Change Risk Perception." *Environment: Science and Policy for Sustainable Development*, vol. 46, 2004, pp. 22–39.
Macdonald, Fiona et al. "Introduction: Envisioning Geopolitics." *Observant States: Geopolitics and Visual Culture*. I. B. Tauris, 2010, pp. 1–19.
Madden, Chris. "The Tree of Life. Cartoon Guide to Biodiversity Loss." *Conservation Bytes*, January 2011, conservationbytes.com/2011/01/14/cartoon-guide-to-biodiversity-loss-x/.
Manzo, Kate. "Imagining Vulnerability: The Iconography of Climate Change." *Area*, vol. 42, 2010, pp. 96–107.
Miles, Malcolm. "Representing Nature: Art and Climate Change." *Cultural Geographies*, vol. 17, 2010, pp. 19–35.
O'Neill, Saffron and Sophie Nicholson-Cole. "Fear Won't Do It: Promoting Positive Engagement with Climate Change through Visual and Iconic Representations." *Science Communication*, vol. 30, 2009, pp. 355–379.
O'Tuathail, Gearóid and Simon Dalby. "Introduction: Rethinking Geopolitics, towards a Critical Geopolitics." *Rethinking Geopolitics*, edited by Gearóid O'Tuathail and Simon Dalby, Routledge, 1998, pp. 1–15.
Ridanpaa, Juha. "Geopolitics of Humour: The Muhammed Cartoon Crisis and the Kaltio Comic Strip Episode in Finland." *Geopolitics*, vol. 14, 2009, pp. 729–749.
Schmidt, Gavin and Joshua Wolfe. *Climate Change: Picturing the Science*. W.W. Norton, 2009.
Scott, James C. *Weapons of the Weak: Everyday Forms of Peasant Resistance*. Yale University Press, 1985.
Slocum, Rachel. "Polar Bears and Energy-Efficient Lightbulbs: Strategies to Bring Climate Change Home." *Environment and Planning: Society and Space*, vol. 22, 2004, pp. 413–438.
Smyth, David. "Special Report: Global Warming." *British Journal of Photography*, vol. 15, 2007, pp. 29–32.
Toles, Tom. "Climate Forecast." *Cartoon*. Originally published in The Washington Post, 5 December 2015, nexusmedianews.com/the-madhouse-effect-6f503846e01c/.
Whitmarsh, Lorraine. "What's in a Name? Commonalities and Differences in Public Understanding of 'Climate Change' and 'Global Warming'." *Public Understanding of Science*, vol. 18, 2009, pp. 401–420.

Index

FSC
www.fsc.org
MIX
Papier | Fördert
gute Waldnutzung
FSC® C083411

Kolibri 360 GmbH
Ferdinand-Jühlke-Straße 7
99095 Erfurt, Deutschland
produktsicherheit@kolibri360.de